Michael Schweizer is Head of Yeast Genetics at the Institute of Food Research, Norwich and Professor of the University of Erlangen-Nuremberg. In addition to teaching students in the UK and Germany, his research encompasses lipid and nucleotide metabolism and the functional analysis of *Saccharomyces cerevisiae*.

Molecular fungal biology

RICHARD OLIVER AND MICHAEL SCHWEIZER

CAMBRIDGE
UNIVERSITY PRESS

PUBLISHED BY THE PRESS SYNDICATE OF THE UNIVERSITY OF CAMBRIDGE
The Pitt Building, Trumpington Street, Cambridge, United Kingdom

CAMBRIDGE UNIVERSITY PRESS
The Edinburgh Building, Cambridge CB2 2RU, UK www.cup.cam.ac.uk
40 West 20th Street, New York, NY 10011-4211, USA www.cup.org
10 Stamford Road, Oakleigh, Melbourne 3166, Australia
Ruiz de Alarcón 13, 28014 Madrid, Spain

First published 1999

Printed in the United Kingdom at the University Press, Cambridge

Typeset in Adobe Minion 10.5/14pt, in QuarkXPress™ [SE]

A catalogue record for this book is available from the British Library

Library of Congress Cataloguing in Publication data

Molecular fungal biology / [edited by] Richard P. Oliver, Michael
Schweizer.
       p.   cm.
Includes index.
ISBN 0 521 56116 7 (hb). – ISBN 0 521 56784 X (pb)
1. Fungal molecular biology.   I. Oliver, Richard P. (Richard
Peter), 1958–   .   II. Schweizer, Michael, Professor.
QK604.2.M64M665   1999
572.8'295–dc21   99-10041   CIP

ISBN 0 521 56116 7 hardback
ISBN 0 521 56784 X paperback

Contents

Contents

Contributors

T. H. Adams, Department of Biology, Texas A & M University, College Station, TX 77843, USA

D. B. Archer, Institute of Food Research, Norwich Research Park, Colney, Norwich NR4 7UA, UK

M. L. Berbee, Department of Botany, University of British Columbia, Vancouver, Canada

P. Bowyer, Institute of Arable Crops Research, Long Ashton Research Station, Long Ashton, Bristol BS18 9AF, UK

S.-W. Chiu, Department of Biology, The Chinese University of Hong Kong, Shatin, N. T., Hong Kong

J. Doonan, John Innes Centre, Norwich Research Park, Colney, Norwich NR4 7UA, UK

H. Feldmann, Adolf-Butenandt-Institut für Physiologische Chemie, Physikalische Biochemie und Zellbiologie der Universität, Schillerstrasse 44, D-80336 München, Germany

A. R. Hawkins, Department of Biochemistry and Genetics, Medical School, Framlington Place, University of Newcastle Upon Tyne, Newcastle NE2 4HH, UK

S. Hosking, School of Biological Sciences, The University of Manchester, 1.800 Stopford Building, Oxford Road, Manchester M13 9PT, UK

M. Jacobs, Institut für Biotechnologie, T. U. Berlin Sekret. TIB 4/4-1, Mikrobiologie und Genetik, Gustav-Meyer Allee 25, D-13355 Berlin, Germany

H. K. Lamb, Department of Biochemistry and Genetics, Medical School, Framlington Place, University of Newcastle Upon Tyne, Newcastle NE2 4HH, UK

Contributors

L. J. Levitt, Department of Biochemistry and Genetics, Medical School, Framlington Place, University of Newcastle Upon Tyne, Newcastle NE2 4HH, UK

D. Moore, School of Biological Sciences, The University of Manchester, 1.800 Stopford Building, Oxford Road, Manchester M13 9PT, UK

G. H. Newton, Department of Biochemistry and Genetics, Medical School, Framlington Place, University of Newcastle Upon Tyne, Newcastle NE2 4HH, UK. (Present address: Current Biology, Middlesex House, 34–42 Cleveland Street, London W1P 5FP)

M. Penttilä, VTT Biotechnology and Food Research, Espoo, PO Box 1500, FIN-02044 VTT, Finland

C. Pitt, John Innes Centre, Norwich Research Park, Colney, Norwich NR4 7UA, UK

G. Robson, School of Biological Sciences, The University of Manchester, 1.800 Stopford Building, Oxford Road, Manchester M13 9PT, UK

M. Saloheimo, VTT Biotechnology and Food Research, Espoo, PO Box 1500, FIN-02044 VTT, Finland

U. Stahl, Institut für Biotechnologie, T. U. Berlin Sekret, TIB 4/4-1, Mikrobiologie und Genetik, Gustav-Meyer Allee 25, D-13355 Berlin, Germany

J. W. Taylor, Department of Plant and Microbial Biology, University of California at Berkeley, 111 Koshland Hall, Berkeley CA 94720-3102, USA

M. Wedde, Institut für Biotechnologie, T. U. Berlin Sekret. TIB 4/4-1, Mikrobiologie und Genetik, Gustav-Meyer Allee 25, D-13355 Berlin, Germany

K. A. Wheeler, Department of Biochemistry and Genetics, Medical School, Framlington Place, University of Newcastle Upon Tyne, Newcastle NE2 4HH, UK

J. K. Wieser, Department of Biology, Texas A & M University, College Station, TX 77843, USA

Preface

'Fungi are often found in damp places; that is why they are shaped like umbrellas.'

This schoolboy howler, possibly aprocryphal, summarizes the parlous state of knowledge about fungi that all too often exists in university-level students, and it is with the aim of counteracting the perceived low status of fungal biology, in contrast to the biology of animals, plant and bacteria, that this book was conceived. It is our goal to demonstrate that fungal research is dynamic, active and exciting. Fungi have played major roles in developing central concepts in biology (see Chapter 1) and continue to be vital model organisms in many areas of research. They have vital roles in animal and plant disease, ecology and biotechnology.

The key to the resurgence of fungal biology has undoubtedly been the development of molecular biology tools for these organisms. A theme of this book is that molecular biology is almost always a vital tool in current research. Part of the reason for the difficulty in studying fungi is their diversity. This has made a succinct, useful definition of fungi impossible. The definition of fungi as the organisms studied by mycologists is an aphorism designed to confuse. Such confusion is now readily explained since, largely through the use of molecular tools (see Chapter 2) the relationship between the diverse fungal groups has become much clearer.

The rest of this book concentrates on the area of mycology where research is substantial and, largely using molecular biology, a detailed functional understanding is emerging. The yeast genome project has undoubtedly put fungi back at the forefront of molecular and cell biology and its conclusions are described in Chapter 3. The highly detailed understanding of gene regulation and metabolic flux control is described in Chapter 4. The emerging consensus on the molecular basis of development is covered in Chapters 5–8. Our varied levels of understanding of the role of fungi as saprotrophes, phytopathogens and animal pathogens are described in Chapters 9–11. Finally, the uses, limitations and prospects of fungi for some aspects of biotechnology are described in Chapter 12. Thus this book is not intended to be a comprehensive treatise, nor is it intended to be a field-guide, rather it is an introduction to some of the more interesting fungal research carried out in universities and research institutes. It is hoped that this book will enhance undergraduate and graduate courses on mycology, provide a suitable primer for scientists entering fungal research and thereby stimulate an interest to study fungi further.

The treatment of each subject in each chapter is necessarily and unapologetically diverse. In some chapters, reference is made to the primary literature whilst in others, the reader is directed mainly to recent reviews and books. These differences in approach were dictated by a number of factors including the relative maturity of the subject and its familiarity to the intended audience as well as a desire to keep the book from becoming unwieldy. Thus some authors were asked not to refer to too many primary references and we solicit the understanding of the research community in this regard.

The editors thank Lilian Schweizer for compiling the index.

RICHARD OLIVER
MICHAEL SCHWEIZER

1 Fungi: important organisms in history and today

M. WEDDE, M. JACOBS AND U. STAHL

Introduction

Fungal species have been at the forefront of developments in microbiology in particular and biology in general ever since the middle of the
sixteenth century. Fungi have vital roles in agriculture, medicine and biotechnology. This chapter provides an overview of the role of microbiology in history with emphasis on eukaryotes. After a short introduction
into what microbiology encompasses, a historical survey of some mile-
stones in the development of modern microbiology, especially mycology,
is given. This also describes where other scientific fields were influenced
by fungal research.

Microorganisms and microbiology

Microorganisms are the most widely distributed group of living
beings. They are capable of occupying nearly every ecological niche due
to the fact that individual microorganisms can grow under extreme conditions, such as temperatures ranging from $-15\,°C$ to more than $100\,°C$,
or in the absence of oxygen or light. Neither animals nor plants are so
adaptable; this is also true as regards nutrition. Some microorganisms
can even utilize exotic energy sources such as alkanes, inorganic salts, or
substituted aromatic compounds.

Generally, the term 'microorganism' encompasses all living beings
too small to be seen by the human eye without magnification, i.e. smaller
than about 1 mm. They are either prokaryotic, e.g. bacteria, or eukaryotic, e.g. hyphal fungi, yeasts and protozoa. We focus here on fungi and
yeasts. For historical reasons the term 'fungus' is used for filamentously
growing microorganisms, whereas yeasts are single-celled eukaryotes.
However, taxonomically both are fungi (see Chapter 2). It has recently

1

become evident that some yeasts are also able to grow filamentously and *vice versa,* either in response to nutritional starvation or during certain stages of their development. This morphological difference is therefore not rigid.

Fungi have several impacts on the life of man. First, they are responsible for breaking down most of the organic carbon produced by plants. In this respect they not only destroy dead material, but also are able to infect living plants. A number of economically important plants are attacked by fungi, leading to crop losses, either by decomposing plants or plant products, or by producing certain toxins. However, there are also examples of favourable coexistence, i.e. symbiosis of plant and fungus, namely the mycorrhiza. Here a fungus grows in close contact to or within the root system of a plant, providing the plant with vitamins or organic nitrogen, while itself is supported with a desired carbon source. Both fungus and plant take advantage of this cohabitation. Second, fungi are a serious problem in medicine. Apart from obligate pathogens such as *Histoplasma,* there are a number of opportunistic germs such as *Candida* or *Aspergillus* which cause infections in compromised hosts. The numbers of such reported cases have increased over past years. Finally, fungi are used in many biotechnological applications. Examples are traditional processes such as the production of beer and wine and the industrial production of organic acids, antibiotics or enzymes. Moreover, molecular genetics have been used as a tool to optimize the yield of production strains and to express foreign genes in fungi in order to produce corresponding gene products.

Historical overview

The discovery of microorganisms and the understanding of their metabolism and life cycles has influenced human civilization in several respects. Due to their small size, microorganisms were originally discovered only after the invention of the microscope in the seventeenth century by a Dutch merchant, Antonius van Leeuwenhoek (1632–1723). Despite their limited magnification power, early microscopes provided for the first time an insight into the submacroscopic world. Amongst other things, van Leuewenhoek described a large number of microorganisms, which were published by the Royal Society in London. His work began the elucidation of the microbial world.

After van Leeuwenhoek nearly a century elapsed before the next substantial progress in the field of microbiology, mainly due to the lack of microscopes of sufficient quality and microbiological technique. During this time there was serious debate about the origin of microorganisms. It centred on the question whether microorganisms grow from airborne 'germs' or whether they develop spontaneously. Ultimately, it could be shown that organic material could be preserved by airtight conditions and subsequent heating. This preservation was paralleled by the absence of microbial growth. In this way spontaneous generation of microorganisms could be ruled out. Louis Pasteur (1822–1895) concluded these studies by demonstrating that air does indeed contain microorganisms and that chemical changes within nutrients support their growth. During his work on the yeast *Saccharomyces cerevisiae* he established another milestone of microbiology: the theory that all fermentative processes are caused by microorganisms, and that each type of fermentation depends on the presence of a certain organism.

In 1897 Buchner discovered by chance that even a cell-free extract of ground cells of *S. cerevisiae* is able to convert glucose to ethanol and carbon dioxide. This demonstrated that metabolic energy is not produced by a living cell itself, but by some of its constituents. This finding was the basis for modern biochemistry.

At the same time it became evident that microorganisms can cause diseases in plants and animals. Robert Koch (1843–1910) discovered that a bacterium, *Bacillus anthracis*, was the causative agent of anthrax. Moreover, he showed that a related organism, i.e. the hay bacillus, was not able to transmit this disease. He thus concluded that a specific disease has only one causative agent. This observation initiated a rapid characterization of various pathogenic bacteria or fungi and formed the basis of medical microbiology. During the first half of the twentieth century knowledge about harmful microorganisms grew. An appropriate treatment of such diseases became possible with the discovery of penicillin by Fleming in 1928 and other antibiotics which were subsequently identified.

The next step in modern microbiology was taken by Beadle and Tatum, who isolated biochemical mutants of *Neurospora crassa* in 1941.

The one-gene one-protein hypothesis was developed partially on the basis of these mutants. After DNA was identified as the substance responsible for heredity and the techniques for *in vitro* recombination and gene transfer were developed, rapid progress was made in the understanding of microorganisms and higher eukaryotes.

Currently one of the highlights of modern research is the elucidation of eukaryotic gene organization and function. Here, *Saccharomyces cerevisiae* (amongst other organisms) serves as a model system. Its genome is 1.2×10^7 basepairs and hence only four times larger than that of *Escherichia coli*. This sequence information is important not only for optimizing biotechnological processes but also for research of diseases such as cancer.

Fungal metabolism

During the course of human history microorganisms have been exploited in many ways. First, there are a great number of examples of fungi and yeasts being used in the production of food. Even before microorganisms were recognized, they were used in the making of beer, wine, bread, cheese, milk products, and even for simply eating (e.g. edible mushrooms). Fundamental techniques did not change in principle from early civilizations until the beginning of the twentieth century. The Babylonians used the yeast *S. cerevisiae* in the brewing of beer in 6000 BC. In addition, this yeast was used for leavening bread two thousand years later in Egypt and was mentioned at the same time in the fermentation of mashed grapes to yield wine. A common factor in all these uses is that *S. cerevisiae* (known as 'leaven') is able to convert very rapidly glucose to ethanol and carbon dioxide. Both products can be useful, e.g. in the case of alcoholic beverages ethanol is of interest not only as a stimulant but sometimes also as a preservative. Furthermore, carbon dioxide is necessary to raise dough during the leavening of bread, while the ethanol produced is removed during the baking process. Furthermore, ethanol can be concentrated by distillation and roughly 25 per cent of worldwide production is from biological sources.

In the eastern world rice is used instead of malt or mashed grapes for fermentation. Here a mould, usually *Aspergillus oryzae*, initiates the

4

fermentation process by hydrolysing rice starch to fermentable sugars. Later on the sugar is converted to ethanol by spontaneous fermentation by either yeast or bacteria leading to products such as sake.

Another example of the traditional use of fungi in food production is in the making of cheese. Moulds such as *Penicillium camemberti* or *P. roqueforti* excrete different proteases and lipases catalysing the decay of caseine or milk fat, resulting in the softening and flavouring of the cheese. Moreover, a number of *Penicillum* species are used to conserve meat-yielding products such as certain hams. In Asia moulds, mainly *Rhizopus*, *Mucor* and *Aspergillus*, are widely used to ferment soya beans to make, for instance, *tempeh* or *miso*.

Second, microorganisms have been used to make chemical substances. Although it was known that they are capable of promoting different chemical changes whilst growing, it took until the first world war to force their industrial implementation. During this period, by influencing its metabolism during fermentation, *S. cerevisiae* was used to produce glyccrol. Later, yeasts such as *Candida utilis*, which can use pentoses as carbon sources, or *Yarrowia lipolytica*, which can metabolize aliphatic alkanes with 10 to 16 carbon atoms, were grown in large amounts both to feed animals and to increase the protein content of human food.

Another important field of the early industrial application of microorganisms was the production of antibiotics. After penicillin was discovered it was produced in appreciable amounts during the second world war, initially with *Penicillium notatum* and later on with *P. chrysogenum*. This was followed by other antibiotics such as streptomycin (described by Schatz *et al.* in 1944). At present there are several thousand efficacious substances of which less than a hundred are produced in large quantities.

Nowadays, there is a rapidly growing number of fungal metabolites, e.g. cytostatica or hormones. Different fungal strains are incubated with chemical precursors which become covalently modified ('biotransformation') in order to produce secondary metabolites of as yet unknown structure and impact. Moreover, secondary metabolites are screened as a pool because sometimes the efficacy is due to the combination of different substances.

Classical genetics

Fungi are very useful in the study of genetics. First, they are eukaryotic, which means they can be used as model systems for higher and therefore more complex organisms. Furthermore, they are easy to handle due to their relatively low nutritional demands, they grow vegetatively within hours and the progeny of sexual crosses develop within days or weeks. During growth they have only a few differentiational stages (i.e. hyphae, conidia and sexual organs), which facilitates analysis. Finally, mutants can be readily obtained as microorganisms having a mainly haploid life cycle. They are susceptible to different chemical or physical methods such as treatment with nitrosoguanidine or ultraviolet (UV) light. There are ways to positively select for mutants, i.e. only the mutants of interest are able to grow, or to identify mutants negatively, i.e. mutagenized cells do not grow. In the latter case mutant cells can be enriched, for example by using nystatin which kills all wild-type cells growing under given conditions (Snow, 1966). A number of different mutants have been isolated, e.g. nutritional mutants, spore colour mutants, where inheritance can be detected visually and developmental mutants, which require permissive conditions to grow.

Discoveries in the field of classical fungal genetics were established using ascomycetes, since these are well-known and well-documented organisms. Therefore, all of the model organisms described below belong to this group of fungi. They are characterized by the fact that their sexual spores are found within a so-called ascus which contains all products of a meiotic division. Usually the ascus is protected by an outer envelope, e.g. perithecia or cleistothecia.

Two intensively studied hyphal fungi are *Neurospora crassa* and *Aspergillus nidulans*. Both have a haploid vegetative phase and produce haploid conidiospores permitting conclusions directly from the phenotype of a mycelium to its genotype. Moreover, each of their asci contains eight spores, since in both cases the four meiotically produced nuclei are mitotically duplicated afterwards. Therefore, each single strand of originally diploid nuclear DNA leads to one ascospore, allowing a detailed analysis of the initial DNA. That is to say that even heteroduplex DNA characterized by a partial non-identity of both strands of a DNA double helix is accessible. The asci of *Neurospora* represents so-called ordered

6

tetrads, i.e. the combination of genes found in one spore corresponds to the distribution of the two parental chromosomes in meiosis I and the two sister chromatids in meiosis II. Thus, recombination events can be detected directly by analysing the asci formed and thereby the distance of a certain gene from its centromere can be determined giving rise to linkage groups, i.e. chromosomes.

A milestone in fungal genetic research was X-ray experiments in 1941 by Beadle and Tatum who isolated a number of auxotrophic mutants of *Neurospora crassa*. In contrast to the wild-type strain, these mutants were unable to grow on certain media due to defects in individual genes. When single compounds were added to the medium, growth was re-established. Thus, a specific dependence of a certain gene product and an individual metabolic step could be shown. Partially on the basis of these results, the one-gene one-protein hypothesis already mentioned was developed.

Aspergillus nidulans was described genetically by Pontecorvo *et al.* (1953) and later on by Clutterbuck (1974). They isolated a large number of mutant strains, which were derived from one initial isolate. These mutants were subsequently used in several fields of research, e.g. mitotic or meiotic recombination or cellular development.

An important step towards handling imperfect fungi was the discovery of the parasexual cycle of *Aspergillus nidulans* by Pontecorvo and Roper (1952). This process is important because most industrially used fungi have no sexual cycle at all, e.g. *Aspergillus* or *Penicillium*, or the strains used for fermentation have lost their breeding ability and their fermentation efficiency can therefore no longer be improved by sexual crosses.

The most important fungus in both basic research and applied genetics is the yeast *Saccharomyces cerevisiae*. It is a single-celled organism and therefore easier to handle in comparison with filamentous fungi. It reproduces itself via budding both as a haploid or as a diploid cell. There are two mating types, namely a and α. Strains of the opposite mating type can be crossed enabling easy genetic analysis. The asci of *S. cerevisiae* contain four spores which are non-ordered. No specialized containment for the ascus is formed because the yeast cell forms the ascus itself.

Protoplast fusion can be used to make hybrids between cells of the same mating type, as well as of diploid or aneuploid cells or even between yeasts of different species or genera. To do this, the cell walls are enzymatically degraded. The resulting cells are called spheroplasts when the cell wall is only partially removed, or protoplasts when the removal is complete. Both cell types can be fused under appropriate conditions and usually karyogamy occurs. A special case is the use of *kar1* mutants which are hindered in nuclear fusion. The fusion products of these mutants keep the original nuclei. After cell division the nuclei are separated again and only the cytoplasm is newly recombined. This process is called 'cytoduction' and it is used in research on nuclear–mitochondrial interaction or mitochondrial function.

In contrast to the moulds described above, budding yeast is a facultative anaerobic organism which needs no oxygen for the generation of energy. It is therefore possible to induce mitochondrial mutants which are not able to gain adenosine triphosphate (ATP) via the respiration chain. Such mutants were initially described by Ephrussi and are called 'petite mutants', because they form only tiny colonies on solid media containing glucose. They do not grow at all on non-fermentable carbon sources such as glycerol or ethanol. With such types of mutants, an analysis of mitochondrial function and biogenesis became possible.

The last microorganism which should be mentioned as a model system for higher eukaryotes is the fission yeast *Schizosaccharomyces pombe*. On the evolutionary scale it is only distantly related to both budding yeast and mammals. The cell division cycle of *S. pombe* more resembles the higher eukaryotes than does *S. cerevisiae*, in respect to both timing and control (Forsburg and Nurse, 1991; Chang and Nurse, 1996). Moreover, homologues of mammalian or vertebrate genes have been identified which participate in cell cycle control. It could be shown that these genes overcome certain developmental defects of *S. pombe*. Since some of these genes are known to be oncogenes, impairment leads to uncontrolled cell proliferation, thus the understanding of the cell cycle of fission yeast could be helpful in cancer research (Pelech *et al.*, 1990; MacNeill *et al.*, 1991).

The pros and cons of fungi

Today there is an enormous diversity of fungal research and fungal applications. Fungi, however, have both pros and cons. Discussed below are examples of this ambivalence, casting a glance on the fields of agriculture, medicine, biotechnology and basic research.

Fungi and agriculture

In most cases, fungi have a negative impact on agriculture. They infect both living plants and plant products and thereby cause damage and crop loss. Hence, plant-pathogenic fungi are of great economic importance and thus there is intensive research. This research concerns different factors involved in plant infections, e.g. phytoalexins (plant defence compounds), fungal pathogenicity genes such as degrading enzymes, signal transduction suppressors and phytotoxins, and avirulence genes conferring strain specificity.

On one hand, there are fungal genera of a relatively broad host spectrum such as *Alternaria, Sclerotinia, Fusarium* and *Penicillium*. On the other hand, some fungi have only a narrow host range such as *Ustilago maydis* (corn smut fungus) and *Cladosporium fulvum* (leaf mould of tomato). The latter group was particularly intensively investigated to understand mechanisms of pathogenicity and host specificity.

An approach to understanding the development of pathogenicity is typified by the work on *Ustilago maydis*. It is a dimorphic basidiomycete having a haploid, yeast-like growing form (sporidia) which is non-pathogenic. The filamentously growing dikaryotic phase is generated after the mating of compatible strains. The dikaryon is able to invade the corn plant. Mating and pathogenicity are controlled by two gene loci (termed *a* and *b*) whose exact interaction remains to be elucidated. However, there is good evidence that the *a* locus encodes a pheromone/pheromone receptor system which resembles the budding yeast mating system. The *b* locus encodes different DNA-binding proteins involved in transcription regulation that may regulate development-specific genes involved in pathogenicity (Banuett, 1992).

An example of a certain interaction between fungus and plant is *Cladosporium fulvum*. *Cladosporium* strains show race-specific colonization of tomato, i.e. strains which are pathogenic for a certain cultivar of tomato are not able to cause disease in other cultivars. This is due to the interaction between strain-specific peptides ('elicitors') and plant receptors. Resistance is achieved if the elicitor is positively recognized by the plant. The interaction is highly specific, as even a point mutation in an elicitor gene leads to virulence (de Wit *et al.*, 1994).

These efforts in understanding plant colonization by fungi may finally provide means to confer resistance on plants and decrease the economic damage caused by fungi. However, they are impressive examples for the interaction between organisms raising questions such as development-specific gene expression, recognition of cells or cell products, and defence mechanisms.

Fungi and medicine

Infections caused by fungi are a serious problem with an increasing impact. In general, fungal pathogens can be divided into two groups. The first group comprises obligate pathogens, such as *Histoplasma capsulatum* or *Coccidioidis immitis*, which are able to infect even healthy individuals. The second and larger group comprises opportunistic fungi, which need a certain predisposition in the host to be infective. A predisposition may take many forms. It may be an imbalanced immune response, the presence of another infection or even an accidentally acquired injury. The most prominent examples of opportunistic fungi are *Candida albicans, Cryptococcus neoformans* and *Aspergillus fumigatus.*

Both *Histoplasma capsulatum* and *Coccidioidis immitis* are soil inhabitants which are widely distributed. They show a dimorphism as they grow in nature in a saprophytic mycelial form which produces conidia. These conidia are dispersed in the air and usually enter the body by inhalation. They grow within the respiratory tract in a parasitic yeast-like form which is capable of multiplying itself within tissue (Dimorphic fungi in biology and medicine, eds. H. Vanden Bossche *et al.*, Plenum Press, New York, 1993). In most cases symptoms are non-specific, such as fever, cough, or weakness in varying degrees.

To infect the host, fungi have to overcome the human defence system. The first and very effective barriers are human skin and mucus membranes, which hinder microorganisms entering the body. Successive systems are the humoral defence, i.e. antibodies and the complement system, and cell-mediated defence, e.g. tissue macrophages or neutrophils. Besides these there are other defence mechanisms, e.g. iron-binding proteins or membrane-damaging peptides (Lehrer *et al.*, 1993).

10

Together they are sufficient in most cases to prevent the body from being infected. However, obligate pathogens are able to overcome this host defence or parts of it. For instance, *Histoplasma* prevents itself from being attacked by growing within mononuclear phagocytes, hence it is in fact an intracellular parasite (e.g. Wu-Hsieh and Howard, 1993).

In contrast, opportunistic pathogens need particular circumstances to be infective. *Candida albicans*, for instance, is a commensal yeast which inhabits the human gastrointestinal tract and the vaginal mucosa. Usually *Candida* shares the given resources with other microorganisms present and thus they are balanced. If this balance is disturbed, e.g. by the intake of antibiotics or during immunosuppression in the case of organ transplantation or chemotherapy, *Candida* sometimes invades the host tissue and causes either local or even systemic infections. Under similar conditions *Cryptococcus* and *Aspergillus* are known to be pathogens. For instance, *Cryptococcus neoformans* often generates superficial infections in AIDS patients, whereas *Aspergillus fumigatus* is, after *Candida*, the most important infectious agent in the immunocompromised host. Here, it causes life-threatening infections, since *Aspergillus* not only colonizes the lung, but afterwards penetrates blood vessels and becomes disseminated to other parts of the body such as the brain.

Fungal research contributes to the understanding of medical questions, e.g. in cancer development. This field is closely connected to work on the cell division cycle. This cycle consists of different phases which a cell passes through before it divides. There are several so-called checkpoints at which the decisions about cell cycle progress are made (Forsburg and Nurse, 1991; Cross, 1995). The cell has to take into account its nutritional status, but also responds to stress factors such as oxidative damage, UV light or proliferation signals, e.g. cytokines or mitogens (Waskiewicz and Cooper, 1995). If cell cycle control fails cells may continuously proliferate or, in contrast, fail to divide at all.

A number of genes important in cell cycle control have been described. They are putative oncogenes, namely genes associated with uncontrolled cell proliferation. Examples are the genes encoding the cdc2, ras, or p53 proteins. The last is a negative regulator of cell proliferation with multiple functions. Defects of p53 are linked to tumour formation and other diseases (Elledge and Lee, 1995; Meyn, 1995). Within the past few years fungal research has provided insights into the properties of p53, as expressed in fission and in budding yeast, respectively (Bischoff *et al.*, 1992, Nigro *et al.*, 1992; Koerte *et al.*, 1995). In addition, several other cell cycle-related genes from humans (e.g. Paris *et al.*, 1994; Davey and Beach, 1995) or even from *Drosophila* (Campbell *et al.*, 1995) and from tomato (Ach and Gruissem, 1994) have been expressed in *Schizosaccharomyces pombe*. Moreover, there have been attempts at understanding the cell cycle and development in *Aspergillus nidulans* (reviewed by Fry and Nigg, 1995) in which a homologue of the human *ras* gene was identified (Som and Kolaparthi, 1994). In conclusion, microorganisms not only cause diseases in human and in animals, they may also help progress in medical research.

Fungi and biotechnology

There are some biotechnological fields in which fungi play an important role, namely the production of food, of extracellular enzymes, of secondary metabolites and of organic chemicals. In each case fungi compete with other biological systems such as bacteria or cell cultures or with the chemical synthesis of the respective compounds.

The area of food production comprises the main classical fields of the making of fermented foodstuffs such as cheese (mostly *Penicillium* spp.) and alcoholic beverages (mostly *S. cerevisiae*), and the planting of edible fungi such as *Agaricus bisporus* (mushrooms) or *Lentinus edodes* (shiitake mushroom). Moreover, yeasts such as *Candida* or *Kluyveromyces* and filmentous fungi such as *Fusarium* can be used to produce single-cell protein, i.e. the production of biomass by growing microorganisms on relatively poor substrates.

Several extracellular enzymes are used to produce food. They are either naturally synthesized or recombinantly introduced into foreign hosts (e.g. Devchand and Gwynne, 1991; Gelissen *et al.*, 1992; Kinghorn and Unkles, 1994). For instance, the milk-clotting enzyme chymosin from calves is recombinantly produced in *Trichoderma* and *Aspergillus* (Uusitalo *et al.*, 1991; Tsuchiya *et al.*, 1993) and subsequently used for cheese production. Other examples for biotechnologically interesting proteins are amylases (starch degradation, *Aspergillus* spp.), cellulases (hydrolysis of cellulose, *Trichoderma reesei*), and different proteases, lipases or nucleases

12

for clearing beverages, as digestion aids or for food flavouring (for further reading see e.g. Fogarty, 1994). Furthermore, proteins of pharmaceutical relevance are expressed in fungi. For instance, human lactoferrin, an iron-binding protein that may be used as an antimicrobial agent, or human tissue plasminogen activator, which degrades blood clots, were expressed in *A. nidulans* (Ward *et al.*, 1992 Upshall *et al.*, 1987).

Secondary metabolites are those substances considered not to be essential for cell growth *per se*, e.g. antibiotics and steroids. Economically important fungi are *Penicillium* spp. and *Cephalosporium* spp. which produce β-lactam antibiotics inhibiting the synthesis of the bacterial cell wall. Both antibiotics (penicillin G and cephalosporin C) can be modified by introducing different side chains to appropriate precursor molecules excreted by mutant strains. These semi-synthetic β-lactam antibiotics are characterized by an increased chemical stability or a decreased spectrum of resistant target organisms. Furthermore, steroidal immunosuppressants are used in the case of organ transplantations to avoid graft rejection. A well-known example of such a compound is cyclosporin A, which is synthesized by *Tolypocladium inflatum*.

Sometimes, a complete *de novo* synthesis of interesting compounds is not possible, but precursors provided become covalently modified. This is called biotransformation. Possible reactions are hydroxylations, (de)hydrogenations and cleavage of covalent bonds (Berger, 1995). Biotransformations can be used for the development of as yet unknown products. Here, individual fungi have their own spectrum of reactions that they can perform (e.g. Mahato and Majumdar, 1993).

Another area of biotechnological use of fungal metabolism is the production of organic substances. The most prominent example is that of citric acid made by *Aspergillus niger*, with a worldwide production of 500 000 tonnes per year (Bu'Lock, 1990). It is used in the fields of food production, chemical synthesis and in pharmaceuticals (e.g. Kapoor *et al.*, 1982; Harvey and McNeil, 1994). Furthermore, industrial ethanol is produced mainly by *S. cerevisiae* and can be used for chemical synthesis or as a fuel substitute. The latter use is interesting for countries with a high production capacity of fermentable plant material. Appropriate fungi are generally able to produce an enormous number of chemicals which can be derived from fungal cultures, such as vitamins, different organic acids and flavour components. This is due to the fact that fungi have varying metabolisms which can, in turn, be industrially exploited (e.g. Berger, 1995).

Basic research using fungi

Important organisms in basic research are by and large the same as mentioned in the field of classical genetics, i.e. *Saccharomyces cerevisiae* is a major research object covering different fields, *Aspergillus nidulans* is the most prominent representative of hyphal fungi, as is *Schizosaccharomyces pombe* with regard to cell cycle control (see above).

Research on *S. cerevisiae* comprises various topics. For instance, gene expression is a central field encompassing the regulation of yeast genes in respect to the biosynthesis of cellular compounds (e.g. Braus, 1991; Vergeres *et al.*, 1993), translation of mRNA (Kozak, 1992; Linder, 1992) or the post-translational processing of proteins (e.g. Bourbonnais *et al.*, 1991; Omer and Gibbs, 1994) including secretion of proteins (e.g. Moir and Mao, 1990; Caldwell *et al.*, 1995). Other main fields of research are those of mitochondrial biogenesis (Grivell, 1995) and of the expression of foreign genes (e.g. Gelissen *et al.*, 1992; Sleep *et al.*, 1991). Moreover, budding yeast is the model system for higher eukaryotes. Its morphogenesis, including cell cycle regulation and signal transduction, is thus of high interest (Forsburg and Nurse, 1991; Kron and Gow, 1995). Finally, two systems which are now widely used should be mentioned. The first is the so-called yeast two-hybrid system, which can be used for the detection of protein–protein interactions. It is used to analyse yeast proteins and also heterologous proteins which are expressed in yeast for this purpose (Chien *et al.*, 1991). The second system is the development of yeast artificial chromosomes (YACs; Burke *et al.*, 1987). These chromosomes are reliably transmitted during mitosis in both yeast and higher eukaryotes. They are used to introduce large fragments of heterologous DNA into yeast, which enables an analysis of the genes included, e.g. genes which are linked to diseases (e.g. Chaffanet *et al.*, 1996). Furthermore, they can be transferred into an appropriate mutant or cell type in order to study possible effects (e.g. Cabin *et al.*, 1995; Lamb and Gearhart, 1995). YACs are used to study different organisms, e.g. tomato, *Arabidopsis* and mouse. (Bonnema *et al.*, 1996; Creusot *et al.*, 1995;

14

Dietrich *et al.*, 1995). This is just a sample of current research on budding yeasts.

Basic research using *Aspergillus nidulans* mainly comprises the fields of gene regulation, mitosis, and morphogenesis. Gene regulation is studied in response to factors such as the ambient pH (e.g. Espeso *et al.*, 1993; Denison *et al.*, 1995), nitrogen (Caddick *et al.*, 1994) and carbon sources (e.g. Fillinger *et al.*, 1995). Mitosis is linked to cell cycle control (described above). The work on morphogenesis deals with the formation of conidiogenous cells and finally the conidia itself; there is a well-regulated system which controls the conidiogenesis (e.g. Timberlake, 1991; Aguirre, 1993; Sewall, 1994). Furthermore, it has been shown in *Neurospora crassa* that formation of conidia is regulated by light (Lauter and Russo, 1991). Thus, the knowledge of this system may lead to the understanding of the day/night cycle on which higher eukaryotes also rely.

Conclusion

This chapter has shown how multi-faceted fungi and fungal research are. Although our knowledge in this field used to be relatively poor, it has become tremendously enlarged within the past few decades. Fungi are now objects of wide research which should result in numerous applications, not only in classical fields but also in the fields of recombinant DNA techniques and as model systems for higher eukaryotes. In the following chapters there are detailed descriptions of the fields of study mentioned in this introduction.

References

Ach, R. A. and Gruissem, W. (1994). A small nuclear GTP-binding protein from tomato suppresses a *Schizosaccharomyces pombe* cell-cycle mutant. *Proc. Natl. Acad. Sci. USA* 91:5863–5867.

Aguirre, J. (1993). Spatial and temporal controls of the *Aspergillus* brlA developmental regulatory gene. *Mol. Microbiol.* 8:211–218.

Banuett, F. (1992). *Ustilago maydis*, the delightful blight. *Trends Genet.* 8:174–180.

Berger, R. G. (1995). *Aroma biotechnology.* Springer, Berlin, Heidelberg, New York.

Bischoff, J. R., Casso, D. and Beach, D. (1992). Human p53 inhibits growth in *Schizosaccharomyces pombe. Mol. Cell. Biol.* 12:1405–1411.

Bonnema, G., Hontelez, J., Verkerk, R., Zhang, Y. Q., van Daelen, R., van Kammen, A. and Zabel, P. (1996). An improved method of partially digesting plant megabase DNA suitable for YAC cloning: application to the construction of a 5.5 genome equivalent YAC library of tomato. *Plant J.* 9 125–133.

Bourbonnais, Y., Germain, D., Latchinian-Sadek, L., Boileau, G. and Thomas, D. Y. (1991). Prohormone processing by yeast proteases. *Enzyme* 45:244–256.

Braus, G. H. (1991). Aromatic amino acid biosynthesis in the yeast *Saccharomyces cerevisiae*: a model system for the regulation of a eukaryotic biosynthetic pathway. *Microbiol. Rev.* 55:349–370.

Bu'Lock, J. D. (1990). Swings and roundabouts for citric acid producers. *Biotechnol. Insight* 84:5–6.

Burke, D. T., Carle, G. F. and Olson, M. V. (1987). Cloning of large segments of exogenous DNA into yeast by means of artificial chromosome vectors. *Science* 236:806–812.

Cabin, D. E., Hawkins, A., Griffin, C. and Reeves, R. H. (1995). YAC transgenic mice in the study of the genetic basis of Down syndrome. *Prog. Clin. Biol. Res.* 393:213–226.

Caddick, M. X., Peters, D. and Platt, A. (1994). Nitrogen regulation in fungi. *Antonie van Leeuwenhoek* 65:169–177.

Caldwell, G. A., Naider, F. and Becker, J. M. (1995). Fungal lipopeptide mating pheromones: a model system for the study of protein prenylation. *Microbiol. Rev.* 59:406–422.

Campbell, S. D., Sprenger, F., Edgar, B. A. and O'Farrell, P. H. (1995). *Drosophila* Wee1 kinase rescues fission yeast from mitotic catastrophe and phosphorylates *Drosophila* Cdc2 *in vitro. Mol. Biol. Cell* 6:1333–1347.

Chaffanet, M., Imbert, A., Adelaide, J., Le Paslier, D., Wagner, M. J., Wells, D. E., Birnbaum, D. and Pebusque, M. J. (1996). A 3.1-Mb YAC contig within the Werner syndrome region, on the short arm of human chromosome 8. *Cytogenet. Cell Genet.* 72:63–68.

Chang, F. and Nurse, P. (1996). How fission yeast fission in the middle. *Cell* 84:191-194.

Chien, C. T., Bartel, P. L., Sternglanz, R. and Fields, S. (1991). The two-hybrid system: a method to identify and clone genes for proteins that interact with a protein of interest. *Proc. Natl. Acad. Sci. USA* 88:9578–9582.

Clutterbuck, J. A. (1974). *Aspergillus nidulans.* In: *Handbook of Genetics*, ed. King, R.C., Plenum Press, New York, pp. 447–510.

16

Creusot, F., Fouilloux, E., Dron, M., Lafleuriel, J., Picard, G., Billault, A., Le Paslier, D., Cohen, D., Chabouté, M.-E., Durr, A., Fleck, J., Gigot, C., Camilleri, C., Bellini, C., Caboche, M. and Bouchez, D. (1995). The CIC library: a large insert YAC library for genome mapping in *Arabidopsis thaliana*. *Plant J.* 8:763–770.

Cross, F. R. (1995). Starting the cell cycle: what's the point? *Curr. Opin. Cell Biol.* 7:790–797.

Davey, S. and Beach, D. (1995). RACH2, a novel human gene that complements a fission yeast cell cycle checkpoint mutation. *Mol. Biol. Cell* 6:1411-1421.

de Wit, P. J., Joosten, M. H., Honee, G., Wubben, J. P., van den Ackerveken, G. F. and van den Broek, H. W. (1994). Molecular communication between host plant and the fungal tomato pathogen *Caldosporium fulvum*. *Antonie van Leeuwenhoek* 65:257–262.

Denison S. H., Orejas M. and Arst H. N. Jr. (1995). Signaling of ambient pH in *Aspergillus* involves a cysteine protease. *J. Biol. Chem.* 270:19–22.

Devchand, M. and Gwynne, D. I. (1991). Expression of heterologous proteins in *Aspergillus*. *J. Biotechnol.* 17:3–9.

Dietrich, W. F., Copeland, N. G., Gilbert, D. J., Miller, J. C., Jenkins, N. A. and Lander, E. S. (1995). Mapping the mouse genome: current status and future prospects. *Proc. Natl. Acad. Sci. USA* 92:10849–10853.

Elledge, R. M. and Lee W. H. (1995). Life and death by p53. *BioEssays* 17:923–930.

Espeso, E. A., Tilburn; J., Arst, H. N. and Penalva, M. A. (1993). pH regulation is a major determinant in expression of a fungul penicillin biosynthetic gene. *EMBO J.* 12:3947–3956.

Fillinger, S., Panozzo, C., Mathieu, M. and Felenbok, B. (1995). The basal level of trancription of the alc genes in the ethanol regulon in *Aspergillus nidulans* is controlled both by the specifc transactivator AlcR and the general carbon catabolite repressor CreA. *FEBS Lett.* 368:547–550.

Fogarty, W. M. (1994). Enzymes of the genus *Aspergillus*. In: *Aspergillus*, ed. Smith, J. E., Plenum Press, New York, pp. 177–218.

Forsburg, S. L. and Nurse, P. (1991). Cell cycle regulation in the yeasts *Saccharomyces cerevisiae* and *Schizosaccharomyces pombe*. *Annu. Rev. Cell Biol.* 7:227–256.

Fry, A. M. and Nigg, E. A. (1995). The NIMA kinase joins forces with Cdc2. *Curr. Biol.* 5:1122–1125.

Gellissen, G., Melber, K., Janowicz, Z. A., Dahlems, U. M., Weydemann, U., Piontek, M., Strasser, A. W. and Hollenberg, C. P. (1992). Heterologous protein production in yeast. *Antonie van Leeuwenhoek* 62:79–93.

17

Grivell, L. A. (1995). Nucleo-mitochondrial interactions in mitochondrial gene expression. *Crit. Rev. Biochem. Mol. Biol.* 30:121-164.

Harvey, L. M. and McNeil, B. (1994). Liquid fermentation systems and product recovery of *Aspergillus*. In: *Aspergillus*, ed. Smith, J. E., Plenum Press, New York, pp. 141-176.

Kapoor, K. K., Chaudhary, K. and Tauro, P. (1982). Citric acid. In: Industrial Microbiology, ed. Reed, G., AVI Publishing, Westport, pp. 709–747.

Kinghorn J. R. and Unkles, S. E. (1994). Molecular genetics and expression of foreign proteins in the genus *Aspergillus*. In: *Aspergillus*, ed. Smith, J. E., Plenum Press, New York and London 65–100.

Koerte, A., Chong, T., Li, X., Wahane, K. and Cai, M. (1995). Suppression of the yeast mutation rft1-1 by human p53. *J. Biol. Chem.* 270:22556–22564.

Kozak, M. (1992). Regulation of translation in eukaryotic systems. *Annu. Rev. Cell. Biol.* 8:197–225.

Kron, S. J. and Gow, A. R. (1995). Budding yeast morphogenesis: signalling, cytoskeleton and cell cycle. *Curr. Opin. Cell Biol.* 7:845–855.

Lamb, B. T. and Gearhart, J. D. (1995). YAC transgenics and the study of genetics and human disease. *Curr. Opin. Genet. Dev.* 5:342–348.

Lauter, F. R. and Russo, V. E. (1991). Blue light induction of conidiation-specific genes in *Neurospora crassa*. *Nucl. Acids Res.* 19:6883–6886.

Lehrer, R. I., Lichtenstein, A. K. and Ganz, T. (1993). Defensins: antimicrobial and cytotoxic peptides of mammalian cells. *Annu. Rev. Immunol.* 11:105–128.

Linder, P. (1992). Molecular biology of translation in yeast. *Antonie van Leeuwenhoek* 62:47–62.

MacNeill, S. A., Warbrick, E. and Fantes, P. A. (1991). Controlling cell cycle progress in the fission yeast *Schizosaccharomyces pombe*. *Curr. Opin. Genet. Dev.* 1:307–312.

Mahato, S. B. and Majumdar, I. (1993). Current trends in microbial steroid bio-transformation. *Phytochemistry* 34:883–898.

Meyn, M. S. (1995). Ataxia-telangiectasia and cellular responses to DNA damage. *Cancer Res.* 55:5991-6001.

Moir, D. T. and Mao, J. I. (1990). Protein secretion systems in microbial and mammalian cells. *Bioprocess Technol.* 9:67–94.

Nigro, J. M., Sikorski, R., Reed, S. I. and Vogelstein, B.(1992). Human p53 and CDC2Hs genes combine to inhibit the proliferation of *Saccharomyces cerevisiae*. *Mol. Cell. Biol.* 12:1357–1365.

Omer, C. A. and Gibbs, J. B. (1994). Protein prenylation in eukaryotic microorganisms: genetics, biology and biochemistry. *Mol. Microbiol.* 11:219–225.

18

Fungi: important organisms in history and today

Paris, J. , Leplatois, P. and Nurse, P. (1994). Study of higher eukaryotic gene function CDK2 using fission yeast. *J. Cell Sci.* 107:615–623.

Pelech, S. L., Sanghera, J. S. and Daya-Makin, M. (1990). Protein kinase cascades in meiotic and mitotic cell cycle control. *Biochem. Cell. Biol.* 68:1297–1330.

Pontecorvo, G. and Roper, J. A. (1952). Genetic analysis without sexual reproduction by means of polyploidy in *Aspergillus nidulans. J. Gen. Microbiol.* 6:VII.

Pontecorvo, G., Roper, J. A., Hemmons, L. M., MacDonald, K. D. and Bufton, A. W. J. (1953). The genetics of *Aspergillus nidulans. Adv. Genet.* 5:141-238.

Sewall, T. C. (1994). Cellular effects of misscheduled brlA, abaA and wetA expression in Aspergillus nidulans. *Can. J. Microbiol.* 40:1035–1042.

Schatz, A., Bugie, E. and Waksman S. A.(1944). Streptomycin, a substance exhibiting antibiotic activity against Gram-positive and Gram-negative bacteria. *Proc. Soc. Exp. Biol. Med.* 55:66–69.

Sleep, D., Belfield, G. P., Ballance, D. J., Steven, J., Jones, S., Evans, L. R., Moir, P. D. and Goodey, A. R. (1991). *Saccharomyces cerevisiae* strains that overexpress heterologous proteins. *Biotechnology* 9:183–187.

Snow, R. (1966). An enrichment method for auxotrophic yeast mutants using the antibiotic "nystatin". *Nature* 211:206–207.

Som, T. and Kolaparthi,V. S. R. (1994). Developmental decisions in *Aspergillus nidulans* are modulated by Ras acitvity. *Mol. Cell. Biol.* 14:5333–5348.

Timberlake, W. E. (1991). Temporal and spatial controls of *Aspergillus* development. *Curr. Opin. Genet. Dev.* 1:351–357.

Tsuchiya, K., Gomi, K., Kitamoto, K., Kumagai, C. and Tamura, G. (1993). Secretion of calf chymosin from the filamentous fungus *Aspergillus oryzae. Appl. Microbiol. Biotechnol.* 40:327–332.

Upshall, A., Kumar, A. A., Bailey, M. C., Parker, M. D., Favreau, M. A., Lewison, K. P., Joseph, M. L., Maraganore, J. M. and McKnight, G. L. (1987). Secretion of active human tissue plasminogen from the filamentous fungus *Aspergillus nidulans. Biotechnology* 5:1301-1304.

Uusitalo, J. M., Nevalainen, K. M., Harkki, A. M., Knowles, J. K. and Penttilä, M. E. (1991). Enzyme production by recombinant *Trichoderma reesei* strains. *J. Biotechnol.* 17:35–49.

Vanden Bossche, H., Odds, F. C. and Kerridge, D. (eds.) (1993). *Dimorphic Fungi in Biology and Medicine.* Plenum Press, New York, p. 129.

Vergeres, G., Yen, T. S., Aggeler, J., Lausier, J. and Waskell, L. (1993). A model system for studying membrane biogenesis. Overexpression of cytochrome b5 in yeast results in marked proliferation of the intracellular membranes. *J. Cell Sci.* 106:249–259.

Ward, P. P., May, G. S., Headon, D. R. and Conneely, O. M. (1992). An inducible expression system for the production of human lactoferrin in *Aspergillus nidulans*. *Gene* 122:219–223.

Waskiewicz, A. J. and Cooper, J. A. (1995). Mitogen and stress response pathways: MAP kinase cascades and phosphatase regulation in mammals and yeast. *Curr. Opin. Cell Biol.* 7:798–805.

Wu-Hsieh, B. and Howard, D. H. (1993). Histoplasmosis. In: *Fungal Infections and Immune Responses*, eds. Murphy, J. W. *et al.*, Plenum Press, New York, pp. 213–250.

2 Fungal phylogeny

M. L. BERBEE AND J. W. TAYLOR

In less than two decades, beginning when Walker and Doolittle (1982) examined relationships among aquatic fungi based on RNA nucleotide sequence, comparative studies of nucleic acid variation have revolutionized evolutionary mycology. Although there is far more remaining to be done than has been done, the big picture of fungal evolution is coming into focus. In this chapter we combine a review of the main points found in the new phylogenetic trees of fungi with a look at the methods used to make the trees. The challenge is not just to learn the current trees, but to learn how they are made and evaluated. The concept of testing alternative evolutionary histories is as important as the new insights that have been gained on fungal evolution; specific trees will change as new data become available but the need to evaluate alternative trees will remain.

Phylogenetic patterns among fungi

To appreciate the phylogeny of fungi, it is necessary to consider them in relation to the rest of life on earth, which is believed to be a continuum beginning several billions of years (Gyr) ago and leading to the present. We will break the continuum into three levels, the 'Big Picture' of fungal evolution or how the organisms studied by mycologists relate to the rest of life, the four phyla that make the kingdom Fungi, and individual members of the four phyla.

Big picture of fungal evolution

The most basic question is, how are fungi related to other biota? This is a more complicated question than you might guess, because mycologists study a very broad range of organisms, and have suspected for well over a century that some of their charges were not closely

21

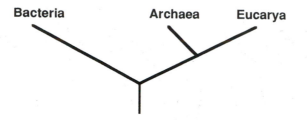

Figure 2.1 The three known Domains of life (Woese *et al.*, 1990). Fungi are among the Eucarya, the nuclear genome of which shares a common ancestor with the Archaea (archaebacteria). Mitochondrial genomes, including those of fungi, originated in the Bacteria.

related. Concern about the distant relationships among fungi arose from reproductive morphology and details of flagellation, and was reinforced by studies of lysine synthesis and cell wall polysaccharides. Comparative studies of nucleic acid sequences from organisms studied by mycologists have confirmed these suspicions (Bruns *et al.*, 1991), marking a clear distinction between a *monophyletic clade* of true fungi (the kingdom Fungi with its four phyla) and the separate clades of other organisms studied by mycologists, i.e. the two groups of cellular slime moulds (Acrasiomycota and Dictyosteliomycota), the plasmodial slime moulds (Plasmodiophoromycetes) and the fungal Stramenopiles (Oomycota, Hyphochytriomycota, and Labyrinthulomycota) (see Figure 2.2).

The earth formed about 4.5 Gyr ago, and geochemical and microfossil evidence indicates that life may have begun as early as 4.2 to 4.3 Gyr ago. The *root* of the *phylogenetic* tree of all extant life is in the prokaryotes, and it defines two main lineages, the domain Bacteria and the ancestor to the domains Archea and Eucarya (Woese *et al.*, 1990) (Figure 2.1). These two branches are commonly thought to have diverged approximately 4 Gyr ago. Based on comparison of nuclear ribosomal genes, eukaryotes arose on the Archaea lineage about 1.8 Gyr ago, followed later by the endosymbiotic acquisition of the mitochondrion from the Bacteria (Figure 2.1). The evolution of the eukaryotes appears to involve the sequential divergence of many 'protist' taxa culminating in a radiation of more complex eukaryotes about 1 Gyr ago, a radiation of branches that resembles the crown of a tree (Figure 2.2).

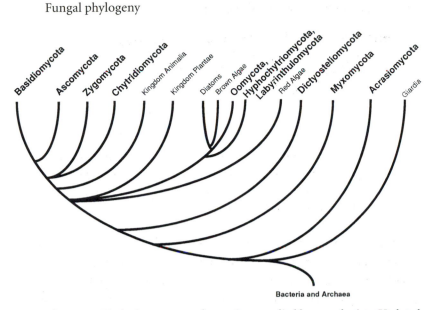

Figure 2.2 Evolutionary tree of organisms studied by mycologists. Updated and redrawn from Bruns *et al.*, 1991.

At the base of the domain Eukarya are found branches leading to organisms whose divergences may predate the acquisition of mitochondria (such as the intestinal parasite *Giardia* and microsporidia) followed by branches leading to the acrasid slime moulds (phylum Acrasiomycota) the plasmodial slime moulds (phylum Myxomycota) and the dictyostelid slime moulds (Dictyosteliomycota) (Figure 2.2). In the case of *Dictyostelium*, phylogenetic analyses of protein coding genes and ribosomal genes are in conflict, but, for the present, we will side with the ribosomal genes. All of these groups are amoeboid and can engulf small food items (e.g. bacteria) or invade and extract the cytoplasm from larger ones (e.g. yeast cells). In wet conditions, the amoebae of some species may develop flagella. Larger, multicellular or multinucleate structures can form through the aggregation or fusion of amoebae, often culminating in spore formation, sometimes involving meiosis. Relatives of these groups of slime moulds include the Euglenoids and other groups of amoeboflagellates. The Plasmodiophoromycetes, a group that includes plant pathogens such as *Plasmodiophora* and *Spongospora*, probably also branches among these and other amoeboflagellates.

Climbing further up the tree, we come to the 'crown' commemorating the radiation of complex eukaryotes (stramenopiles, alveolates, red algae, plants and green algae, animals and choanoflagellates and fungi), all of which form the 'eukaryotic crown group' (Wainright *et al.*, 1993; Baldauf and Palmer, 1993) (Figure 2.2). Among the members of the 'crown group' is the Stramenopila (Leipe *et al.*, 1994 [=Chromista of Cavalier-Smith, 1981]) home to three taxa studied by mycologists: the Oomycota, the Hyphochytriomycota and the Labyrinthulomycota, (Leipe *et al.*, 1994), along with the diatoms, brown and golden brown algae. The stramenopiles have tubular mitochondrial cristae, a shared ancestral character, and tripartite tubular flagellar hairs, a shared derived character. The Oomycota, Hyphochytriomycota and Labyrinthulomycota are basal to the algae in the stramenopiles, supporting the idea that they diverged before one of their relatives acquired a photosynthetic cell with chlorophylls a + c, which was the endosymbiotic event that led to the radiation of the stramenopile algae. Although these organisms are not close relatives of the kingdom Fungi, as shown by the details of flagella structure and small subunit (SSU) rDNA comparison they can seem indistinguishable from true fungi in the form of their thalli (both determinate and filamentous) and their ecological roles (parasitism and decomposition). These features, combined with the profound economic effects of plant-parasitic Oomycota, ensure their continuing study by mycologists.

The Labyrinthulomycota includes some organisms whose morphology resembles the Chytridiomycota of the kingdom Fungi (Thraustochytriaceae), as well as others that form a net-like plasmodium superficially reminiscent of amoeboflagellates (Labyrinthulaceae). One species, *Labyrinthula zosterae*, is parasitic on marine eel grass and has caused serious disease (Muehlstein *et al.*, 1991).

Somatic structures of species in the Hyphochytriomycota and Oomycota closely resemble members of the kingdom Fungi, some having determinate thalli and others making hyphae and complex mycelia. Hyphochytrids do not cause economic problems for humans, but the same cannot be said of the Oomycota, which include plant pathogens responsible for seedling damping off (*Pythium* spp.), innumerable *Phytophthora* blights and rots, including the one responsible for the Irish potato famine,

24

and the downy mildews (Peronosporaceae) (Agrios, 1988; Alexopoulos *et al.* 1996). All oomycetes can produce flagellated cells (zoospores), but many are terrestrial plant pathogens that no longer rely on zoospores.

Kingdom Fungi

The kingdom Fungi is also part of the 'eukaryotic crown group', the radiation of complex eukaryotes that occurred about 1.0 Gyr ago. As with any ancient radiation that happened over a relatively short period of time, determining the relative order of divergences, and therefore the sister taxa, is difficult. In the case of fungi, comparisons of nuclear rDNA, elongation factor and cytoskeletal proteins suggest that fungi are the *sister group* of animals (plus choanoflagellates), but this result is controversial (Figure 2.2). Additional support for a common ancestor of fungi and animals about 1.0 Gyr ago comes from shared biochemical features (storage of glycogen, extracellular chitin matrix, mitochondrial codon UGA specifying tryptophan instead of termination) and shared phenotype of the motile stage (unicellular with a single posterior flagellum, as found in fungal zoospores, animal male gametes, and choanoflagellates). The number of species in the kingdom Fungi is unknown, but potentially very large. Fewer than 100 000 species have been described, but as many as 1 to 1.5 million may exist (Hawksworth, 1991; Hawksworth *et al.*, 1996).

Chytridiomycota and *Zygomycota*, the 'coenomycetes'

From the branch that unites the animals plus choanoflagellates with the kingdom Fungi, we now move into the fungal kingdom. Among the groups that diverge at the base of the fungal clade are those with posterior, whiplash flagella. These fungi have been classified in the Chytridiomycota, and their inclusion in the Fungi has been controversial because they have flagella and the other fungal groups do not (Margulis and Schwartz, 1988). Putting the Chytridiomycota in the protists ignored other phenotypic features shared by Chytridiomycota and other fungi, such as the pathway of lysine biosynthesis and the chitinous cell wall. Comparison of SSU rDNA sequence (Bowman *et al.*, 1992) showed that chytrids are part of a monophyletic kingdom Fungi.

Just as flagella misled biologists about the exclusion of Chytridiomycota from the fungi, they also have misled mycologists about the

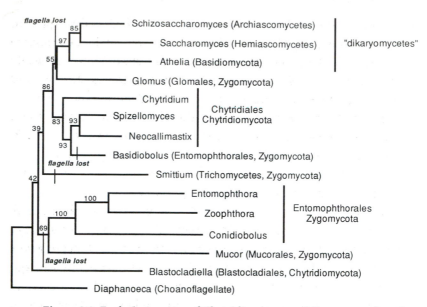

Figure 2.3 Evolutionary tree of Chytridiomycota and Zygomycota based on nuclear SSU rDNA nucleotide sequence (redrawn from Nagahama *et al.* 1995). Note the multiple independent losses of flagella implied by this tree, showing that presence of flagella is not a good character for the classification of fungi found in these two Divisions. In fact, neither Division is monophyletic, but orders in both divisions are monophyletic.

monophyly of the Chytridiomycota. This problem was foreshadowed by phylogenetic analysis of a few representatives of chytrids and zygomycetes, which revealed difficulties in separating species in the two phyla (Bruns *et al.*, 1992). It has been made clear by more thorough comparison of nuclear rDNA sequence which showed the Chytridiomycota and Zygomycota to be non-monophyletic due to the multiple independent losses of flagella (Nagahama *et al.*, 1995; (Figure 2.3). Intertwined as they are, the names Chytridiomycota and Zygomycota have lost utility, and it is becoming popular to refer to both groups as the coenomycetes, an informal name commemorating their scarcity of hyphal cross walls and resulting coenocytic cytoplasm. Of course, there are monophyletic groups (orders of these fungi) within both the Chytridiomycota and Zygomycota, and these are what we shall consider, after making a few generalizations about the two phyla. The thalli of chytrids and zygomycetes are haploid, and sexual reproduction results in a thick-walled spore in which meiosis occurs. Many chytrids are developmentally determinate with most of the thallus becoming a sporangium, although they typically produce filaments (called rhizoids) that anchor the thallus in food sources and provide nutrition. Other chytrids and most zygomycetes are developmentally indeterminate and produce hyphae and mycelia, albeit without many cross walls. Some zygomycetes can also grow as yeasts. In both phyla, mitospores typically are produced in a sporangium, but neither mitospores nor meiospores are produced in a fruiting body, with the exception of the Endogonales in the Zygomycota.

Based upon comparison of SSU rDNA sequence by Nagahama *et al.* (1995), the Blastocladiales (Chytridiomycota) lie at the base of the fungi. One of the oldest fossil fungi is a specimen of the blastocladiaceous genus *Allomyces* from the Devonian, which was living about 400 million years ago (Mya) (T. Taylor *et al.*, 1994). The next higher node leads to a clade of Mucorales (Zygomycota), followed by Entomophthorales (Zygomycota). Next come the Trichomycetes (Zygomycota), followed by a node leading to Chytridiales, Spizellomycetales and Neocallimasticales (Chytridiomycota) plus Basidiobolus (Zygomycota) (Figure 2.3).

The Blastocladiales are saprobic (now that fungi are seen as distinct from plants, we must drop the word saprophyte for fungi) on small bits of organic matter in water and wet soil, and parasitize algae, pollen grains, and small aquatic animals, including mosquito larvae. The Mucorales are home to many spoilage fungi, a famous example being *Rhizopus stolonifer*, the black bread mould, a species that all of us have inadvertently cultivated at one time or another. The Entomophthorales, as their name suggests, destroy insects through pathogenesis. Trichomycetes are found in the guts of arthropods; they attach to the exoskeleton and must reinfect with each moult (Lichtwardt 1986). Chytridiales and Spizellomycetales parasitize the same sorts of small plants such as algae and pollen grains, and are saprobic as well. One group of chytrids, the Neocallimasticales, are symbiotic in the guts of ruminants and other large herbivores, aiding in the animal's digestion of plant material (Wubah *et al.*, 1991). These fungi are obligate anaerobes, and are remarkable for the short duration of their life cycle (between meals taken by the sheep or cow) and the aggressive activity shown by their carbohydrate-degrading enzymes (Wood *et al.*, 1995).

27

Basidiobolus is a fascinating case, because on the one hand it lacks flagella and phenotypically had been classified in the Entomophthorales (Zygomycota), but on the other, its SSU rDNA places it right in the middle of the flagellated chytrids (Nagahama *et al.*, 1995). Could it have lost its flagellum 'recently'? A decade before the nucleic acid comparison, McKerracher and Heath (1985) found an organelle that looked like the base of a chytrid flagellum (basal body or kinetosome, also acting as a centriole at nuclear division) in the cytoplasm of *Basidiobolus*. No study of any other zygomycete has uncovered any organelle similar to a basal body or centriole, so it seems likely that *Basidiobolus* has had a relatively recent loss of its flagellum. The phylogenetic moral of the story of flagella and Chytridiomycota is not to base monophyletic groups on the presence or absence of shared ancestral characters, in this case flagella.

Continuing up the tree, we come to the most recently diverged group of the coenomycetes, the Glomales (Zygomycota); in this position the Glomales is the sister group to the most complex fungi, the Ascomycota plus Basidiomycota (Figure 2.3). The Glomales are also an old group of fungi judging by their apparent preservation in the rhizomes of plants fossilized in the Rhynie chert, *c.* 400 Mya (Pirozynski and Malloch 1975). Together, the chytrids and zygomycetes account for no more than 3 per cent of described fungal species, and as far as human health is concerned, neither group is a major player, although *Basidiobolus* and *Rhizopus* species do infect humans (Kwon-Chung and Bennett, 1992). However, ecologically, some of the most important fungi of all are in the Glomales, which is home to the arbuscular mycorrhizal fungi, which are endomycorrhizal partners with 70 per cent or more of the world's higher plants (Smith and Read, 1997).

Basidiomycota plus *Ascomycota*; the dikaryomycetes

Leaving the coenomycetes behind, we find ourselves on a branch supporting the final two phyla of fungi, the Ascomycota and Basidiomycota (Figure 2.3). Members of each group can grow as yeasts or as hyphae, or both. Mitotic and meiotic spores are produced abundantly, and many species make a fruiting body which protects the developing meiotic spores and aids in their dispersal. Most species are haploid, but following mating they delay nuclear fusion, forming a binuclear or

28

dikaryotic stage, a feature commemorated by the informal name for the combined Ascomycota and Basidiomycota, dikaryomycetes. The hyphae of the dikaryomycetes also have regularly spaced cross walls, or septa. These septa are useful for distinguishing hyphae of coenomycetes from those of dikaryomycetes; their absence helped to identify hyphae as coenomycetes in the Rhynie chert, fossilized some 390 Mya (Kidston and Lang, 1921).

Basidiomycota

The life cycles of Basidiomycota typically contain a long dikaryotic phase, making it necessary for the mitotic divisions of the two haploid nuclei to be synchronized. Often, a specialized hyphal branch, called either a clamp connection or a hook cell, is associated with synchronized mitosis. A fossil of this distinctive structure is known from the woody tissue of a Carboniferous fern, preserved *c.* 290 Mya (Dennis 1969).

In terms of number of species, the Basidiomycota is much larger than the phyla that we have already considered, comprising almost 35 per cent of described fungi. The shared derived morphological character defining the group is the basidium, a meiocyte produced at the end of a hyphal filament. The basidium gives rise to externally produced meiospores (basidiospores) which are forcibly discharged into the air. If the basidiomycete makes a fruiting body, the basidia will be formed at the surface in a fertile layer called a hymenium. There is great variation in the shape of fruiting bodies and in the form of basidia, especially the presence or absence of septa that develop in basidia following meiotic divisions. Historically, the form of the basidium and of the fruiting body have been the most important systematic characters. Comparison of SSU rDNA sequences has defined three lineages, two without fruiting bodies, Urediniomycetes and Ustilaginomycetes, and one with fruiting bodies, Hymenomycetes (Swann and Taylor, 1993) (Figure 2.4). For the present, relationships among the three classes are unresolved, suggesting that a rapid radiation accounted for their origins. The SSU rDNA trees challenge previous classifications based on basidium form (Swann and Taylor, 1995b), and imply that basidial septation is not useful in delimiting the main basidiomycete lineages.

The Urediniomycetes embraces one of the most renowned groups of plant pathogenic fungi, the rusts (Uredinales) as well as the

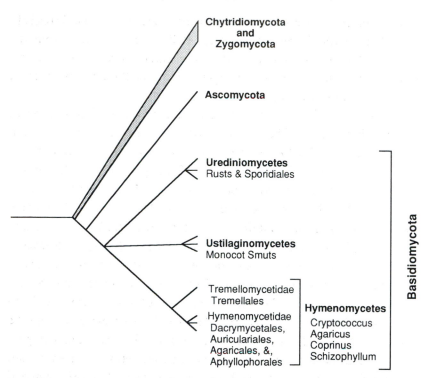

Figure 2.4 Evolutionary tree of Basidiomycota based on nuclear SSU rDNA genes based on Swann and Taylor (1993). The three major lineages, Urediniomycetes, Ustilaginomycetes and Hymenomycetes are mono-phyletic, but resolving their interrelationships is problematic.

Sporidiales, which grow as yeasts (Swann and Taylor 1995a). Also included in this clade are less well-known taxa, including *Septobasidium*, parasite of scale insects, and *Mixia osmunda* (Nishida *et al.*, 1995), recently shown to be a basidiomycete after decades of classification in the ascomycetes. Urediniomycetes lack a fruiting body, but many have a thick-walled spore, the teliospore, which serves as a dispersal agent, a resistant spore, and the cell in which nuclear fusion occurs. When the teliospore germinates, the external basidium and teliospores are formed.

Ustilaginomycetes includes the smuts (Ustilaginales), which is another group of renowned plant pathogens, and the Exobasidiales, less renowned, but no less phytopathogenic. Oddly, one smut-like fungus, *Microbotryum violaceum*, which causes anther smut of Caryophyllaceae,

30

is found on the Urediniomycete clade (Boekhout *et al.*, 1995; Gottschalk and Blanz, 1985; Swann and Taylor, 1995b), well separated from the Ustilaginales. *Microbotryum* is a reminder that even complex morphology and life histories can arise independently. Like the Urediniomycetes, the Ustilaginomycetes have teliospores.

The third class, Hymenomycetes, comprises the basidiomycetes with fruiting bodies and basidia arranged in hymenia. There are two basal branches, one leading to the Tremellales (jelly fungi) and the other to the Dacrymycetales (also having gelatinous fruiting bodies (Wells, 1994)), Auriculariales (tree ear fungi (Berres *et al.*, 1995)), Agaricales (mushrooms) and Aphyllophorales (shelf fungi (Hibbett and Donoghue, 1995)). The Agaricales contain many ectomycorrhizal fungi and, along with the Aphyllophorales, many plant pathogens. Although ectomycorrhizae are less common than endomycorrhizae, they are extremely important for hardwood and coniferous trees in temperatre forests (Smith and Read, 1997). The Aphyllophorales is home to the wood decay fungi, both the brown rots that specialize in cellulose degradation and the white rots that can decompose both cellulose and lignin. Recent comparison of nucleic acids shows that the boundaries between the Agaricales and Aphyllophorales are not real, and that fungi with hymenia arranged into gills, tubes and other forms have arisen independently many times (Hibbett and Vilgalys, 1993; Hibbett *et al.* 1997). The most severe basidiomycete human pathogen, *Filobasidiella neoformans* (= *Cryptococcus neoformans*) is found in the Tremellales (Swann and Taylor, 1995b).

Ascomycota

The last fungal phylum is the Ascomycota, sister group to the Basidiomycota and the other member of the dikaryomycetes. The Ascomycota is the largest phylum of fungi, having almost 45 per cent of described species and also laying claim to most of the asexual dikaryomycetes (which have been classified in a form taxon, the deuteromycetes), another 20 per cent or so of described fungi. Many of the most destructive plant pathogens are ascomycetes, as are nearly all of the severe animal pathogens. Although there are several groups of mycorrhizal ascomycetes (LoBuglio *et al.*, 1996), none is as important as the Glomales (Zygomycota) or Agaricales (Basidiomycota). Compensating for the lack of root symbionts are about 13 500 species symbiotic with algae, the lichenized fungi, accounting for about one-third of the Ascomycota. The morphological character that defines the Ascomycota is the ascus, a meiocyte that differentiates at or near the ends of highly branched dikaryotic hyphae or within yeast cells, and in which are formed the meiospores (ascospores). All asci lack internal septa and almost all asci have at least four spores and usually eight (due to a post-meiotic mitosis). Ascus shape varies considerably and is correlated with the ability of the ascus to forcibly eject its spores: ascospores are typically ejected from elongated, fusiform asci and are passively released from globose asci.

Near the base of the Ascomycota lies the first group of ascomycetes to diverge, the Archiascomycetes, a non-monophyletic group discovered through comparison of SSU rDNA sequences (Berbee and Taylor, 1993; Nishida and Sugiyama, 1994) (Figure 2.5). There are some surprises in this group. A prominent member, *Schizosaccharomyces pombe*, the fission yeast, turns out not to be closely related to *Saccharomyces cerevisiae*, the other model yeast (Taylor *et al.*, 1993). Second, *Pneumocystis carinii*, formerly thought to be a protozoan, but having the life cycle and morphology of a yeast, lies here (J. W. Taylor *et al.*, 1994). Third, *Neolecta vitellina*, a filamentous ascomycete that makes a large fruiting body (apothecium) with the hymenium of asci broadly exposed, also lies in the Archiascomycota (Landvik *et al.*, 1993; Figure 2.5). Not so surprising is the presence in the Archiascomycetes of *Taphrina* and *Protomyces* spp., plant pathogenic fungi have often been noted to possess basidiomycete-like character states (Berbee and Taylor, 1993; Nishida and Sugiyama, 1993; Savile, 1955).

Moving along the main ascomycete clade, we come to a dichotomy between the Hemiascomycetes and the euascomycetes (Figure 2.5). The Hemiascomycetes lack fruiting bodies and most of them exist as yeasts. The Hemiascomycetes is home to the most famous fungus, *Saccharomyces cerevisiae*, as well as *c.* 300 other described yeast species including the widespread human pathogen, *Candida albicans* (Kurtzman 1994). Not all Hemiascomycetes are exclusively yeasts, e.g. even *Saccharomyces cerevisiae* can produce a filament of sorts, and some species are predominately filamentous, i.e. *Galactomyces geotrichum* and *Dipodascus albidus*.

The sister group to the Hemiascomycetes is the euascomycetes, an

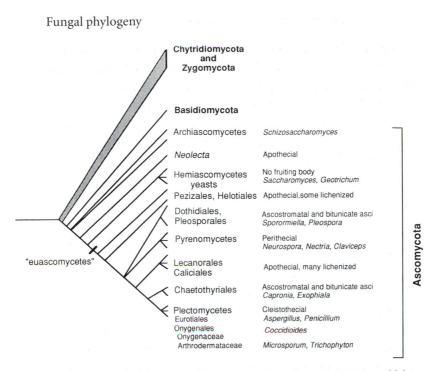

Figure 2.5 Evolutionary tree of Ascomycota based on work in several labs (e.g. Gargas *et al.*, 1995; Gargas and Taylor, 1995; LoBuglio *et al.*, 1996; Spatafora and Blackwell, 1993; Spatafora *et al.*, 1995). Some traditional Ascomycota divisions have proved to be monophyletic (Plectomycetes, Pyrenomycetes), but others have not (Discomycetes and Loculoascomycetes) and are better treated as monophyletic orders. Fruiting body type (if applicable) and some regenerative genera are given.

informal name for the ascomycetes that are filamentous and possess fruiting bodies. However, filamentous growth is clearly an ancestral character, and at least one ascomycete, *Neolecta*, developed a fruiting body before the divergence of the euascomycetes. *Neolecta* is a puzzle, because it suggests that fruiting bodies developed prior to the divergence of the Hemiascomycetes, making it likely that fruiting bodies were lost in this group of yeasts. Of course, it is possible that fruiting bodies arose twice, once with *Neolecta* and again in the euascomycetes. Perhaps careful comparison of *Neolecta* and euascomycete fruiting bodies will provide the answer. Within the euascomycetes, classes traditionally have been based on fruiting body morphology; comparison of SSU and large subunit (LSU) rDNA has shown some of the traditional classes of filamentous ascomycetes to be monophyletic, and others not.

Species diverging from the base of the euascomycetes have their asci arranged in exposed hymenia covering parts of open fruiting bodies, or apothecia. These fruiting bodies provide more evidence that the first type of ascomycete fruiting body was the apothecium. The asci are thin walled and under pressure, so that the ascospores are forcibly discharged as a group of eight into the air above the apothecium. Synchronization of discharge creates a draught that increases the height of discharge (Ingold 1965). There are several clades of apothecial taxa, which correspond to morphologically based orders and families (Pezizales, Helotiales, Lecanorales, Caliciales) (Gargas et al., 1995; Gargas and Taylor, 1995). Although individual clades may be well supported, relationships among them have not been resolved with rDNA sequences, as might be the case if they represented a rapid, early radiation of euascomycetes. Amongst the apothecial clades are other clades whose members have fruiting bodies that are more closed than apothecia. For example, the Pyrenomycetes have flask shaped fruiting bodies with narrow openings in the neck (perithecia), and the Plectomycetes have completely closed apothecia (cleistothecia) (Spatafora and Blackwell, 1993). Members of the Pyrenomycetes have thin-walled asci that forcibly eject the ascospores, but asci of the Plectomycetes are no longer elongated and the ascospores are not expelled from the globose asci. In addition, there are at least two clades of ascomycetes with partially closed fruiting bodies and two-walled asci. The brittle outer ascus wall breaks to permit the elastic inner one to extend to the opening in the neck of the fruiting body, whereupon the ascospores are discharged one by one. These clades, (1) Dothidiales plus Pleosporales and (2) Chaetothyriales, belong to the morphologically based Loculoascomycetes, but they are not sister groups, so the traditional Loculoascomycetes is not monophyletic (Berbee, 1996; Bowen et al., 1992; Spatafora et al., 1995) (Figure 2.5). The basic concept of an early radiation of apothecial euascomycetes followed by independent closings of the apothecium leading to subsequent radiations may hold sway for quite a while, but the exact relationships of species within well-supported clades, and among the clades, will undoubtedly change regularly as more

34

rDNA sequence becomes available for as yet unstudied taxa, and as sequence is obtained from other nucleic acid regions, i.e. genes coding for proteins. Euascomycetes have distinctive ascospores and mitospores, and a diversity of these spores have been found in the lower Cretaceous, c. 170 Mya, indicating that these fungi were well established by that time (Pirozynski and Weresub, 1979). Now would be a good time to point out that fungal fossils and comparison of fungal nucleic acid sequence have the potential to provide dates for major events in the evolution of fungi. Preliminary attempts have been made (Berbee and Taylor, 1993; Simon *et al.*, 1993), and one resulting tree is shown in Figure 2.6.

When introducing the Ascomycota, we mentioned that sexual reproduction has not been observed for almost 20 per cent of described fungi, most of them closely related to sexual Ascomycota, with a few having basidiomycetous affinities. When morphological features of sexual reproduction were the sole characters for systematics of dikaryomycetes, i.e. before comparative studies of rDNA sequence were commonplace, a separate classification for asexual dikaryomycetes was necessary (Reynolds and Taylor, 1993). Because all fungi have nucleic acids, and because comparison of nucleic acid sequence has proved very useful in fungal systematics, there is no longer any need to classify asexual fungi separately, and we have not done so (Taylor, 1995). For example, fungi such as *Candida albicans*, *Penicillium chrysogenum*, *Bipolaris maydis* or *Exophiala dermatitidis* are all asexual, but all can be, and have been, placed on trees and in higher taxa along with sexual Ascomycota (i.e. Saccharomycetales, Eurotiales, Pleosporales and Chaetothyriales, respectively).

The future

This concludes our brief tour up the tree of life, which covered just the main branches that harbour fungi in the loose sense, i.e. those organisms studied by mycologists. We started at the divergence of the domains Bacteria, and Archae plus Eukarya, and then moved along the Eukarya past the slime moulds (Acrasiomycota, Myxomycota, Dictyosteliomycota and Plasmodiophoromycetes) arriving at the 'crown group'. In one branch of the crown, the Stramenopiles, we visited the phyla Labyrinthulomycota, Hyphochytriomycota and Oomycota, all relatives of the

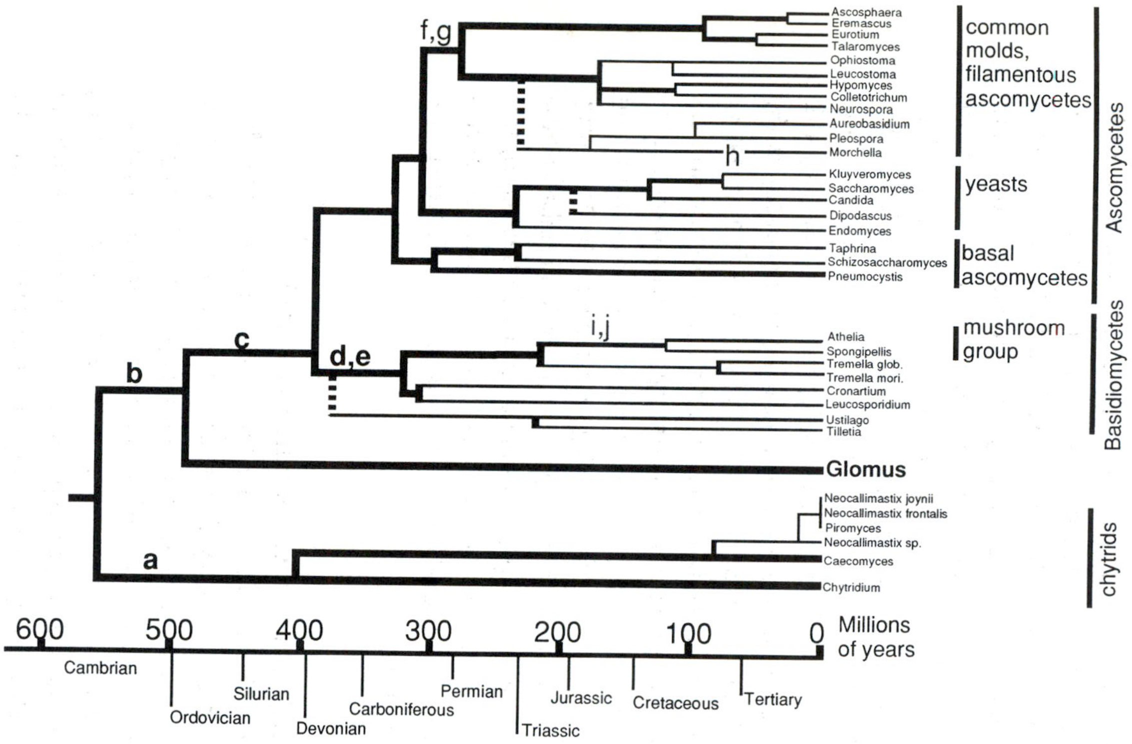

Figure 2.6 Evolutionary tree based on nuclear SSU rDNA sequences fitted to the geological time scale (Berbee and Taylor, 1993). Branch lengths are proportional to the average rate of nucleotide substitution (1 per cent per 100 million years) corrected for substitution rate variation among lineages. The average substitution rate was calibrated with fossil fungi, or fossil fungal hosts or symbionts. Letters on branches correspond to the origins of fungal morphological features (a, zoospores; b, hyphae; c, hyphal septa; d, clamp connections; e, basidia; f, Ascomycota mitospores; g, cylindrical asci that forcibly discharge their ascospores; h, complex Ascomycota fruiting bodies; i, basidia of Agaricales and Aphyllophorales; j, mushrooms).

brown and golden-brown algae. Finally, we travelled through the kingdom Fungi, home to the phyla Chytridiomycota, Zygomycota, Basidiomycota and Ascomycota. At the end of the tour, you may have the impression that comparative molecular biology has a lot more work to do if the fewer than 100 SSU rDNA fungal sequences that now exist are to be expanded to the 60000 fungi that have been described (Hawksworth *et al.*, 1996), or to the 1 to 1.5 million fungi that are postulated to exist (Hawksworth, 1991). The situation is not as dire as it may appear because many smaller clades of fungi are being studied, particularly at the genus level. As a result many of the economically important genera (*Penicillium* [LoBuglio *et al.*, 1993], *Fusarium* [O'Donnell *et al.* 1998], *Aspergillus* [Geiser *et al.*, 1996; 1998], etc.) have developing trees. Accessing this information, which as you can imagine is changing very quickly, is the new challenge. Textbooks such as this cannot be expected to keep pace with the rapid pace of fungal molecular evolution, and as more taxa are added to comparisons, even major branches may change position. For the most current information, the Internet is probably the best source. Two sites that are currently useful for mycologists are the NCBI Taxonomy database <http://www3.ncbi.nlm.nih.gov/Taxonomy/tax.html> and the Tree of Life Project <http://phylogeny.arizona.edu/tree/phylogeny.html>, and new ones should be found at Mycology Resources <http://www.keil.ukans.edu/~fungi/> or by searching the Internet with appropriate key words.

Phylogenetics: how trees are made and tested

Phylogenetics is the study of genealogical relationships and of historical patterns of evolutionary change among organisms. One of the underlying assumptions in phylogenetics is that patterns of change are recorded in the characters inherited by organisms from earlier generations. The record of change can be embedded in characters ranging from the morphological to genotypic features of DNA and protein sequences. Typically, phylogenetic relationships are diagrammed in the form of dichotomously branching trees generated, usually, by one of several computer algorithms. In view of the revolution that the incorporation of molecular data is having on our understanding of fungal phylogenetics, this section aims to explain how trees are generated for such data.

Why bother with phylogenetics?

Both practical concerns and pure curiosity drive phylogenetic studies. Curiosity has led to research on topics varying from identifying the closest relatives of the morphological oddities of the fungus world to evaluating coevolution of ants and basidiomycetes. Most phylogenetic studies, though, address practical concerns including classification of economically important fungi or detection of fungi of medical or environmental importance.

Examples of curiosity-driven phylogenetics   Is sexual reproduction necessary for long-term evolutionary survival in the fungi? Roughly 5000 species of fungi are known only from their asexual states. Some genera, including *Penicillium*, include species reproducing sexually as well as numerous species known only from their asexual states. If sexual reproduction is unnecessary, then the asexual lineages may have originated long ago and perhaps radiated to produce new, clonal species. Alternatively, if sexual reproduction is necessary, then asexual species would have high rates of extinction and extant asexual species would be of recent origin. In the case of *Penicillium*, LoBuglio *et al.* (1993) found evidence for recent and repeated origin of asexual species in phylogenetic trees from ribosomal DNA sequence data. The lack of ancient, radiating asexual lineages supported the hypothesis that sexual reproduction is necessary for long-term evolutionary survival.

How do parasites originate? The Laboulbeniales constitute an order of minute parasitic fungi that attach themselves to the outside surfaces of exoskeletons of arthropods. Unlike most fungi that grow until nutrients are exhausted, the Laboulbeniales produce a fixed number of cells, a complex little fruiting body, and then growth stops. Close morphological study (Blackwell and Malloch, 1989) and phylogenetic analysis of DNA sequence data (Blackwell, 1994) suggested that the Laboulbeniales may be related to *Pyxidiophora*, another odd genus relying on mites and bark beetles for transportation to its nutrient source of dung or dung fungi. Blackwell (1994) suggested that the Laboulbeniales may have evolved from fungi like *Pyxidiophora*, switching from using arthropods as agents of dispersal to sources of food.

Practical reasons for studying phylogeny   In taking a comparative approach, phylogenetic studies contribute to intelligent transfer of information from the three or four very well known 'model systems' fungal species to the hundreds of thousands of other less tractable, less well-known fungi. Experimental research has focused particularly on a single fungus, the famous baker's yeast, *Saccharomyces cerevisiae*. Not only has the whole genome of baker's yeast been sequenced, but in addition, a great deal is known about the function of many of the genes and the proteins. In contrast to the thousands of available gene sequences for *S. cerevisiae*, the international sequence database GenBank in July 1996 contained 153 nucleotide sequences for *Pneumocystis carinii*. *P. carinii* is an unculturable obligate pathogen and a major cause of death by lung infection in AIDS patients. Originally believed to be a protist, phylogenetic trees from ribosomal RNA sequence data revealed *P. carinii* to be an ascomycetous fungus (Edman *et al.*, 1988). Taking advantage of the phylogenetic information, Ludewig *et al.* (1994) investigated the mode of action of pentamidine, an antifungal drug used in treating *P. carinii* infections in *S. cerevisiae*. In *S. cerevisiae*, pentamidine disrupted respiratory growth. *P. carinii* and *S. cerevisiae* are both ascomycetes and the chances are good that the drug would function the same way in both organisms but conducting the experiments was far easier in *S. cerevisiae*.

Even more closely related to *S. cerevisiae* is *Candida albicans*, the fungus that causes candidiasis including vaginal yeast infections. Genetic analysis is difficult in *Candida albicans* since it does not produce a sexual state. *S. cerevisiae*, in contrast, reproduces sexually on command and is an ideal organism for genetic analysis. Heidler and Radding (1995) cloned and sequenced a gene associated with resistance to an antifungal drug, aureobasidin A. Instead of investigating gene function in *C. albicans*, Heidler and Radding (1995) searched the international computer genetic databases to find the *S. cerevisiae* homologue to the *C. albicans* gene. They demonstrated that gene deletion was fatal in *S. cerevisiae* and they concluded that the drug aureobasidin A may work by targeting the homologous gene in *C. albicans*.

39

Phylogenetic approaches using fungal DNA sequences have contributed to ecological studies through species identification in studies of mycorrhizal fungi, the fungi involved in mutually beneficial symbiotic associations with plant roots. Since mycorrhizal fungi usually produce only hyphae when colonizing plant roots and these hyphae are often unculturable, identification based on morphology has been difficult or impossible. To infer the identity of mycorrhizal fungi, Bruns and Gardes (1993) isolated DNA from colonized roots and sequenced the internal transcribed spacer regions of the fungal ribosomal RNA genes and generated phylogenetic trees showing the relationships between the gene sequences of the unidentified fungus associated with the roots and the sequences of identifiable, sporulating mushrooms. Even when the fungal sequences associated with the roots did not match sequences of known species, the phylogenetic trees revealed the mycorrhizal species as members of familiar families or genera.

Testing theories of coevolution

For the past 50 years (more or less), biologists have invoked coevolution to explain, for example, why related mutualists or parasites associate with related hosts. Phylogenetics allows a critical test of coevolutionary theory because, if the organisms are really engaging in evolutionary tracking, speciation in one group of organisms should result in speciation in the coevolving partner. In a perfect case of coevolution, taking a tree showing the relationships of one group of organisms and replacing the name of each organism with the name of its coevolving species would produce an accurate phylogenetic tree for all the coevolving species.

Coevolution: the case of the fungus-farming ants Tropical leaf cutter ants in several genera cultivate fungi that transform leaves into a digestible, proteinaceous food source. The ants require the fungus for survival. To ensure a supply of fungus for her brood, the queen ants in some species carry a fungus pellet to start a new fungus garden as they head out to found a new nest. Have ants and fungi coevolved, with the fungus dependent on the ants for propagation, so that ant speciation resulted in fungal divergence and vice versa? Or is the fungus/ant

40

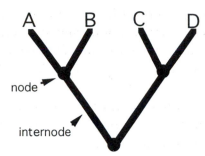

Figure 2.7 A, B, C and D represent taxa, joined at nodes and connected by internodes.

association more casual, with ants 'adopting' a variety of different fungi? Chapela *et al.* (1994) compared a fungus phylogeny from 28S rRNA gene sequences with an ant phylogeny based on morphological characters. In several cases, the branching order of the tree of fungi differed from the branching order of the associated ants. Conclusion? Coevolution among ants and fungi was not absolute. Although most of the leaf cutter ants preferred fungi in the genus *Lepiota*, at least some species of ants switched *Lepiota* species during their evolutionary history.

The structure of phylogenetic relationships

Phylogenetic trees show the ancestor descendant relationships among genes, species, or other taxonomic groups. Trees consist of 'terminal taxa', usually living species, connected by a network of internodes and nodes. The 'terminal taxa' may be taxonomic groups other than species, such as genera or families, or they may be genes or proteins. Since they are not necessarily taxa, alternative names for the terminal units include 'operational taxonomic units'.

Species trees: nodes and internodes (Figure 2.7) Branches in a tree join to form nodes. The nodes represent the transition from a recombining ancestral population to two divergent species with independent evolutionary fates. Even if the terminal taxa in a tree are supraspecific groups like genera or even kingdoms, each node can be interpreted as a point of divergence between two species.

The internodes in a tree, the parts of the branches between the nodes, represent lineages evolving through time. Internode lengths may be arbitrary, or they may be proportional to the number of reconstructed changes that occurred in the evolving lineage.

Trees are completely resolved and dichotomizing when each node is attached to three branches, one representing the ancestral species and two representing diverging species with independent evolutionary histories. Not all trees are completely resolved, or completely dichotomous, however. Polytomies, consisting of four or more internodes radiating from one node most frequently indicate uncertainty about branching order due to insufficient or contradictory information. Polytomies may also indicate that a single, interbreeding ancestral population gave rise to more than two new species in a short period of time.

Gene trees If the terminal taxa are genes, a node represents a transition from a single population of the gene to two diverging lineages of the gene, either because the organisms carrying the gene have speciated, or because the gene has duplicated within a species and the two copies are evolving independently (Figure 2.8). Copies of genes that are found at a homologous locus in different organisms are termed orthologous genes. Duplicated genes occupying separate loci within a species are termed paralogous genes.

Rooting and hierarchical relationships Trees may be either rooted or unrooted (Figure 2.9). In unrooted trees, the direction of evolution is unspecified. Relationships among taxa in unrooted trees are only partially specified because a pair of taxa arising from a common node may be either closely related (if the node represents a recent split) or highly divergent (if the node represents an ancient split). Although rooted and unrooted trees can be topologically identical, rooted trees have much higher information content because they specify the direction of transfer of genetic information.

Rooting to specify the direction of evolution and common ancestry for taxa often comes after establishing the topological relationships among taxa. Imagine that branches in the tree are pieces of rope and nodes are knots. Mechanically, rooting is analogous to choosing one of the knots and dragging it to the bottom of the tree. The knot at the

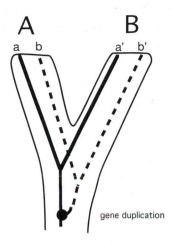

Figure 2.8 Origin of orthologous and paralogous genes. Paralogous genes begin with a gene duplication in an ancestral species. Gene duplication is indicated in the diagram by the split between the solid line and the dotted line. Taxa A and B each bear two copies of the gene. The two copies in A, labelled a and b, are paralogues. Similarly, a' and b' in B are paralogues. The genes a and a' are orthologues that diverged from one another when species A and B diverged. b and b' constitute a second pair of orthologues.

bottom of the tree corresponds to the first divergence among the taxa in the tree. The most distantly related taxa in the tree are the two groups that originated from the first divergence in the tree. The most recent divergences are represented by nodes near the tips of the tree and the closest relatives in the tree are pairs of terminal taxa coming from the same node. In rooted trees, one internode represents the evolving ancestral lineage, and the remaining two internodes represent its two descendent 'sister taxa'. Because they evolved from the same ancestor, sister taxa are equal in age. The equal age of sister taxa can be useful in inferring the geological ages of taxa. Given a reliable phylogeny, a fossil representing one taxon automatically provides a minimum age for the sister taxon and all of the evolving lineages that preceded it. For example, only basidiomycetes have clamp connections and so Dennis (1969), on finding clamp connections in Carboniferous coals, inferred that basidiomycetes were at least 300 million years old. Since ascomycetes are the sister taxon to the basidiomycetes, ascomycetes must also be at least 300 million years old. Chytrids

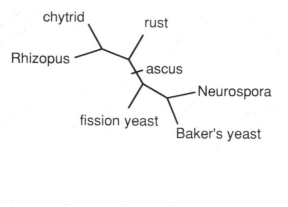

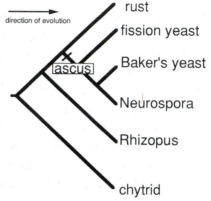

Figure 2.9 Rooted and unrooted trees. The top, unrooted tree does not reveal the direction of evolution, define monophyletic groups, or suggest the direction of character state changes. In this tree, baker's yeast could be more closely related either to the fission yeast or to *Neurospora*. The presence of an ascus could be either a primitive character lost from the rust, Chytrid and *Rhizopus*, or an advanced character state that originated in the ancestor to *Neurospora* and the two yeasts. In the tree at the bottom, the chytrid was chosen as an outgroup to root the tree. The rooted tree specifies that the baker's yeast is more closely related to *Neurospora* than to the fission yeast. Further, the direction of evolution of the ascus is clear. The ascus is a derived character that evolved in the common ancestor to the three ascomycetes, the fission yeast, baker's yeast and *Neurospora*.

and zygomycetes diverged before the split between ascomycetes and basidiomycetes and so they must also have originated more than 300 million years ago (Berbee and Taylor, 1993).

Identifying the root   Determining which divergence occurred first often involves an element of subjectivity. Even in molecular phylogenetic studies, initial assumptions about rooting are often based on morphology. One strategy involves choosing a rooting consistent with an *a priori* assumption about the direction and cost of character evolution (Figure 2.10).

A second strategy involves rooting using one or more outgroups. An 'ingroup' is the monophyletic group of taxa under consideration. 'Outgroups' are other taxa, ideally near relatives of the ingroup but not part of the ingroup. For the kingdom Fungi, possible outgroups would include an animal, plant, or protist (Figure 2.10). The basal node of the rooted tree would be point of divergence of the outgroup from the fungi. Initial assumptions about an outgroup can sometimes be tested by adding another outgroup with an even more distant phenotype. The initial outgroup should then appear as the second most divergent lineage in the tree. For example, the second divergence in an animal and fungal tree with bacteria as the outgroup should fall between the fungi and the animals, supporting the choice of animals as an outgroup for fungi.

A third strategy, applicable to nucleotide and amino acid data, involves assuming that sequences change at approximately the same rate in the organisms in a study. This is the assumption of a molecular clock. If the rate of change is approximately constant, then the taxa that differ by few substitutions must have diverged from a common ancestor recently. The taxa that differ by the most substitutions must have diverged from a common ancestor long ago. Placing the deepest divergence in terms of per cent substitution at the base of the tree effectively roots the tree (see Figure 2.14).

Rooting gene trees   Trees of a duplicated gene can be rooted at the point of gene duplication, with paralogous copies of the gene in the same organism serving as outgroups for orthologous copies of the gene (Figure 2.11). Using this approach Iwabe *et al.* (1989) inferred that the first split

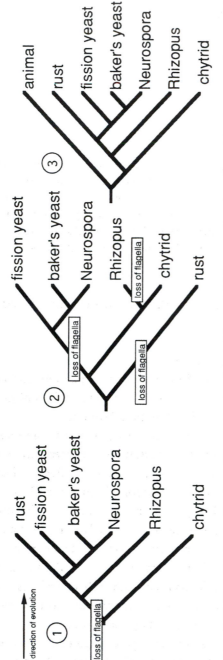

Figure 2.10 Strategies for rooting trees. One strategy involves making an initial assumption about how characters can evolve. Assume, for example, that flagella can be lost but not gained and that trees minimizing the number of losses should be preferred. Since the chytrids are fungi and have flagella, the ancestor to all fungi must also have had flagella. Because it requires only one flagellar loss, rooting between the chytrid and the other taxa (tree 1) is better than rooting between the rust and the other taxa (tree 2), which requires three flagellar losses. Tree 3 shows rooting of a fungal tree using an animal as an outgroup.

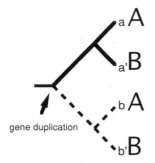

Figure 2.11 Rooting using a duplicated gene. As in Figure 2.2, the diagram at the left shows taxa A and B. Genes a and b are paralogues, genes a and a' constitute the first pair of orthologues and genes b and b' constitute the second pair of orthologues. Since taxa A and B both appear twice in the tree, gene duplication must have preceded speciation. The tree can be rooted at the most ancient node, which must be the point of gene duplication.

in the tree of living organisms divides the Bacteria from the Archaea plus Eukarya.

Trees, groups and classification

To maximize their power of prediction, classifications should be phylogenetic. The next three paragraphs discuss three kinds of phylogenetic groups and the significance of each in classification.

Monophyletic groups. A monophyletic group is an evolving ancestral species and all of the taxa descended from it. In a rooted tree, a monophyletic group begins with a node or an internode and includes all of the nodes, internodes and terminal taxa originating from that node or internode (Figure 2.12). Monophyletic groups are the basic units of taxonomy and in this chapter, we are using the words 'clusters' of taxa or 'clades' to specify monophyletic groups. The predictive value of a taxonomic system is based on the assumption that members of monophyletic groups, sharing a common genetic origin, will share other characteristics as well. The kingdom Fungi, including chytrids, zygomycetes, ascomycetes and basidiomycetes, is an example of a monophyletic group.

Paraphyletic groups. A paraphyletic group includes an ancestral species and some, but not all of its descendants (Figure 2.12). In a rooted

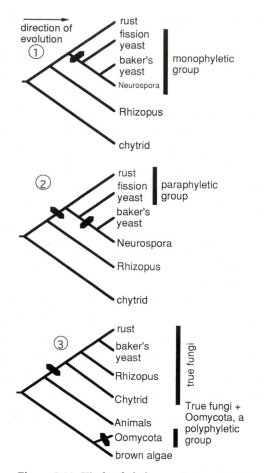

Figure 2.12 Kinds of phylogenetic groups. Monophyletic groups include an ancestor and *all* of its descendants. In tree 1, the fission yeast, baker's yeast and *Neurospora* form a monophyletic group. Paraphyletic groups include an ancestor but not all of its descendants. In the middle tree, the rust plus the fission yeast form a paraphyletic group that includes the ancestor to both taxa but not baker's yeast and *Neurospora*, the other two taxa descended from the same ancestor. Polyphyletic groups, diagrammed in the bottom tree, do not include the common ancestor to all the taxa in the group. The true fungi and Oomycota were at one time all considered fungi. However, the common ancestor to the true fungi and the Oomycota is represented by the bottom node in the tree. Since the organism at the base of the tree would be an ancient protozoan and not a fungus, the grouping of true fungi plus the Oomycota is polyphyletic.

tree, a paraphyletic group begins at a node or an internode and includes some, but not all, of the branches and terminal taxa extending from the ancestral node. Typically, the taxa that are excluded from the paraphyletic group evolved enough distinctive characters so that they no longer resemble the common ancestor. Fishes, for example, are paraphyletic because they are all derived from the same ancestral fish, but the amphibians, reptiles, birds and mammals derived from the same fish ancestor are not considered to be fish. Similarly, water moulds in the Chytridiomycota are paraphyletic. The Chytridiomycota share a common ancestor which also gave rise to the terrestrial zygomycetes, ascomycetes and basidiomycetes, groups that are not included in the Chytridiomycota. Paraphyletic groups do not reveal which species are most closely related to the derived taxa. As a result, most taxonomists, when revising classification systems, reduce the numbers of paraphyletic groups and increase the numbers of monophyletic groups where possible.

Polyphyletic groups. A group is polyphyletic if it does not include the ancestor to all group members (Figure 2.12). In a rooted tree, the taxa represented by the most recent node common to all members of the polyphyletic group would not be included in the group. Polyphyletic groups are often defined based on similar characteristics that arose independently in different groups. Because polyphyletic groups include unrelated taxa, they are usually eliminated from classification systems when they are detected.

From groups to classification

Standard classification involves clustering fungi into hierarchical, ranked groups: species, genera, families, orders, classes, phyla and kingdoms and domains. Ideally, the ranked groups in a classification should all be monophyletic so that the genetic relationships of all included organisms are specified. In reality, classification systems are far from ideal, largely because the genetic relationships among organisms are incompletely known. New characters and techniques for inference of relationships often reveal paraphyly or even polyphyly in groups originally considered monophyletic.

To maximize efficiency of information storage and retrieval, classifications should also be stable and the names should remain constant over time. Not too surprisingly, the goals of stability and of a phylogenetically based system often conflict. Most older taxonomies, including Whittaker's (1969) five kingdom system, classify the Oomycota, the slime moulds and the true fungi together in one phylum and a single kingdom. As you now know, more recent classifications, reflecting an influx of phylogenetic information, have reduced the kingdom Fungi by transferring the slime moulds to the protist kingdom and the Oomycota to the Stramenopila. Changes to classifications driven by progress in phylogenetic understanding will ultimately improve the taxonomic system although the transitional period of taxonomic plasticity will be confusing. Phylogenetically based systems of classification available over the Internet may prove helpful in tracking recent taxonomic changes

Making and testing phylogenetic trees
Phenotypic characters as evidence for relationships

For hundreds of years, fungal phylogeny and taxonomy were based almost entirely on morphology. Morphological characters remain defining features for many fungal groups. Characters can have one or more alternative character states. To be phylogenetically useful, the various states of a character must be homologous, sharing a common evolutionary origin. In addition, the characters must be variable so that state changes provide a record of phylogenetic events. An example of a particularly useful morphological character is the fungal meiocyte, with two alternative character states: (1) the ascus with internal spore production and (2) the basidium with external spore production.

For formal phylogenetic analysis, character states are usually coded in a uniform way recognized by the computer software. Presence of asci might be coded as 1 for example, presence of basidia as 2 and the absence of either asci or basidia as 0. Polarity, or the evolutionary direction of character state changes, can be specified or not, depending on the kind of analysis and the intended interpretation of the results. Initial specification of polarity of character state changes is reasonably common in parsimony analysis of morphological characters. For example, assuming that the presence of asci is an irreversible, derived character state $0 = >1$ but not $1 = >0$ changes are permitted in the tree. Character state changes can also be ordered without specifying the

50

direction or polarity of change. Specifying order or polarity of character change increases the information content in data sets and resolution levels in trees, but the tree cannot be used later as evidence to evaluate whether the assumed character changes actually occurred without introducing circular reasoning.

DNA sequence-based studies support the traditional view that the ascus is a derived character that evolved in the ancestor to the ascomycetes. Similarly, the basidium is a derived character marking the monophyletic group, the basidiomycetes. Interpretation of many other morphological characters has been more difficult. Convergence and homology are difficult to distinguish in simple characters like the presence of a yeast state or production of asexual spores by budding. Most of the available characters are useful over a limited taxonomic range. Further, the total number of morphological characters for the fungi is small.

Sequence characters as evidence for relationships

DNA and protein sequences support highly resolved, statistically testable phylogenetic hypotheses. The virtues of sequences in phylogenetic reconstruction stem from the abundance and the predictability of available data. For example, the *Neurospora crassa* genome includes about 47 million nucleotides (Orbach *et al.*, 1988). Sequence characters, just because they are numerous, are likely to provide a reconstructable record of changes revealing ancestry not only among species, but also at higher taxonomic levels.

Since all cellular organisms have the same basic molecular genetic components, understanding of sequence evolution in one organism can often be generalized to others. Homologous sequences often evolved at similar rates in diverse organisms. For example, because they evolve quickly, sequences of the internal transcribed spacer regions in ribosomal RNA genes have proved useful for reconstructing phylogenies of closely related species from a wide variety of fungi. The same regions are also useful for species of red algae and flowering plants. Further, DNA and amino acid substitutions are relatively easy to model based on experience with many sequences in numerous organisms. Morphological changes, in contrast, were often rare events restricted to a few lineages of organisms and therefore difficult to model. A combination of high numbers of

51

characters and availability of models of change, as is found in nucleic acids, facilitate design of tests of statistical support in sequence-based trees.

Databases as repositories of sequence data

The initial cost of sequencing DNA or proteins remains high. However, once determined, sequences are usually deposited in international databases and then used repeatedly in various studies by different authors. Many journals will not publish a sequence-based study until the authors have submitted their sequences to a database. The international databases, National (United States) Center for Biotechnology Information (NCBI) GenBank, European Molecular Biology Laboratory (EMBL) and the DNA Database of Japan (DDBJ), SWISS-PROT and PROTEIN SEQUENCE DATABASE of PIR-International exchange data regularly so that all sequences are available from each database.

Finding sequences in the databases

Sequences can be found in any of the international databases by searching for an organism name, for a gene or for the unique code assigned to each sequence by the databases (e.g. NCBI http://www.ncbi.nlm.nih.gov/). In addition NCBI allows BLAST searches through its 'Entrez browser' world wide web site at http://www3.ncbi.nlm.nih.gov/Entrez/index.html. With a BLAST search, researchers submit one sequence or a region of a sequence to the nucleic acid or proteir databases. The NCBI computers then search through the databases for other sequences that are more similar to the original query sequence than expected by chance. The output of a BLAST search is a list of sequences, arranged in order of similarity to the original sequence. Information associated with the similar sequences often provides insight into the function or identity of the query sequence.

Generating trees

Data matrices for phylogenetic analysis

Most analyses of phylogenetic information involves generating a data matrix and submitting the data matrix to an appropriate computer program for inference of a tree. A typical data matrix for phenotypic

STATE LABELS
    meiocyte: 0=thick walled resting spore, 1=ascus, 2=basidium
    dikaryon: 0=absent, 1=present
    growth: 0=thallus, 1=yeast 2=hyphal
    habitat: 0=aquatic, 1=terrestrial
    perithecium: 0=absent, 1=present

	meiocyte	dikaryon	growth	habitat	perithecium
chytrid	0	0	0	0	0
baker's yeast	1	0	1	1	0
Neurospora	1	1	2	1	1
fission yeast	1	0	1	1	0
rust	2	1	2	1	0

Data matrix of aligned DNA sequence data

	1 1111111112 2222222223 33333 1234567890 1234567890 1234567890 12345
chytrid	AATTGTCAGA GGTGAAATTC TTGGATTTAA TTAAG
baker's yeast	AATTGTC?G? GGTGAAATTC TTGGATTTAT TGAAG
Neurospora	AATTGTCAGA GGTGAAATTC TTGGATTTAT TGAAG
fission yeast	AATTGTCAG? GGTGAAATTC TTGGATTTAT TGAAG
rust	CATCGTCAGA GGTGAAATTC TTGGATTGAT GTAAG

Figure 2.13 Data matrices. For morphological data, the possible character states are usually defined, coded and entered into a table. In a data matrix of DNA nucleotides or amino acids, sequences from diverse organisms are aligned so that homologous characters form the vertical columns.

characters consists of a column of taxon names to the left, followed by columns of coded character states for a succession of characters (Figure 2.13).

If the characters are nucleic acid or amino acid sequence characters, the sequences are aligned so that each site, or column in the alignment corresponds to a homologous position in the sequences of different taxa. A variety of computer software packages varying in power and price are available to help with this critical task. Among alignment programs, ClustalW is free, works well on a variety of kinds of computers and can be obtained at http://iubio.bio.indiana.edu. In analyses of DNA sequences, the 'cost' assigned to some changes may be higher than the 'cost' of other

UPGMA, distance matrix from 18S rRNA genes from five fungal species.

	sacch	neuro	schizo	chytrid	rust
sacch	0	0.0826	0.0513	0.0931	0.1231
neuro		0	0.0787	0.1182	0.1324
Schizo			0	0.0746	0.1126
chytrid				0	0.1321
rust					0

UPGMA, recalculated distance matrix from first cycle

	sacch/schizo	neuro	chytrid	rust
sacch/schizo	0	0.0807	0.0839	0.1179
neuro		0	0.1182	0.1324
chytrid			0	0.1321
rust				0

UPGMA, recalculated distance matrix from second cycle

	sacch/schiz/neuro	chytrid	rust
sacch/schiz/neuro	0	0.0953	0.1227
chytrid		0	0.1321
rust			0

Last cycle

	sacch/schiz/neuro/chyt	rust
sacch/schiz/neuro/chyt	0	0.1251
rust		0

UPGMA tree

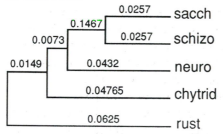

Figure 2.14 Constructing a tree using UPGMA. The top box shows a distance matrix from 18S rRNA gene sequences for five taxa. The distances have received a Kimura correction for multiple hits. UPGMA begins by clustering the most similar pair of taxa in the data matrix. The length of the branch from the node to each taxon is equal to half of the difference between the two taxa. In this example, the lowest distance is 0.0513,

54

changes, but no assumption is made about the polarity or order of character state change.

Analysis of data and generating phylogenetic trees

Methods of inferring trees fall into two classes, distance methods and discrete character methods (Hillis *et al.*, 1996). Distance methods include UPGMA, Fitch-Margoliash and Neighbour-joining. These methods use a distance matrix (Figure 2.14), rather than a data matrix, as their input. Some distance methods, like UPGMA and Neighbour-joining, use an exact algorithm to infer a tree from the distance matrix. Others, like Fitch-Margoliash, use an exact algorithm to infer a tree from a matrix but then explore the fit of the data matrix to alternative tree topologies.

Discrete character methods including parsimony and maximum likelihood use a data matrix as input. The data matrix can be a matrix of morphological character states or a DNA or amino acid sequence alignment. The discrete character methods superimpose each character state change on as many trees as possible.

Figure 2.14 (*cont.*)
between *Saccharomyces cerevisiae* (sacch) and *Schizosaccharomyces pombe* (schizo). 'Sacch' and 'schizo' become the first pair of terminal taxa and the branch length to each taxon is 0.0513/2 or 0.0257 units.

After the initial pairing, the distance matrix is recalculated. Distances between each of the other taxa and sacch and schizo are replaced by the average of the distance of each taxon from sacch and from schizo. For example, the distance from sacch to *Neurospora crassa* (neuro) is 0.0826. The distance from schizo to neuro is 0.0787. Averaging the two distances gives 0.0807, the recalculated distance from neuro to the new 'conglomerate' taxon schizo/sacch. Dividing 0.0807 in half gives 0.0432, the distance from the new node to neuro. Subtracting the branch length to sacch (0.0257) from 0.0432 gives 0.1467, or the distance from the new node to the node for schizo and sacch. In the second cycle, the recalculation of the data matrix is repeated, calculating average distances of each taxon from sacch, from schizo and from neuro. Each recalculation produces a new data matrix that is one row and one column shorter than the original matrix. The process of clustering and recalculating the data matrix continues until only a single pair-wise comparison remains in the data set and all taxa are included in the tree.

Table 2.1. *Corrections for multiple hits*

Observed %			Corrected %	
Transitions	Transversions	Total substitution	Jukes-Cantor	Kimura's
2.50%	1.25%	3.75%	3.85%	3.85%
1.25%	2.50%	3.75%	3.85%	3.86%
20.00%	30.00%	50.00%	82.40%	93.24%
30.00%	20.00%	50.00%	82.40%	83.11%
40.00%	20.00%	60.00%	120.71%	120.71%

In general, if the data set contains the signal or information necessary to correctly resolve branching order, the particular algorithm used to generate a tree has little influence on the outcome. Differences in outcome related to differences in the algorithm used to generate the tree sometimes reflect insufficient information in the original data set. Infrequently conflicts between trees generated by different methods result from more serious failure of the data to match the assumptions implicit in a particular method.

Distance methods in phylogenetic analysis

Calculating a distance matrix and correcting for multiple hits
Before applying a distance algorithm, a distance matrix must be generated from the initial data matrix. The 'distances' can be calculated as the proportion of differences between pairs of taxa relative to the total number of characters. Distances for DNA sequence data or protein data are usually corrected for multiple hits, or repeated substitutions at the same site. Jukes and Cantor (1969) suggested a simple and widely used correction for DNA sequence data based on the assumptions that substitutions follow a Poisson distribution and that all kinds of substitutions are equally likely. In Table 2.1, the 'observed substitutions' columns represent the proportion of differences directly counted between two hypothetical DNA sequences. The corrected percentages are estimates of the total numbers of substitutions that occurred, including the historical substitutions that cannot be counted directly because they had been concealed by new substitutions at the same site.

56

When observed percentage differences between taxon pairs are low, the probabilities of repeat substitutions at the same sites are also low and corrections for multiple hits make essentially no difference (Table 2.1). However, as the per cent substitution between a pair of sequences increases, a higher and higher proportion of the new substitutions conceal older substitutions at the same sites. When the per cent substitution exceeds 50 per cent, the corrected per cent substitution becomes much higher than the observed percentage. The inferred proportion of multiple hits makes up the difference between the observed proportion of substitution of '50 per cent' and the Jukes and Cantor corrected proportion of '82.4 per cent'.

Other corrections for multiple hits use essentially the same logic, but assume more complex (and sometimes more realistic) evolutionary models. The observed differences between sequences include transitions, or replacements of purines by other purines ($A \leftrightarrow G$ substitutions) and pyrimidines by pyrimidines ($C \leftrightarrow T$ substitutions). All other substitutions are transversions or replacements of purines by pyrimidines and vice versa. Typically, the transitions are most common. Unlike the Jukes and Cantor correction, the Kimura (1980) correction can take into account different transition/transversion frequencies. For the Kimura correction in Table 2.1, we assumed that transitions were twice as probable as transversions. When the observed percentage of transitions is less than twice the percentage of transversions, the Kimura correction assumes that the missing transitions were hidden by the transversions. As a result, the reconstructed percentage is sometimes higher with the Kimura correction than with the Jukes and Cantor correction.

Computer software for generating *corrected* distance matrices from aligned DNA or protein sequences are available in the PHYLIP (Felsenstein 1993), MEGA (Kumar *et al.*, 1993) and PAUP* (Swofford, 1998) packages.

UPGMA Conceptually, UPGMA, or 'unweighted pair group method with arithmetic means' is the simplest of the distance methods and it clusters taxa successively based on similarity (Figure 2.14). UPGMA assumes that an evolutionary clock operated, or in other words, that changes accumulated at a constant rate among all lineages. Averaging distances in the recalculation of the data matrix reflects this assumption of constant evolutionary rates. A side-effect of the clock assumption is polarization of the tree. The most similar taxa are assumed to have diverged most recently and are clustered together at the tips of the tree. The most divergent taxa originated earliest and emerge from the base of the tree. In this way, unlike most tree-building algorithms, UPGMA produces rooted trees.

If rates of evolutionary change are approximately the same for all lineages, UPGMA will recover the correct tree and accurate branch lengths. However, in reality, evolutionary rates often vary. When rates vary, UPGMA reconstructs the wrong branch lengths and possibly the wrong rooting and wrong branching order. In the example in Figure 2.14, the UPGMA tree roots the fungal tree incorrectly with a rust rather than with the flagellated chytrid. Because the rust lineage evolved more quickly than the others, UPGMA placed it at the base of the tree.

UPGMA does not take into account all available information about branch lengths. In Figure 2.14, the UPGMA branch length between the baker's yeast (sacch) and the fission yeast *Schizosaccharomyces pombe* (schizo) comes from a single distance in the distance matrix. However, information on the length of the branch to 'sacch' is also embedded in the distance between 'neuro' and 'sacch', the chytrid and 'sacch', and the rust and 'sacch' since each of these pairs of distances covers the distance from 'sacch' to its node. Considering all of the sources of information, as in the next two distance methods, leads to more accurate phylogenetic reconstruction under a broader range of evolutionary conditions.

Fitch-Margoliash   The Fitch-Margoliash algorithm (Fitch and Margoliash, 1967) is like UPGMA in its approach to finding a tree topology, but a molecular clock is not assumed. Branch lengths are calculated based on all contributing information embedded in the data set and the resulting tree is not rooted. After finding a tree, an optimization routine proceeds, 'trying' different topologies until the branch lengths calculated from the algorithm are as close as possible to the numbers calculated in the distance matrix. For a worked example of the Fitch-Margoliash algorithm and optimization, see Weir (1996). The computer package PHYLIP includes FITCH, a program implementing the Fitch-Margoliash algorithm. Because it includes an optimization routine, FITCH is slow compared with other distance programs. However, FITCH performs well in terms of recovering the correct tree in computer simulation experiments.

Neighbour-joining The neighbour-joining algorithm (Saitou and Nei, 1987), like UPGMA and Fitch-Margoliash algorithms, begins with a distance matrix. Neighbour-joining works by clustering taxa to minimize the total distance in the tree. First, a star phylogeny shows each taxon as equally distant from every other taxon (Figure 2.15). The initial distance in the star is calculated. Then, the neighbour-joining algorithm temporarily removes a pair of taxa from the star, clustering the pair at a new node. The total distance in the tree is recalculated for the tree that now includes two nodes (Figure 2.15). The first pair of taxa is returned to the star and the next pair of taxa is clustered at a new node. This process continues until the total distance in the tree has been recalculated with every possible pair of taxa forming the new node. To infer a tree for five taxa, this first round of recalculation would involve 10 distance recalculations.

If a pair of taxa are closely related, then placing them together decreases the total distance in the tree because the distance between the pair is small and the distance from their shared node to the centre of the star is large but is counted only once. The neighbour-joining algorithm searches through the recalculated distances and permanently joins the pair of taxa that produced the lowest total distance. The process of removing each pair of taxa in succession from the central star and then permanently joining the pair that most reduces the distance across the whole tree continues until all of the taxa are joined in a completely resolved tree. Branch lengths are calculated as with the Fitch-Margoliash algorithm. For a worked example of application of the neighbour-joining algorithm, see Swofford et al. (1996).

The neighbour-joining algorithm can be implemented using programs in the PHYLIP, MEGA, and PAUP* computer packages. Neighbour-joining does not necessarily find the tree that minimizes the total distance among taxa since it only explores a few of the possible tree

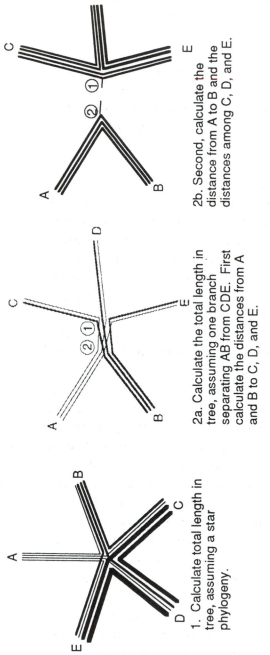

1. Calculate total length in tree, assuming a star phylogeny.

2a. Calculate the total length in tree, assuming one branch separating AB from CDE. First calculate the distances from A and B to C, D, and E.

2b. Second, calculate the distance from A to B and the distances among C, D, and E.

Figure 2.15 Calculating tree length in neighbour-joining. The neighbour-joining algorithm first calculates the total distance in a tree, assuming a star phylogeny. The letters in this diagram designate taxa and the lines indicate distances measured between taxa. The fine lines, for example, indicate the distances from taxon A to each of the other four taxa. The next coarsest lines indicate the distance from B to the remaining three taxa. To find the total distance in the star phylogeny, sum all pair-wise distances, e.g. $AB + AC + AD + AE + BC + BD + BE + CD + CE + DE$. Note that four 'lines' lead to each taxon, indicating that adding together all possible pairs of distances has covered the total distance in the tree four times (the number of taxa minus one). Dividing the summed distance by four therefore gives the total distance in the star. After calculating the total distance for the star phylogeny, the next step is to introduce a second node connecting a pair of taxa and recalculate the total length of the tree. In this example, in 2a, taxa A and B share a new node (2) while C, D, and E remain with the original node (1). Finding the total distance in the tree now involves first (2a) adding together the distance from A to C, from A to D, from A to E and from B to C, B to D and B to E. In this way, the distance from A to B is measured three times, from node 1 to node 2 is measured six times, and the distance from node 2 to each of the other three taxa is measured two times. So that each distance in the tree is measured six times (2b), three times the distance from A to B and twice the distance from C to D, from C to E and from D to E is added into the total. Dividing the total by six gives the length of the new tree.

topologies. However, it does not assume an evolutionary clock, it is very fast and it performs well on many data sets under varied conditions.

Discrete character methods

Two discrete character methods, parsimony and maximum likelihood, both superimpose data on trees and choose the tree that best represents the data. Parsimony selects the tree that minimizes the number of character state changes. Maximum likelihood selects the tree that maximizes the probability of the data.

Parsimony Parsimony analysis is commonly applied to both morphological and sequence data. As illustrated in Figure 2.16, the minimum number of character state changes needed to explain the original data is calculated for possible trees. The trees requiring the fewest character state changes are considered optimal (i.e. the most parsimonious). Individual characters are termed 'informative' in parsimony analysis when they require different numbers of changes on different trees. To be informative, an unpolarized character must have at least two character states and each state must be present in at least two taxa. Character state changes are often called steps and the total number of character state changes required for the tree is the length of the tree.

As the number of taxa increases, the number of possible trees increases even faster (e.g. there are three possible unrooted trees for four taxa, but more than 2.2×10^{20} for 20 taxa (Felsenstein, 1978)) and finding the most parsimonious trees for a given data set can be a computationally intensive task. No exact algorithm exists for uniting taxa so that the most parsimonious tree results. Instead, finding the most parsimonious trees involves superimposing data on as many trees as possible and, as in Figure 2.16, choosing the tree that results in fewest character state changes. Branch and Bound searches maximize the efficiency of finding all equally parsimonious trees with a quick, initial determination of a length for a reasonably parsimonious tree from a particular data set, followed by a much slower search for better or equally good trees. The Branch and Bound search systematically examines all possible combinations of taxa that do not exceed the length of the shortest tree found up to that point. Although the number of taxa that can reasonably be included in a search

61

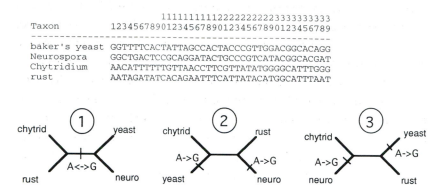

```
                         1111111111222222222233333333333
 Taxon                   12345678901234567890123456789
 -------------------------------------------------------
 baker's yeast  GGTTTTCACTATTAGCCACTACCCGTTGGACGGCACAGG
 Neurospora     GGCTGACTCCGCAGGATACTGCCCGTCATACGGCACGAT
 Chytridium     AACATTTTTTGTTAACCTTCGTTATATGGGGCATTTGGG
 rust           AATAGATATCACAGAATTTCATTATACATGGCATTTAAT
```

Figure 2.16 Parsimony analysis. This figure shows a data matrix including only the informative sites extracted from an alignment of 18S rRNA sequence data. The three, topologically distinct unrooted trees that can be drawn for the four taxa in the alignment are diagrammed below the matrix. To explain the data in the first column on the first unrooted tree, a single nucleotide substitution would be required. The substitution would have occurred in the evolving lineage represented by the branch between the two nodes. The ancestor to the whole group could equally parsimoniously have had a 'G' or an 'A' at this position. If, for example the ancestor had a A in position one, then an A to G substitution occurred in the lineage that gave rise to the yeast and the *Neurospora*. To explain the data at the first site on the second tree, a minimum of two substitutions would have been required. Either two independent substitutions of G for A occurred in the yeast and the *Neurospora* lineages, or two independent substitutions of A for G occurred in the chytrid and rust lineage. Similarly, tree 3 requires a minimum of two substitutions. Based on the information content in the first site, tree 1 would be the most parsimonious tree.

However, not all of the sites in the same data matrix support the same tree. Of the 39 informative sites in the data matrix, 21 support tree 1, requiring a single substitution on that tree and a minimum of two substitutions on tree 2 and tree 3. On the other hand, five of the 39 informative sites support tree 3 and require more substitutions on each of the other two trees. When the information conflicts, the tree requiring the fewest changes overall is considered the most parsimonious. Adding up all of the changes required to explain the data, tree 1 requires 57 substitutions or steps; tree 2 requires 65 steps and tree 3 requires 73 steps. Tree 1 is the most parsimonious tree over all the sites in the data matrix. It also is in agreement with prevailing, morphological taxonomy as the two ascomycetes, the yeast and the *Neurospora*, cluster together as compared with the chytrid and the rust.

keeps increasing as computers become more powerful, Branch and Bound searches remain computationally impractical for numbers of taxa included in many standard phylogenetic analyses. Heuristic searches are much faster, but they do not necessarily find all equally parsimonious trees or even the most parsimonious tree (Maddison 1991). Heuristic searches find a reasonably parsimonious tree quickly and then search for other equally or more parsimonious trees by switching the position of branches. As a result, heuristic searches may produce a tree that is shorter than neighbouring trees (a local optimum), but that is longer than a more distant tree with a fundamentally different topology.

Parsimony may be used as a criterion to map changes onto a tree. If two sister taxa have a derived character state and all of the other taxa in the tree have the primitive state of the same character, it is most parsimonious to assume that the change occurred only once, and that the change took place in the evolving ancestor of the two sister taxa. Maddison and Maddison (1992) provide a straightforward discussion of some of the potential uses and considerations in character state reconstruction. It is also possible to map phenotypic characters onto a tree made from nucleic acid data to investigate the evolution of particular characters, e.g. ascospore discharge (Berbee and Taylor, 1992a) or meiotic reproduction (LoBuglio *et al.*, 1993).

Maximum likelihood Maximum likelihood is concerned with the probability of a character changing state along a branch of a tree. Because there may be many characters and many branches, the formal description is more complex. As Felsenstein (1984) explains:

likelihood turns out to be:

- [i] The product over all characters, of
- [ii] the sum, over all ways that character states could be assigned to the unobserved forks on the phylogeny, of
- [iii] the product over all segments of the phylogeny, of
- [iv] the probability, for the given character, of changing from the state specified at the start of the segment to the state specified at the end of the segment.

Calculating maximum likelihood requires an initial estimate for part [iv] above of the probability of all possible character state changes.

Data matrix

taxon	A	B	C
character 1	0	1	1
character 2	0	1	0

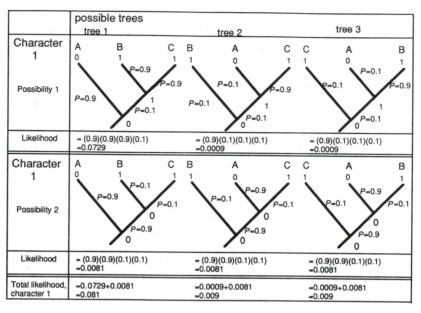

	possible trees		
	tree 1	tree 2	tree 3
Character 1 Possibility 1	(tree diagram)	(tree diagram)	(tree diagram)
Likelihood	= (0.9)(0.9)(0.9)(0.1) =0.0729	= (0.9)(0.1)(0.1)(0.1) =0.0009	= (0.9)(0.1)(0.1)(0.1) =0.0009
Character 1 Possibility 2	(tree diagram)	(tree diagram)	(tree diagram)
Likelihood	= (0.9)(0.9)(0.1)(0.1) =0.0081	= (0.9)(0.9)(0.1)(0.1) =0.0081	= (0.9)(0.9)(0.1)(0.1) =0.0081
Total likelihood, character 1	=0.0729+0.0081 =0.081	=0.0009+0.0081 =0.009	=0.0009+0.0081 =0.009

	possible trees		
	tree 1	tree 2	tree 3
Character 2 Possibility 1	(tree diagram)	(tree diagram)	(tree diagram)
Likelihood	= (0.9)(0.9)(0.1)(0.1) =0.0081	= (0.1)(0.1)(0.1)(0.1) =0.0001	= (0.9)(0.9)(0.1)(0.1) =0.0081
Character 2 Possibility 2	(tree diagram)	(tree diagram)	(tree diagram)
Likelihood	= (0.9)(0.9)(0.9)(0.1) =0.0729	= (0.1)(0.9)(0.9)(0.9) =0.0729	= (0.9)(0.9)(0.9)(0.1) =0.0729
Total likelihood, character 2	=0.0081+0.0729 =0.081	=0.0001+0.0729 =0.073	=0.0081+0.0729 =0.081
Total likelihood, both characters	=(0.081)(0.081) =0.00656	=(0.009)(0.073) =0.000656	=(0.009)(0.081) =0.000729
Ln likelihood	−5.0266	−7.328	−7.224

Fungal phylogeny

Applied to DNA sequences, maximum likelihood methods determine the probability of substitution based on the level of substitution in the data, the relative frequencies of the four nucleotides, and the different probabilities of transitions and transversions.

Likelihood analysis tends to be computationally intensive because it involves calculation of the probabilities of all possible ways that the observed data could have been generated (in parts [i], [ii] and [iii] above) given a particular tree (Figure 2.17). As in parsimony, there is no exact algorithm for assembling the most likely tree. To find an optimal tree, the

Figure 2.17 How maximum likelihood works: a simple example (thanks to David Maddison, personal communication, August 1996). In this example, the model for evolution specifies two character states. 0 is the ancestral character but reversals are permitted. The probability of change along each branch of the tree is the same. The probability of a single character state change is 0.1. The probability of a character state remaining the same is 0.9. Likelihood is calculated by drawing all three possible trees and then drawing all possible combinations of character state changes, consistent with the data matrix and the constraints of the model. For the first character on the first tree, the first node must have the character state 0, because the model specifies that 0 is ancestral. The second node could be either 0 or 1. Assuming that the second node is 1, the first tree would require a single character state change between the first and second node, which would have a probability of 0.1. No character state changes need be assumed on the branches to the terminal taxa. The probability of no character state change on each of the terminal branches is 0.9. The likelihood for the tree, given a 1 at the second node is $(0.9)(0.9)(0.9)(0.1) = 0.0729$. Of course, instead of a 1 at the first node, a 0 could have been present. If so, character state changes would have had to take place along two of the four branches of the tree, giving a likelihood of $(0.9)(0.9)(0.1)(0.1) = 0.0081$. The likelihood for the first character, given the first tree, would be the sum of the likelihoods of the two possible ways the character could have come about, equal to 0.0729 + 0.0081, or 0.081. A similar calculation gives the likelihood for all three possible trees and for both characters. The likelihood of both character 1 and character 2, given tree 1 is the product of the individual likelihoods, or 0.081×0.081, is equal to 0.00656. Because the number of decimal places in likelihoods becomes difficult to manage, log likelihoods are usually presented. The natural logarithm of 0.00656 is -5.0266.

calculation of the likelihood must be repeated for all possible trees. As in heuristic searches in parsimony analysis, computer programs that calculate likelihood, including PAUP*, DNAML in PHYLIP (Felsenstein, 1993) and fastDNAml 1.0 (Olsen *et al.*, 1993) calculate and present the most likely tree that it was feasible to find. Where parsimony analysis programs will give the 'length' or the number of steps required in the most parsimonious tree, likelihood programs will give the natural logarithm of the likelihood of the tree. The higher the log likelihood (the less negative the log likelihood), the more likely are the data given the tree.

Evaluation of trees

Any of the algorithms used in phylogenetic analysis can generate a dichotomously branching tree from completely random data. But, do the nodes and internodes in a tree reflect chance events, or do they result from shared evolutionary history? Tests of trees evaluate how strongly the data support a tree and distinguish a random pattern of data distribution from a pattern resulting from phylogenetic information. Four approaches to testing trees are explained briefly in the next section.

Bootstrapping

Bootstrap percentages (Felsenstein, 1985; Hillis and Bull, 1993) often accompany published phylogenetic trees. The percentages provide a sense of whether more data similar to the original data would support the same branching order (Figure 2.18). Bootstrapping works by creating replicate data sets through sampling from among the columns of characters in the original data set randomly and with replacement. After having created usually 500–1000 replicate data sets, one or more trees are inferred from each replicate data set. The trees can be inferred using any available algorithm. The percentage associated with a branch in a tree is the percentage of times that the cluster of taxa at the end of the branch appeared among all the trees generated from all the replicate data sets.

Since not all of the characters from the original data set are used in each replicated data set, the particular characters that support a branch from the original data may or may not be included in the new data set (Figure 2.18). When many uncontradicted characters support a branch, the chances are good that almost all the replicated data sets will include at

66

```
Original Data:

        site #    1 2 3 4 5 6 7 8 9 10
BAKER'S YEAST     G G T T T T C A C  T
NEUROSPORA        G G C T G A C T C  C
CHYTRID           A A C A T T T T T  T
RUST              A A T A G A T A T  C
```

```
Two bootstrapped data sets:
          site # 1 2 3 3 3 4 4 6 9 9
BAKER'S YEAST     G G T T T T T T C C
NEUROSPORA        G G C C C T T A C C
CHYTRID           A A C C C A A T T T
RUST              A A T T T A A A T T

          site # 1 1 1 2 2 2 3 3 4 4
BAKER'S YEAST     G G G G G G T T T T
NEUROSPORA        G G G G G G C C T T
CHYTRID           A A A A A A C C A A
RUST              A A A A A A T T A A
```

Partitions found in one or more
trees and frequency of occurrence:

```
1234    Freq
------------
..**    466.50
.*.*     33.50
```

Bootstrap 50% majority-rule consensus tree

```
/----------------- BAKER'S YEAST(1)
+----------------- NEUROSPORA(2)
|          /--------- CHYTRID(3)
\--93---+--------- RUST(4)
```

Figure 2.18 Bootstrapping. This figure shows the original data set at the top and then two of the 500 bootstrap data sets created from the original data. All three data sets contain the same number of columns, but each boot-strapped set may include any given column from the original data once, more than once, or not at all. The second data set, for example, contains the third column three times but lacks the fifth column from the original data set. Using parsimony, trees were inferred from each bootstrap data set. The tree at the bottom of the figure is a consensus of all the trees drawn from all 500 bootstrap data sets. The 93 in the figure indicates that baker's yeast and *Neurospora* clustered together in 93 per cent of 500 bootstrap replicates. The partition table shows that in 33.5 trees, or the remainder of the 500 repli-cates, *Neurospora* and the rust clustered together instead.

least some of these supporting characters and the branch will receive a high bootstrap percentage. Bootstrapping is a rough, non-statistical measure of the extent of support from the data for many branches in a tree. Percentages over 95 per cent indicate that the data strongly support the branch. Bootstrap percentages under 50 per cent indicate little support for the branch; in fact, most of the trees from the replicate data sets lacked the branch.

Bremer decay indices

Bremer decay indices (Bremer 1988) can only be applied to parsimony analysis and are not applicable in a statistical sense to establish confidence intervals on branches. Like bootstrap percentages, decay indices are useful in providing a general sense of support for branches in a tree. In the most parsimonious trees, taxa form the cluster that minimizes the total number of character state changes. The decay index for a branch gives the number of additional character state changes needed to collapse the branch in a consensus because some of the equally parsimonious trees of the same length have either added taxa to or removed taxa from the cluster at the end of the branch. In other words, by increasing the length of the most parsimonious tree, it is possible to reposition taxa using the additional character state changes or steps. A cluster receiving little support from the data will have a low decay index because adding taxa to it or removing taxa from it would require only a few extra character state changes. Alternatively, a cluster that is strongly supported by the data will have a high decay index because adding to or removing its taxa would require many extra steps. Calculating decay indices for branches involves successively searching for trees one, two and three or more character state changes longer than the most parsimonious tree. The minimum number of extra character state changes associated with the collapse of a branch into a polytomy is the decay index for the branch.

Archie's randomization test

Archie's randomization test is representative of a class of tests that ask whether the data set contains more hierarchical phylogenetic information than expected by chance. Although Archie's test is described for use with parsimony, related tests could be applied with other phylogenetic methods. Replicate data sets are created by randomizing the character states for each character. This maintains the original character state frequencies but removes all phylogenetic structure. The length of the most parsimonious tree from each randomized data set is recorded. To establish the distribution of trees from randomized data sets, the randomization and the tree building procedures are repeated 100 or more times. The length of the most parsimonious tree from the original data set can then be compared with the lengths of trees from the randomized data sets. Archie (1989) suggested testing whether the length of the most parsimonious tree from the original data exceeds $100(1-a)$ per cent of the lengths of most parsimonious trees from randomized data sets, where a is the accepted type I error, or the probability of mistaking noise for signal. Setting a to 0.05 for example, the most parsimonious tree should be shorter than at least 95 per cent of the random trees. If the most parsimonious tree is not shorter than at least 95 per cent of random trees, the data set probably lacks phylogenetic structure. Note that randomization tests detect overall structure in the data set, unlike bootstrapping and decay indices that show support for each branch. Phylogenetic structure could result from strong evidence for close relationship of two taxa in an otherwise noisy data set, or for overall, weak support for the whole phylogenetic tree.

The Kishino-Hasegawa test

The Kishino and Hasegawa (1989) test takes a maximum likelihood approach to statistical comparison of the fit of the data to alternative trees. Kishino and Hasegawa initially used the test to demonstrate statistical support from DNA sequence data for clustering humans with African great apes rather than with orangutans. Mechanically, the test works by calculating the mean and variance of the log likelihood for each site in the alignment and for each tree. If the mean log likelihoods for trees differ by more than two standard deviations, then the fit of the data to the trees is significantly different at the 95 per cent confidence level. To implement the test fairly, the most likely trees consistent with each hypothesis must first be found, to avoid testing aspects of the tree topology irrelevant to the hypothesis. We usually use a slightly less fair implementation, finding instead the most parsimonious tree that satisfies the

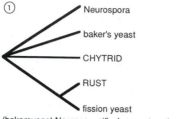

(bakersyeast,Neurospora(fissionyeast,rust),chytrid)

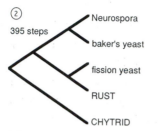

(bakersyeast,(NEURO,((fissionyeast,RUST),CHYTRID)));

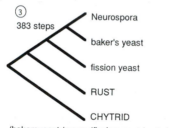

(bakersyeast,(neuro,(fissionyeast,(rust,chytrid))))

Figure 2.19 Testing alternative trees. Tree 1 is a constraint tree for the hypothesis that fission yeast, *Schizosaccharomyces pombe,* is the closest relative of the rust. Representing the constraint, a single branch separates the fission yeast and the rust from the other taxa. Computer programs recognize branching order as shown under each tree, in the form of nested parentheses, where each pair of parentheses represents a cluster in a hierarchical tree. Given a data matrix from 18S rRNA gene sequences and a nested parenthesis constraint tree, the computer package PAUP produced tree 2, the most parsimonious resolution of the constraint. Tree 3 is the unconstrained maximum likelihood tree. Using parsimony to reconstruct changes, the constrained tree 2 required 395 substitutions while the unconstrained tree required only 383 substitutions.

Table 2.2. *Log likelihood of trees 2 and 3 from Figure 2.19*

Tree number	Log	Diff Ln L likelihood	Its S.D.	Significantly worse?
2	−4516.39	−69.56	19.47	Yes
3	−4446.82	←———		Best tree

initial constraint of the hypothesis (Figure 2.19). The computer programs PAUP* DNAML in the PHYLIP package (Felsenstein, 1993) performs the Kishino and Hasegawa tests after the data matrix and the nested parenthesis trees have been inputted into memory. In Table 2.2, we estimated the log likelihood of tree 2 and tree 3 from Figure 2.19. Tree 3 is the most likely tree. In tree 2, the ascomycete fission yeast clusters with a rust instead of with the other ascomycetes. Putting the fission yeast in the 'wrong' place would require a series of nucleotide substitutions that are significantly less probable than the substitutions required to cluster the fission yeast where it belongs, with the other ascomycetes.

References

(Works of general interest are marked with an asterisk)

*Agrios, G. N. (1988). *Plant Pathology*, third edition. Academic Press, San Diego.

*Alexopoulos, C. J., Mims, C. W. and Blackwell, M. (1996). *Introductory Mycology*, fourth edition. Wiley, New York.

Archie, J. W. (1989). A randomization test for phylogenetic information in systematic data. *System. Zool.* 38:239–252.

Baldauf, S. L. and Palmer, J. D. (1993). Animals and fungi are each other's closest relatives: congruent evidence form multiple proteins. *Proc. Natl. Acad. Sci. USA* 90:11558–11562.

Berbee, M. L. (1996). Loculoascomycete origins and evolution of filamentous ascomycete morphology based on 18S rRNA gene sequence data. *Mol. Biol. Evol.* 13:462–470.

Berbee, M. L. and Taylor, J. W. (1992a). Two ascomycete classes based on fruiting-body characters and ribosomal DNA sequence. *Mol. Biol. Evol.* 9:278–284.

Berbee, M. L. and Taylor, J. W. (1992b). Detecting morphological convergence in true fungi, using 18S rRNA gene sequence data. *Biosystems* 28:117–125.

Berbee, M. L. and Taylor, J. W. (1992b). (1993). Dating the evolutionary radiations of the true fungi. *Can. J. Bot.* 71:1114–1127.

Berres, M. E., Szabo, L. J. and McLaughlin D. J. (1995). Phylogenetic relationship in auriculariaceous basidiomycetes based on 25S ribosomal DNA sequences. *Mycologia* 87:821–840.

Blackwell, M. (1994). Minute mycological mysteries: the influence of arthropods on the lives of fungi. *Mycologia* 86:1–17.

Blackwell, M. and Malloch, D. (1989). Pyxidiophora (Pyxidiophoraceae) : a link between the Laboulbeniales and hyphal Ascomycetes. *Mem. N. Y. Bot. Gard.* 49:23–32.

Boekhout, T., Fell, J. W. and O'Donnell, K. (1995). Molecular systematics of some yeast-like anamorphs belonging to the Ustilaginales and Tilletiales. *Stud. Mycol.* 38:175–183.

Bowen, A. R., Chen-Wu, J. L., Momany, M., Young, R., Szaniszlo, P. J. and. Robbins, P. W. (1992). Classification of fungal chitin synthases. *Proc. Natl. Acad. Sci. USA* 89:519–523.

Bowman, B. H., Taylor, J. W., Brownlee, A. G., Lee, J., Lu, S.-D and White, T. J. (1992). Molecular evolution of the fungi: relationship of the Basidiomycetes, Ascomycetes, and Chytridiomycetes. *Mol. Biol. Evol.* 9:285–296.

Bremer, K. (1988). The limits of amino acid sequence data in angiosperm phylogenetic reconstruction. *Evolution* 42:795–803.

Bruns, T. D. and Gardes, M. (1993). Molecular tools for the identification if ectomycorrhizal fungi: taxon specific oligonucleotide probes for Suilloid fungi. *Mol. Ecol.* 2:233–242.

Bruns, T. D., White, T. J. and Taylor, J. W. (1991). Fungal molecular systematics. *Annu. Rev. Ecol. Syst.* 22:525–64.

Bruns, T. D., Vilgalys, R., Barns, S. M., Gonzalez, D., Hibbett, D. S., Lane, D. J., Simon, L., Stickel, S., Szaro, T. M., Weisburg, W. G. and Sogin, M. L. (1992). Evolutionary relationships within the fungi: analyses of nuclear small subunit rRNA sequences. *Mol. Phylog. Evol.* 1:231–241.

Cavalier-Smith, T. (1981). Eukaryote kingdoms: seven or nine? *Biosystems* 14:461–481.

*Chapela, I. H., Rehner, S. A., Schulz, T. R. *et al.* (1994). Evolutionary history of the symbiosis between fungus-growing ants and their fungi. *Science* 266:1691–1694.

Dennis, R. L. (1969). Fossil mycelium with clamp connections from the Middle Pennsylvanian. *Science* 163:670–671.

72

Edman, J. C., Kovacs, J. A., Masur, H., Santi, D. V., Elwood, H. J. and Sogin, M. L. (1988). Ribosomal RNA sequence shows *Pneumocystis carinii* to be a member of the Fungi. *Nature* 334:519–522.

Felsenstein, J. (1978). Cases in which parsimony or compatibility methods will be positively misleading. *Syst. Zool.* 27:401–410.

(ed.) (1984). The statistical approach to inferring evolutionary trees and what it tells us about parsimony and compatibility *In*: Cladistics: perspectives on the reconstruction of evolutionary history.

(1985). Confidence limits on phylogenies: an approach using the bootstrap. *Evolution* 19:783–791.

(1993). PHYLIP (Phylogeny Inference Package). Available from J. Felsenstein through Department of Genetics, University of Washington, Seattle, USA.

Fitch, W. M. and Margoliash, E. (1967). Construction of phylogenetic trees. *Science* 155: 279–284.

Gargas, A., and. Taylor, J. W (1995). Phylogeny of Discomycetes and early radiations of the apothecial Ascomycotina inferred from SSU rDNA sequence data. *J. Exp. Mycol.* 19:7–15.

*Gargas, A., DePriest, P. T., Grube, M. and Tehler, A. (1995). Multiple origins of lichen symbioses in fungi suggested by SSU rDNA phylogeny. *Science* 268:1492–1495.

Geiser, D. M., Timberlake, W. E. and Arnold, M. L. (1996). Loss of meiosis in *Aspergillus. Mol. Biol. Evol.* 13:809–817.

Geiser, D. M., Frisvad, J. C. and Taylor, J. W. (1998). Evolutionary patterns in *Aspergillus* section Fumigati inferred from partial beta-tubulin and hydrophobin sequences. *Mycologia* 90:831–845.

Gottschalk, M. and Blanz, P. A. (1985). Untersuchungen an 5S ribosomalen ribonukleinsauren als beitrag zur Klarung von systematik und phylogenie der Basidiomyceten. *Z. Mykol.* 51:205–243.

*Hawksworth, D. L. (1991). The fungal dimension of biodiversity: magnitude, significance and conservation. *Mycol. Res.* 95:641–655.

Hawksworth, D., Kirk, P., Sutton, B. and Pegler, D. (1996`). *Ainsworth's and Bisby's Dictionary of the Fungi*, eighth edition. CABI, Wallingford, UK.

Heidler, S. A. and Radding, J. A. (1995). The *AUR1* gene in *Saccharomyces cerevisiae* encodes dominant resistance to the antifungal agent aureobasidin A (LY295337). *Antimicrobial Agents Chemother.* 39:2765–2769.

Hibbett, D. S. and Donoghue, M. J. (1995). Progress toward a phylogenetic classification of the polyporaceae through parsimony analysis of mitochondrial ribosomal DNA sequences. *Can. J. Bot.* 73S:S853–S861.

Hibbett, D. S., Pine, E. M., Langer, E., Langer, G. and Donoghue, M. J. (1997). Evolution of gilled mushrooms and puffballs inferred from ribosomal DNA sequences. *Proc. Natl. Acad. Sci. USA* 94:12002–12006.

Hibbett, D. S. and Vilgalys, R. (1993). Phylogenetic relationships of the basidiomycete genus *Lentinus* inferred from molecular and morphological characters. *Syst. Biol.* 18:409–433.

Hillis, D. M. and Bull, J. J. (1993). An empirical test of bootstrapping as a method for assessing confidence in phylogenetic analysis. *Syst. Biol.* 42:182–192.

*Hillis, D. M., Moritz, C. and Mable B. K. (eds.) (1996). *Molecular Systematics* Sinauer Associates, Sunderland, Mass.

*Ingold, C. T. (1965). *Spore Liberation.* Clarendon Press, Oxford.

Iwabe, N., Kuma, K.-I., Hasegawa, M. *et al.* (1989). Evolutionary relationships of archaebacteria eubacteria and eukaryotes inferred from phylogenetic trees of duplicated genes. *Proc. Nat. Acad. Sci.* USA 86: 9355–9359.

Jukes, T. H. and Cantor, C. R. (1969). Evolution of protein molecules. *Mammal. Protein Metab.* 3:21–132.

Kidston, R. and Lang, W. H. (1921). The fungi of the Rhynie Deposit. *Trans. R. Soc. Edin.* 52:855–902.

Kimura, M. (1980). A simple method for estimating evolutionary rates of base substitutions through comparative studies of nucleotide sequences. *J. Mol. Evol.* 16:111–120.

Kishino, H. and Hasegawa, M. (1989). Evaluation of the maximum likelihood estimate of the evolutionary tree topologies from DNA sequence data, and the branching order in Hominoidea. *J. Mol. Evol.* 29:170–179.

Kumar, S., Tamura, K. and Nei, M. (1993). *MEGA: Molecular Evolutionary Genetics Analysis.* The Pennsylvania State University, University Park, PA 16802, USA.

Kurtzman, C. P. (1994). Molecular taxonomy of the yeasts. *Yeast* 10:1727–1740.

*Kwon-Chung, K. J. and Bennett, J. E. (1992). *Medical Mycology.* Lea & Febiger, Philadelphia.

Landvik, S., Eriksson, O. E, Gargas, A. and Gustafsson, P. (1993). Relationships of the genus *Neolecta* (Neolectales ordo nov., Ascomycotina) inferred from 18s rDNA sequences. *Syst. Ascomycetum* 11:107–118.

Leipe, D. D., Wainright, P. O., Gunderson, J. H., Porter, D., Patterson, D. J.,Valois, F., Himmerich, S. and Sogin, M. L. (1994). The stramenopiles from a molecular perspective: 16S-like rRNA sequences from Labyrinthuloides minuta and Cafeteria roenbergensis. *Phycologia* 33:369–377.

Lichtwardt, R. W. (1986). *The Trichomycetes.* Springer-Verlag, New York.

LoBuglio, K. F., Pitt, J. I. and Taylor, J. W. (1993). Phylogenetic analysis of two ribosomal DNA regions indicates multiple independent losses of a sexual *Talaromyces* state among asexual *Penicillium* species in subgenus *Biverticulum*. *Mycologia* 85:592–604.

LoBuglio, K. F., Berbee, M. L. and Taylor, J. W. (1996). Phylogenetic origins of the asexual mycorrhizal symbiont *Cenococcum geophilum* Fr. and other mycorrhizal fungi among the ascomycetes. *Mol. Phylog. Evol.* 6:287–294.

Ludewig, G., Williams, J. M., Li, Y. *et al.* (1994). Effects of pentamidine isethionate on *Saccharomyces cerevisiae*. *Antimicrobial Agents Chemother.* 38: 1123–1128.

Maddison, D. R. (1991). The discovery and importance of multiple islands of most parsimonous trees. *Syst. Zool.* 40:315–328.

Maddison, W. P. and Maddison, D. R. (1992). *MacClade*. Sinauer Associates, Sunderland, Mass., USA.

Margulis, L. and. Schwartz, K. V (1988). *Five Kingdoms*. W. H. Freeman, New York.

McKerracher, L. J. and Heath, I. B. (1985). The structure and cycle of the nucleus-associated organelle in two species of *Basidiobolus*. *Mycologia* 77:412–417.

Muehlstein, L. K., Porter, D. and Short., F. T. (1991). *Labyrinthula zosterae* sp. nov., the causative agent of wasting disease of eelgrass, *Zostera marina*. *Mycologia* 83:180–191.

Nagahama, T., Sato, H., Shimazu, M. and Sugiyama, J. (1995). Phylogenetic divergence of the entomophthoralean fungi: evidence from nuclear 18s ribosomal RNA gene sequence. *Mycologia* 87:203–209.

Nishida, H. and Sugiyama, J. (1993). Phylogenetic relationships among *Taphrina*, *Saitiella*, and other higher fungi. *Mol. Biol. Evol.* 10:431–436.

(1994). Archiascomycetes: detection of a major new linage within the Ascomycota. *Mycoscience* 35:361–366.

Nishida, H., Ando, K., Ando, Y., Hirata, A. and Sugiyama, J. (1995). *Myxia osumndae*: transfer from the Ascomycota to the Basidiomycota based on evidence from molecules and morphology. *Can. J. Bot.* 73:S660–S666.

O'Donnell, K., Cigelnik, E. and Nirenberg, H. I. (1998). Molecular systematics and phylogeography of the *Gibberella fujikuroi* species complex. *Mycologia* 90:465–493.

Olsen, G. J., Matsuda, H. and Hagstrom, R. (1993). fastDNAml. [Computer program available through its authors.] Urbana, Illinois.

Orbach, M. J., Vollrath, D., Davis, R. W. *et al.* (1988). An electrophoretic karyotype of *Neurospora crassa*. *Mol. Cell. Biol.* 8: 1469–1473.

Pirozynski, K. A. and Malloch, D. W. (1975). The origin of land plants: a matter of mycotrophism. *Biosystems* 6:153–164.

Pirozynski, K. A. and Weresub, L. K. (1979). A biogeographic view of the history of ascomycetes and the development of their pleomorphism. Pp. 93–123 in *The Whole Fungus: the Sexual-Asexual Synthesis.* (B. Kendrick, ed.) National Museums of Canada and the Kananaskis Foundation, Ottawa, Canada.

Reynolds, D. R. and Taylor, J. W. (eds.). (1993). *The Fungal Holomorph: Mitotic, Meiotic and Pleomorphic Speciation in Fungal Systematics.* CAB International, Wallingford, U.K.

Saitou, N. and Nei, M. (1987). The neighbour-joining method: a new method for reconstructing phylogenetic trees. *Mol. Biol. Evol.* 4:406–425.

Savile, D. B. O. (1955). A phylogeny of the basidiomycetes. *Can. J. Bot.* 33:60–104.

Simon, L., Bousquet, J. J., Levesque, R. C. and Lalonde, M. (1993). Origin and diversification of endomycorrhizal fungi and coincidence with vascular land plants. *Nature* 363:67–69.

*Smith, S. E. and Read, D. J. (1997). *Mycorrhizal Symbiosis*, second edition Academic Press, San Diego.

Spatafora, J. and Blackwell, M. (1993). Molecular systematics of unitunicate perithecial Ascomycetes. The Clavicipitales–Hypocreales connection. *Mycologia* 85:912–922.

Spatafora, J. W., Mitchell, T. G. and Vilgalys, R. (1995). Analysis of genes coding for small-subunit rRNA sequences in studying phylogenetics of dematiaceous fungal pathogens. *J. Clin. Mycol.* 33:1322–1326.

*Swann, E. C. and Taylor, J. W. (1993). Higher taxa of Basidiomycetes: an 18S rRNA gene perspective. *Mycologia* 85:923–936.

Swann, E. C. and Taylor, J. W. (1995a). Phylogenetic diversity of yeast-producing basidiomycetes. *Mycol. Res.* 99:205–210.

Swann, E. C. and Taylor, J. W. (1995b). Phylogenetic perspectives on basidiomycete systematics: evidence from the 18S rRNA gene. *Can. J. Bot.* 71 (Suppl.): S862–S868.

Swofford, D. L. (1998). Phylogenetic analysis using Parsimony (*and other methods), Version 4. Sinauer, Sunderland, MA.

Swofford, D. L., Olsen, G. J., Waddell, P. J. *et al.* 1996. Phylogeny reconstruction. Pp. 407–514 *in Molecular Systematics,* second edition. D. M. Hillis, C. Moritz and B. K. Mable. Sinauer Associates, Sunderland, MA.

Taylor, J. W. (1995). Making the Deuteromycota redundant: a practical integration of mitosporic and meiosporic fungi. *Can. J. Bot.* 73 (Suppl.): S754–S759.

Taylor, J. W., Bowman, B., Berbee, M. L. and White, T. J. (1993). Fungal model organisms: phylogenetics of *Saccharomyces, Aspergillus* and *Neurospora. Syst. Biol.* 42:440–457.

Taylor, J. W., Swann, E. and Berbee, M. L. (1994). Molecular evolution of ascomycete fungi: phylogeny and conflict. Pp. 201–211 *in First International Workshop on Ascomycete Systematics,* ed. D. L. Hawksworth. NATO Advanced Science Institutes Series. Plenum Press, New York.

Taylor, T. N., Remy, W. and Hass, H. (1994). Allomyces in the Devonian. *Nature* 367:601.

*Wainright, P. O., Hinkle, G., Sogin, M. L. and Stickel, S. K. (1993). Monophyletic origins of the Metazoa: an evolutionary link with fungi. *Science* 260:340–342.

Walker, W. F. and Doolittle, F. W. (1982). Nucleotide sequences of 5S ribosomal RNA from four oomycete and chytrid moulds. *Nucl. Acids Res.* 10:5717–5721.

Weir, B. S. (1996). *Genetic Data Analysis.* II. *Methods for Discrete Population Genetic Data.* Sinauer, Sunderland MA.

Wells, K. (1994). Jelly fungi, then and now! *Mycologia* 86:18–48.

Whittaker, R. H. (1969). New concepts of kingdoms of organisms. *Science* 163: 150–160.

*Woese, C. R., Kandler, O. and Wheelis, M. L. (1990). Towards a natural system of organisms: proposal for the domains archaea, bacteria, and eucarya. *Proc. Natl. Acad. Sci. USA* 87:4576–4579.

*Wolfe, K. H., Gouy, M., Yang, Y.-W., *et al.* 1989. Date of the monocot-dicot divergence estimated from chloroplast DNA sequence data. *Proc. Natl. Acad. Sci. USA* 86: 6201–6205.

Wood, T. M., Wilson, C. A. and McCrae, S. I. (1995). The cellulase system of the anaerobic rumen fungus *Neocallimastix frontalis*: studies on the properties of fractions rich in endo-(1 fwdarw 4)-beta-D-glucanase activity. *Appl. Microbiol. Biotechnol.* 44:177–184.

Wubah, D. A., Fuller, M. S. and Akin, D. E. (1991). *Neocallimastix*: a comparative morphological study. *Can. J. Bot.* 69:835–843.

3 The yeast *Saccharomyces cerevisiae*: insights from the first complete eukaryotic genome sequence

H. FELDMANN

Introduction

The yeast *Saccharomyces cerevisiae* is arguably one of the most important fungal organisms used in biotechnological processes. Making bread and alcoholic beverages, yeast has served mankind for several thousands of years. Many enzymes and biological compounds useful in biochemical research have been produced from yeast cells. In the mid-1930s yeast was introduced as an experimental system for molecular biology (Roman, 1981) and has since received increasing attention. The elegance of yeast genetics and the ease of manipulation of yeast, and finally the technical breakthrough of yeast transformation to be used in reverse genetics, have substantially contributed to the enormous growth in yeast molecular biology (Broach *et al.*, 1981; Strathern *et al.*, 1981; Guthrie and Fink, 1991). This success is also due to the fact, which was not anticipated a couple of years ago, that the extent to which basic biological structures and processes have been conserved throughout eukaryotic life is remarkable.

Yeast: an experimental system for molecular biology
Yeast is a versatile eukaryotic model organism

It is now well established that yeast is an ideal system in which cell architecture and fundamental cellular mechanisms can be successfully investigated. Among all eukaryotic model organisms, *S. cerevisiae* combines several advantages. It is a unicellular organism which, unlike more complex eukaryotes, is amenable to mass production. It can be grown on defined media giving the investigator complete control over environmental parameters. Yeast is tractable to classical genetic techniques and

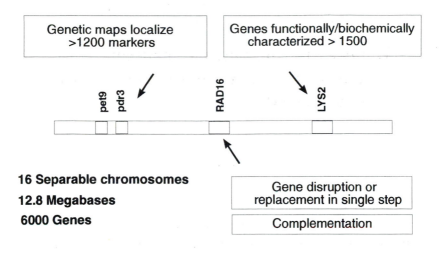

Figure 3.1 Yeast as a versatile model system.

functions in yeast have been studied in great detail by biochemical approaches (Strathern *et al.*, 1981; Broach *et al.*, 1981; Guthrie and Fink, 1991). In fact, a large variety of examples provide evidence that substantial cellular functions are highly conserved from yeast to mammals and that corresponding genes can often complement each other (Figure 3.1).

It is not surprising, therefore, that yeast has again reached the forefront in experimental molecular biology in taking its place as the first eukaryotic organism of which the entire genome sequence has been made available (Dujon, 1996; Goffeau et al., 1996). The wealth of information obtained in the yeast genome project (Goffeau et al., 1997a) will be used as a reference against which sequences of human, animal or plant genes, and those of a multitude of unicellular organisms now under study may be compared (Clayton et al., 1997). Moreover, the ease of genetic manipulation in yeast opens the possibility for functionally dissecting gene products from other eukaryotes in this system.

Experimental approaches in yeast molecular biology

It may be useful to briefly outline some of the characteristics of the yeast system. With its 12.8 Mb, the yeast genome is about 200 times smaller than the human genome but less than four times bigger than that of Escherichia coli. At the onset of the sequencing project, knowledge about some 1200 genes encoding either RNA or protein products had accumulated (Mortimer et al., 1992). The complete genome sequence now defines some 6000 open reading frames (ORFs) most of which are likely to encode specific proteins. A protein-encoding gene is found every two kb in the yeast genome, with approximately 70 per cent of the total sequence being covered (Mewes et al., 1997). In addition to the protein-encoding genes, the yeast genome contains some 120 ribosomal RNA genes in a large tandem array on chromosome XII, 40 genes encoding small nuclear RNAs (sRNAs), 274 tRNA genes (belonging to 42 families) which are scattered throughout the genome, and 51 copies of the yeast retrotransposons (Ty elements). Finally, the sequences of non-chromosomal elements, such as the 6 kb of the 2µ plasmid DNA, the killer plasmids present in some strains, and the yeast mitochondrial genome (c. 80 kb) have to be considered.

The genome of S. cerevisiae is divided up into 16 chromosomes ranging in size between 250 kb and >2500 kb. Choosing appropriate conditions, it is feasible to separate all 16 chromosomes in pulsed field gel electrophoresis (PFGE). This provides definition of 'electrophoretic karyotypes' of strains by sizing the chromosomes (Carle and Olson, 1985). The gels can be utilized for Southern blotting followed by hybridization, or to isolate chromosome-specific DNA.

The first genetic map of S. cerevisiae was published by Lindegren in

Saccharomyces cerevisiae: the first complete genome sequence

1949 (Lindegren, 1949); many revisions and refinements have appeared since (e.g. Mortimer *et al.*, 1992). The latest compilation comparing genetic and physical maps can be found in Cherry *et al.* (1997). Both meiotic and mitotic approaches have been developed to map yeast genes. The life cycle of *S. cerevisiae* normally alternates between diplophase and haplophase. Both ploidies can exist as stable cultures. In heterothallic strains, haploid cells are of two mating types, *a* and *α*. Mating of *a* and *α* cells results in *a/α* diploids that are unable to mate but can undergo meiosis. The four haploid products resulting from meiosis of a diploid cell are contained within the wall of the mother cell (the ascus). Digestion of the ascus and separation of the spores by micromanipulation yield the four haploid meiotic products. Analysis of the segregation patterns of different heterozygous markers among the four spores constitutes tetrad analysis and reveals the linkage between two genes (or between a gene and its centromere) (Mortimer and Schild, 1991). On the whole, genetic distance in yeast appears to be remarkably proportional to physical distance, with a global average of 3 kb/cM.

A large variety of protocols for genetic manipulation in yeast are available (Guthrie and Fink, 1991; Johnston, 1994). Yeast has a generation time of *c.* 80 minutes and mass production of cells is easy. Simple procedures for the isolation of high molecular weight DNA, rDNA, mRNA, and tRNA are at hand. It is possible to isolate intact nuclei or cell organelles such as intact mitochondria (maintaining respiratory competence). High-efficiency transformation of yeast cells is achieved, for example, by the lithium acetate procedure (Ito *et al.*, 1983) or by electroporation. A large variety of vectors have been designed to introduce and to maintain or express recombinant DNA in yeast cells (e.g. Guthrie and Fink, 1991; Johnston, 1994).

Furthermore, a large number of yeast strains carrying auxotrophic markers, drug resistance markers or defined mutations are available. Culture collections are maintained, for example, at the Yeast Genetic Stock Center (YGSC) and the American Type Culture Collection (ATCC). In the near future, mutant strains with defined gene deletions together with clones carrying the corresponding gene cassettes will emerge from the EUROFAN project (see below). A comprehensive library of recombinant lambda clones constructed as part of an *S. cerevisiae* physical mapping project and grouped in contigs (Olson *et al.*, 1986) is maintained and distributed by ATCC. Ordered cosmid libraries using different vectors were constructed during the yeast sequencing project (Riles *et al.*, 1993; Stucka and Feldmann, 1994; Thierry *et al.*, 1995).

Gene disruptions and single-step gene replacements are unique to *S. cerevisiae* and offer an invaluable advantage for experimentation. Yeast genes can functionally be expressed when fused to the green fluorescent protein (GFP) thus allowing to localize gene products in the living cell by fluorescence microscopy (Niedenthal *et al.*, 1996). The yeast system has also proved a useful tool to clone and to maintain large segments of foreign DNA in yeast artificial chromosomes (YACs) being indispensable in other genome projects (Burke *et al.*, 1987), and to monitor protein–protein interactions using the two-hybrid approach (Fields and Song, 1989).

The yeast genome sequencing project

In 1989 the European Commission decided to initiate a yeast sequencing project within the framework of its biotechnology programmes. Based on a network approach, some 35 European laboratories were initially involved in this enterprise (Vassarotti and Goffeau, 1992). Chromosome III was the first chromosome to be completed in 1992 (Oliver *et al.*, 1992). In 1994 chromosomes XI (Dujon *et al.*, 1994) and II (Feldmann *et al.*, 1994) were published from the EU project. Soon after its beginning, several other laboratories joined the project and agreed upon an international collaboration that would enable the whole yeast genome sequence to be finalized in late 1995. Finally, more than 600 scientists in Europe, North America and Japan became involved in this effort (Levy, 1994). Figure 3.2 lists the laboratories involved in the collaborative project from which the entire sequence was released in April 1996. A comprehensive survey on the complete yeast genome has recently appeared (Goffeau *et al.*, 1997a).

As is evident from a variety of reports on analyses of the entire yeast genome, the findings obtained from the analysis of the first completed chromosomes, reflecting chromosome architecture, genome organization and coding capacity, may now be generalized for the yeast chromosomes throughout.

Saccharomyces cerevisiae: the first complete genome sequence

Cloning and mapping procedures

The sequencing of chromosome III started from a collection of overlapping plasmid or phage lambda clones that were distributed by the DNA coordinator to the contracting laboratories (Oliver *et al.*, 1992). However, it soon became evident that cosmid libraries were much more advantageous in large-scale sequencing. Several of these libraries were constructed (e.g. Riles *et al.*, 1993; Stucka and Feldmann, 1994; Thierry *et al.*, 1995) and used for the rest of the yeast chromosomes. Obvious advantages of cloning DNA segments in cosmids, which normally accommodate 35 to 45 kb of DNA, were (i) larger genes could be obtained on a single recombinant clone; (ii) several linked genes could be isolated together with their intergenic regions; (iii) fewer colonies had to be maintained and screened to isolate a clone of interest; (iv) cosmid clones turned out to be stable for many years under usual storage conditions. Not surprisingly, the isolation of sequentially overlapping cosmid clones has enabled physical linkage and extensive characterization of genes from various organisms.

The construction of a library with as complete coverage as possible, with as few clones as possible, implies that the cloned DNA fragments are randomly distributed on the DNA. The number of clones (N) in a library representing each genomic segment with a given probability (P) is

$$N = \ln(1-P)/\ln(1-f)$$

where f is the insert length expressed as fraction of the genome size (Clarke and Carbon, 1976). For example, with the size of 12 800 kb for the yeast genome and assuming an average insert length of 35 kb, a cosmid library containing 4600 random clones would represent the yeast genome at $P = 99.99$ per cent, i.e. about 12 times the genome equivalent. The actual number of cosmid clones obtained by the usual procedures is very high ($>200 000/\mu$g DNA).

The low number of clones was of interest in setting up ordered yeast cosmid libraries or sorting out and mapping chromosome specific sublibraries. For example, a chromosome XI specific sublibrary composed of 138 clones has been generated from an unordered cosmid library by colony hybridization using PFGE purified chromosome XI DNA as a probe. The 'nested chromosomal fragmentation' method (Thierry and

83

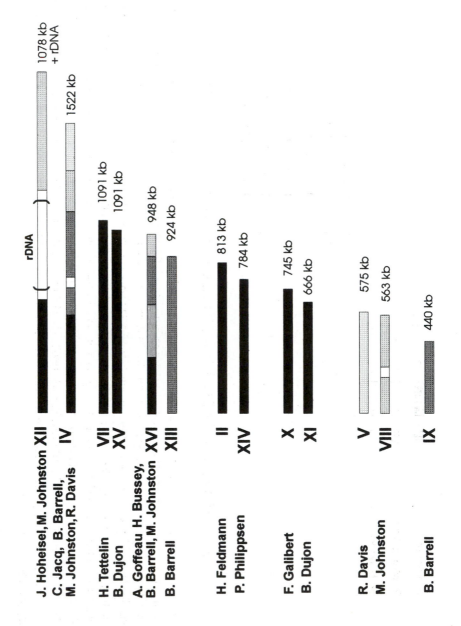

rDNA

J. Hoheisel, M. Johnston XII

C. Jacq, B. Barrell,
M. Johnston, R. Davis IV   1078 kb +rDNA

1522 kb

H. Tettelin VII   1091 kb
B. Dujon XV   1091 kb

A. Goffeau H. Bussey,
B. Barrell, M. Johnston XVI   948 kb

B. Barrell XIII   924 kb

H. Feldmann II   813 kb
P. Philippsen XIV   784 kb

F. Galibert X   745 kb
B. Dujon XI   666 kb

R. Davis V   575 kb
M. Johnston VIII   563 kb

B. Barrell IX   440 kb

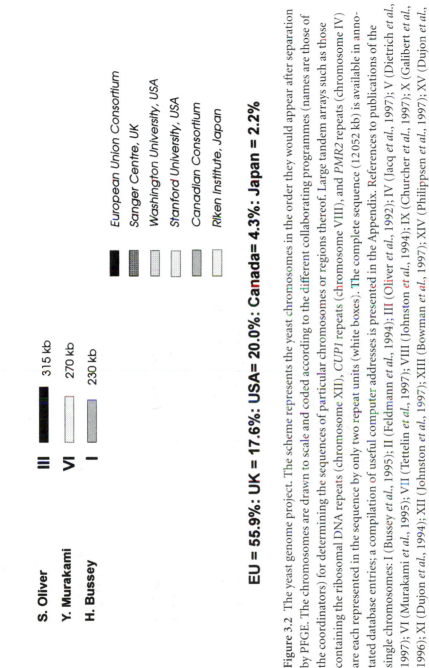

S. Oliver ▮▮▮ 315 kb

Y. Murakami VI 270 kb

H. Bussey I 230 kb

European Union Consortium
Sanger Centre, UK
Washington University, USA
Stanford University, USA
Canadian Consortium
Riken Institute, Japan

EU = 55.9%: UK = 17.6%: USA= 20.0%: Canada= 4.3%: Japan = 2.2%

Figure 3.2 The yeast genome project. The scheme represents the yeast chromosomes in the order they would appear after separation by PFGE. The chromosomes are drawn to scale and coded according to the different collaborating programmes (names are those of the coordinators) for determining the sequences of particular chromosomes or regions thereof. Large tandem arrays such as those containing the ribosomal DNA repeats (chromosome XII), *CUP1* repeats (chromosome VIII), and *PMR2* repeats (chromosome IV) are each represented in the sequence by only two repeat units (white boxes). The complete sequence (12052 kb) is available in annotated database entries; a compilation of useful computer addresses is presented in the Appendix. References to publications of the single chromosomes: I (Bussey *et al*, 1995); II (Feldmann *et al*, 1994); III (Oliver *et al*, 1992); IV (Jacq *et al*, 1997); V (Dietrich *et al*, 1997); VI (Murakami *et al*, 1995); VII (Tettelin *et al*, 1997); VIII (Johnston *et al*, 1994); IX (Churcher *et al*, 1997); X (Galibert *et al*, 1996); XI (Dujon *et al*, 1994); XII (Johnston *et al*, 1997); XIII (Bowman *et al*, 1997); XIV (Philippsen *et al*, 1997); XV (Dujon *et al*, 1997); XVI (Bussey *et al*, 1997).

Dujon, 1992) was then applied to rapid sorting of these clones. Finally, a set of some 30 overlapping cosmids was sufficient to build a contig of chromosome XI. This approach has also been successfully applied to many of the other yeast chromosomes sequenced in the following (e.g. Huang *et al.*, 1994). Connected with the sequencing of chromosome II (Feldmann *et al.*, 1994), we had selected 6000 independent clones from a total yeast cosmid library to prepare the DNA of each single cosmid from minilysates of 5 ml cultures. These samples were numbered and the corresponding cultures kept as glycerol stocks at $-70\,°C$. By chromosomal walking, 43 of these clones had been sorted out to cover yeast chromosome II. We employed a vector which carries a yeast marker and therefore can be used in direct complementation experiments (Stucka and Feldmann, 1994).

To facilitate sequencing and assembly of the sequences, contigs of overlapping cosmids and fine-resolution physical maps of the respective chromosomes were constructed first, by application of classical mapping methods (fingerprints, cross-hybridization) or by novel methods developed for this programme, such as site-specific chromosome fragmentation (Thierry and Dujon, 1992) or the high-resolution cross-hybridization matrix (Scholler *et al.*, 1995). These techniques might also be of interest for other genomes as well and, particularly, for mapping YAC inserts. It may be noted that by convention of all laboratories engaged in sequencing the yeast genome, the strain αS288C or isogenic derivatives thereof were chosen as the source of DNA, as these strains have been fairly well characterized and employed in many genetic analyses.

Sequencing strategies, sequence assembly and quality control
Sequencing strategies

In the European network, clones were distributed to the collaborating laboratories according to a scheme worked out by the DNA coordinators. Each contracting laboratory was free to apply sequencing strategies and techniques of its own provided that the sequences were entirely determined on both strands and unambiguous readings were obtained. Two principal approaches were used to obtain subclones for sequencing: (i) generation of sublibraries by the use of a series of appropriate restriction enzymes or from nested deletions of appropriate subfragments made by exonuclease III; (ii) generation of shotgun libraries from whole cosmids or subcloned fragments by random shearing of the DNA. Sequencing by the Sanger technique was either performed manually, labelling with [^{35}S]dATP being the preferred detection method, or by using automated devices, following the various established protocols. Two types of devices for on-line detection with fluorescence labelling, the Applied Biosystems ABI373A and the Pharmacia ALF, were employed. One laboratory used the direct blotting electrophoresis system from GATC (Konstanz). Similar procedures were applied to the sequencing of chromosomes outside the European network. The American laboratories largely relied on machine-based large-scale sequencing.

Sequencing telomeres

Sequencing of the yeast chromosome telomeres presented a particular problem. Due to their repetitive substructures and the lack of appropriate restriction sites they could be cloned by conventional procedures with only a few exceptions. Largely, telomeres were physically mapped relative to the terminal-most cosmid inserts using the I-*Sce*I chromosome fragmentation procedure (Thierry and Dujon, 1992). The sequences were then determined from plasmid clones obtained by 'telomere trap cloning', an elegant strategy developed by E. Louis at Oxford (Louis, 1994; Louis and Borts, 1995). Particular features of the telomeric sequences will be discussed below.

Sequence assembly and quality control

Within the European network, all original sequences were submitted by the contracting laboratories to the Munich Information Centre for Protein Sequences (MIPS). The sequences were kept in a data library, assembled into progressively growing contigs, and updated during the course of the project by the application of appropriate criteria in a number of quality controls, starting with chromosome XI (Vassarotti et al., 1995). In collaboration with the DNA coordinators, the final chromosome sequences were derived.

In phase 1, sequences were collected that had entirely been determined on both strands with sufficient overlaps between individual readings and experimental solving of all possible uncertainties, and accompanied by detailed sequencing strategies and documentation of the results. Phase 1 sequences were then examined for matching of predicted restriction sites with the physical map, putative frameshifts, and quality of sequence overlaps between neighbours. The results were reported to the contracting laboratories for verification or correction. In phase 2, sequences were then compared to the 'verification' sequences from particular regions. Selected regions were either long fragments (total of 15–20 per cent per chromosome) that were entirely resequenced using the same criteria as for original sequences, or short segments (total of 1–2 per cent per chromosome) chosen from suspected or difficult zones which were resequenced directly from cosmids using designated pairs of oligonucleotides as primers. Discrepancies, if any, were returned to the authors of the original sequence and to the authors of the verification sequence, and both were asked to re-examine their data or to solve the problem experimentally. After appropriate corrections, the sequences became classified as phase 3, ready for final editing of the complete sequence. We were confident, therefore, that a level of 99.97 per cent accuracy had been achieved (Dujon, 1996). At this level, of course, the sequences were not entirely free from errors, but it would have meant an enormous effort to reach a higher accuracy at that time. Meanwhile, numerous corrections have been incorporated in the data collections, thus improving accuracy.

Also in the other yeast chromosomes, automated procedures were employed for sequence assembly, based for example on the programme package developed at Cambridge (e.g. Dear and Staden, 1991) or on the ACeDB programme developed for the *Caenorhabditis elegans* genome project (Thierry-Mieg and Durbin, 1992). In all cases, correct assembly of the sequences was guaranteed by establishing that the order of restriction sites predicted from the sequence was consistent with the physical maps of these sites that had been determined independently and care was taken to perform quality controls that would result in a high accuracy (Riles *et al.*, 1993).

Sequence analysis
Prior to data submission by the single laboratories, and finally when the complete sequences were available, sequences were subjected to analysis by various algorithms. For the European network, special software

developed for the VAX at MIPS was used to locate and translate open reading frames (ORFs), to retrieve non-coding intergenic sequences, and to display various features of the sequence(s) on graphic devices (XCHROMO; an interactive graphics display program). The sequences have been interpreted using the following principles. (i) All intron splice site/branch-point pairs detected by using specially defined patterns (Kalegoropoulos, 1995; Kleine and Feldmann, unpublished) were listed. (ii) All ORFs containing at least 100 contiguous sense codons and not contained entirely in a longer ORF on either DNA strand were listed (this included partially overlapping ORFs). (iii) The two lists were merged and all intron splice site/branch-point pairs occurring inside an ORF but in opposite orientation were disregarded. (iv) Centromere and telomere regions, as well as tRNA genes and Ty elements or remnants thereof were sought by comparison with previously characterized datasets of such elements (Kleine and Feldmann, unpublished) including the database entries provided in a continuously updated library of tRNAs and tRNA genes (Sprinzl *et al.*, 1996). Finally, the various routines to search for tRNA genes were applied as they were developed (Pavesi *et al.*, 1994; El-Mabrouk and Lisacek, 1996; Percudani *et al.*, 1997). Catalogues of the ribosomal genes, the genes for the small RNAs, the tRNA genes, and the Ty elements are available at the MIPS web site (see Appendix).

Searches for similarity of proteins to entries in the databanks were performed by FASTA (Pearson and Lipman, 1988), BLASTX (Altschul *et al.*, 1990), and FLASH (Califano and Rigoutsos, 1993), in combination with the Protein Sequence Database of PIR-International and other public databases. Protein signatures were detected by using the PROSITE dictionary (Bairoch *et al.*, 1995) as well as BLOCKS and PRODOM domains whenever relevant for the interpretation of the query sequence. Compositional analyses of the chromosomes (base composition; nucleotide pattern frequencies, GC profiles; ORF distribution profiles, etc.) were performed by using GCG programmes (Devereux *et al.*, 1984) or the X11 program package (C. Marck, unpublished). For calculations of CAI and GC content of ORFs the algorithm CODONS (Sharp and Li, 1987; Lloyd and Sharp, 1992a,b) was used. At MIPS, comparisons of the chromosome sequences with data bank entries were based on a newly developed algorithm (HPT) (Mewes and Heumann, 1995). Furthermore, particular nucleotide patterns were searched for, which will be mentioned below. Basically, the same strategies were applied by other laboratories to interpret their sequences, again combining well-established routines with special software developed in these laboratories (e.g., Johnston *et al.*, 1994).

The yeast genome
High gene density, few introns

The majority of the open reading frames (ORFs) in yeast vary in size between 100 to more than 4000 codons (Dujon, 1996). Of the total ORFs, some 64 per cent fall into the range <500 codons, 27 per cent into the range 500–1000 codons, 6 per cent into the range 1000–1500 codons, and some 3 per cent have >1500 codons. Only a minority (probably less than 1 per cent) of the ORFs is estimated to be below 100 codons; the smallest mature peptides that have been characterized are the two mating pheromones. Approximately 70 per cent of the yeast DNA is coding sequence, with ORFs rather evenly distributed on both of the strands. On average, one gene is encountered every two kb. These figures were already derived from the analysis of the first three chromosomes but by and large hold true for all yeast chromosomes. The present estimate, after completion of the entire sequence, is that some 5800 genes are encoded by the yeast genome (Mewes *et al.*, 1997). The actual gene number can be precisely defined only when all ORFs have been experimentally examined whether or not they represent truly transcribed genetic entities.

The compact nature of the *S. cerevisiae* genome is apparent when compared with more complex eukaryotic systems. For example, the genome of *C. elegans* contains a potential protein-encoding gene only every 6 kb (Hodgkin *et al.*, 1994) and, in the human genome, gene density might be as low as one gene in 30 kb (Olson, 1993). Current data indicate that even the genome of the fission yeast, *Schizosaccharomyces pombe*, possesses a lower gene density (one gene per 2.3 kb) (Bowman and Barrell, http://www.sanger.ac.uk/yeast/pombe.html) than *S. cerevisiae*. The difference between the two yeast genomes appears to be due to the fact that in the fission yeast *c.* 40 per cent of the genes contain introns, whereas only a minor fraction (less than 5 per cent of the protein-encoding genes in *S. cerevisiae*) are predicted (or already experimentally shown) to be interrupted by introns. To date, only two cases have been encountered where two introns are present: the *MAT* locus on chromosc   II and a ribosomal protein gene, *RPL6A*, on chromosome VII. In one latter, the second intron encodes a small RNA. Generally, the intron is located at the extreme 5'-end of each gene, sometimes even preceding the coding region. The predominant population of intron-containing genes is recruited by the ones encoding ribosomal proteins. The functional significance of the introns is by no means clear, and despite the low number of intron-containing genes yeast maintains a highly complex and sophisticated machinery for splicing (overview: Newman, 1994).

Open reading frames and gene function

How many genes are functional?

In the analysis of chromosome XI, a strategy was developed for the interpretation of sequences (Dujon *et al.*, 1994) which was later applied to the other chromosomes to follow. Each ORF was evaluated using the codon adaptation index (CAI) (Sharp and Li, 1987) and ORF sizes as criteria: ORFs that were both shorter than 150 codons and have a CAI of less than 0.110 were considered 'questionable'. Most of the overlapping ORFs were sorted out by special criteria, and only the remaining ORFs considered 'real' genes.

Meanwhile, a few cases have been found where overlapping ORFs indeed exist and are expressed. In one particular case it was even shown that expression of the two ORFs occurs at different stages of cell growth. An interesting question was, how many pseudogenes might be present in the yeast genome. From earlier studies, it was anticipated that this number in yeast should be low compared with that in mammalian genomes. Generally, this assumption seems to hold true for most of the yeast chromosomes, but chromosome I turned out to be an exception to this rule (Bussey *et al.*, 1995). Chromosome I is the smallest naturally occurring functional eukaryotic nuclear chromosome so far characterized. The central 165 kb resembles other yeast chromosomes in both its high density and distribution of genes. In contrast, the remaining sequences flanking this DNA (i.e. the two 'ends' of this chromosome) have a much lower gene density, are largely not transcribed, contain no

91

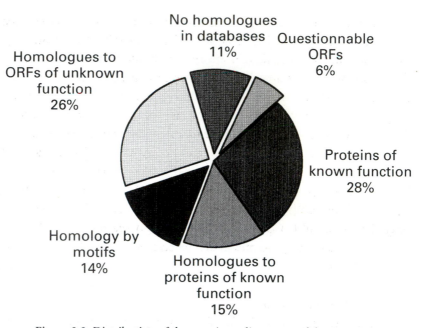

No homologues
in databases
11%

Questionnable
ORFs
6%

Homologues to
ORFs of unknown
function
26%

Proteins of
known function
28%

Homology by
motifs
14%

Homologues to
proteins of known
function
15%

Figure 3.3 Distribution of the protein-coding genes of the yeast genome as predicted from the sequence.

essential genes for vegetative growth, and contain four apparent pseudo-genes, and a stretch of 15 kb of redundant sequence. These terminally repetitive regions consist of a telomeric repeat, flanked by DNA closely related to *FLO1*, a yeast gene involved in cell flocculation and encoding a large serine/threonine-rich cell wall protein with internal repeats. The pseudogenes are related to known yeast genes but have internal stop codons. Extreme care has been taken in such cases to reconfirm the sequences of the regions in question by independent laboratories.

At present, a comparison of the final sequence with public databases reveals (Figure 3.3) that some 28 per cent of the ORFs correspond either to previously known protein-encoding genes or to genes whose functions have been determined previously or during the course of the yeast genome project. An estimated 6 per cent of the total have to be considered 'questionable ORFs', because some of these represent small ORFs partly overlapping with larger ORFs, or small ORFs completely included in larger ORFs located on the opposite strand.

Thus, 66 per cent of the total ORFs represent novel putative yeast genes the majority of which, as far as we can extrapolate from transcriptional mapping of particular chromosomes, should represent 'real' genes, though many of those appear to be transcribed at an extremely low level (Yoshikawa and Isono, 1990; Fairhead and Dujon, 1994). Approximately 15 per cent of the total genes have homologues among gene products from yeast or other organisms whose functions are known, while another 14 per cent of the total have homologues the functions of which are not well defined or show just recognizable motifs. The remaining 37 per cent of the total ORFs have either homologues to ORFs of unknown function on other chromosomes (26 per cent of the total) or no homologues in data libraries at all (11 per cent of the total). Based on the predictable number of yeast genes, this means that we still have to attribute functions to some 2200 to 2700 novel yeast genes.

The 'orphan' paradoxon

It is noteworthy that the number of yeast genes to which functions could be attributed through comparisons with the database entries from all organisms has not substantially increased during the course of the project, although this information has grown exponentially in the past few years. Thus we are left with the conclusion that besides the genes studied by the geneticists, which are most often common to many organisms, a large set of the previously undiscovered genes, sometimes called orphans (Dujon, 1996), make up a major part of the yeast genome. Systematic transcript maps have indicated that orphans represent real genes, which are regulated and as frequently expressed as other genes. Even in absence of homologues, computer programs have provided some clues to some orphans, and with time, structural homologues of known function will appear in the databases allowing a reduction of the number of genes of unpredictable function. Ultimately, however, there is good reason to assume that there will remain a core of ORFs specific to the yeast genome, the functions of which can only be demonstrated by experimentation.

A similar observation of the occurrence of obviously phylum-specific genes was made, when the complete genomes of small prokaryotes, such as *Haemophilus influenzae* (1.83 Mb, 1727 predicted coding regions) (Fleischmann *et al.*, 1995), *Mycoplasma genitalium* (0.6 Mb, 470 predicted coding regions) (Fraser *et al.*, 1995), or the archeon *Methano-coccus jannaschii* (1.7 Mb, 1759 predicted coding regions) (Bult *et al.*, 1996) were determined: a large proportion of the genes have no counterparts in other organisms. While in *H. influenzae* this percentage is comparable to that from yeast (some 40 per cent), the archeon presents the most extreme case: only 40 per cent of the genes could be assigned a putative function with high confidence. In all these organisms, genes for basic cellular functions (DNA replication and repair, transcription and translation, cellular transport, and energy metabolism) could be identified. Apparently, differences in genome content are reflected as profound differences in physiology and metabolic capacity between these organisms (Clayton *et al.*, 1997). Most strikingly, results from the analysis of the *M. jannaschii* genome confirmed the close evolutionary relationship of the archea to the eukaryotes (Olsen and Woese, 1996).

Proteins classified

Once the complete sequence of the yeast genome was available, it was most interesting to compare systematically all of the ORFs which can be classified by homology with proteins of known function or according to the presence of known functional motifs (or protein signatures). Following different principles of categorization, several useful inventory lists of the yeast proteins have been compiled (Wilkins *et al.*, 1996; Garrells, 1996; Mewes *et al.*, 1997) and are available as data libraries on the respective web sites (see Appendix). The scheme in Figure 3.4 is based on a classification of the yeast proteins into 14 major categories (Gazetteer: Mewes *et al.*, 1997).

Putative membrane and mitochondrial proteins

Particular attention has been given to the yeast membrane protein families. Applying the ALOM algorithm (Klein *et al.*, 1985) to predict putative membrane spans, a first estimate for chromosome III revealed that some 38 per cent of the real genes may code for transmembrane proteins containing from 1 to 14 potential membrane transversions (Goffeau *et al.*, 1993). Based on the complete yeast genome sequence, an analysis of approximately 2300 yeast membrane proteins has been conducted

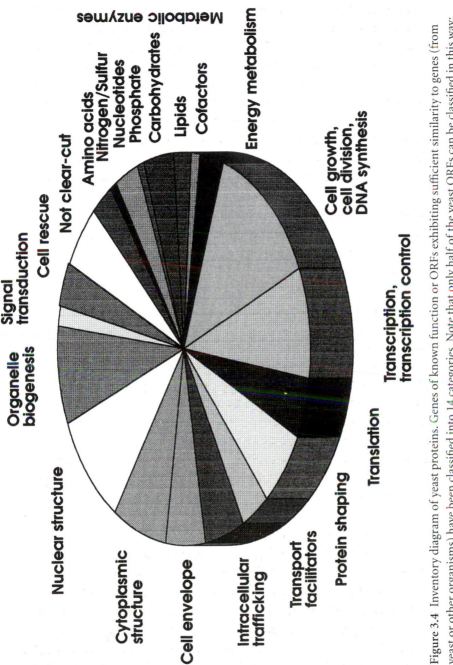

Figure 3.4 Inventory diagram of yeast proteins. Genes of known function or ORFs exhibiting sufficient similarity to genes (from yeast or other organisms) have been classified into 14 categories. Note that only half of the yeast ORFs can be classified in this way; no classification is possible for the rest of approximately 3000 ORFs.

recently (Decottignies and Goffeau, 1997; Goffeau *et al.*, 1997a). A pre-liminary classification of all membrane proteins into five groups was done on the number of their transmembrane spans. Within each group, families were defined on the basis of binary comparisons by means of the PRDF and PRSS program (included in the FASTA software package), and to confirm the composition of the resulting families, BLAST searches were carried out. Alignment of proteins in a family was done by the multiple sequence alignment program DCSE and phylogenetic trees were constructed with the help of TREECON for Windows. This analysis indicates that almost 1600 membrane proteins have no homologues and thus are 'families' with a single member. A total of 704 membrane proteins clustered into 222 families (minimum two members; 40 families with four or more members). Exceptionally large families are the ones with 18 transport P-type ATPases, 33 mitochondrial transporters, 28 ABC proteins (including an impressive set of multidrug pumps), and the nearly 200 major 'facilitators' (such as sugar and amino acid transporters) sub-clustered into some 20 families. It turned out that more than half of the members of these families were unknown before systematic sequencing.

Examination of the ORFs for the occurrence of putative mitochon-drial target signal sequences is difficult due to the complex character of these signatures (Hartl *et al.*, 1989) which has to be achieved mainly by eye inspection. Since not all of the proteins participating in mitochon-drial biogenesis are imported *via* particular signal sequences, the exact number of proteins involved in maintaining mitochondrial function in yeast remains unknown at present. A rough estimate is that some 6–8 per cent of the yeast proteins may be involved in mitochondrial biogenesis.

Other genetic entities

In addition to the genes encoding proteins, we have obtained detailed information on the organization of the genes for tRNAs and other small RNAs, the yeast retrotransposons (Ty elements), as well as the telomeric and centromeric sequences. The genes for the ribosomal RNAs are clustered to some 100 copies on the right arm of chromosome XII, whereas the multiple copies of tRNA genes are found scattered through-out the genome.

Five different types of Ty elements that exhibit substantial homology to retroviruses and retrotransposons from plants and animals are present in the yeast genome: Ty1, Ty2, and Ty4 belong to the 'copia' class of retrotransposons, while Ty3 is a member of the 'gypsy' family (review: Sandmeyer, 1992). αS288C contains 33 Ty1 and 13 Ty2 elements, whereas Ty4 is present at only three locations and Ty3 occurs in two copies. Ty5, a copy of which was found in chromosome III, was characterized as a new class of yeast transposon (Oliver *et al.*, 1992). Altogether, the *S. cerevisiae* strain αS288C carries eight Ty5 insertions, six of which are located near the telomeres, and five are found within 500 bp of autonomously replicating sequences (*ARS*) present in the subtelomeric regions (Zou *et al.*, 1996). While in *S. cerevisiae* the Ty5 elements no longer appear capable of transposition, strains of *S. paradoxus*, a species closely related to *S. cerevisiae*, carry numerous Ty5 insertions, where again they are associated with type X telomeric repeats and flanked by *ARS* (Zou *et al.*, 1996).

Like retroviruses, the Ty elements transpose through an RNA intermediate and by reverse transcription. Transposition rates are low, and the number of elements is kept fairly constant by balancing transposition and excision events (Fink *et al.*, 1986). This is manifest from the presence of 268 solo LTRs or other remnants that are footprints of previous transposition events. Due to the vagabond life style of the retrotransposons, yeast strains differ with respect to the sometimes rather complex 'patterns' formed by these elements resulting from multiple integrations and excisions. However, early data has already revealed that spontaneous transposition events do not appear to occur randomly along the length of individual chromosomes but that the Ty elements are preferably integrated into the upstream regions of tRNA genes (Eigel and Feldmann, 1982; DelRey *et al.*, 1983). In the following, these notions were substantiated through the analysis of a large variety of tRNA gene loci showing that the 5′ flanking regions of tRNA genes were the preferred target sites for Ty transposition (Warmington *et al.*, 1986, 1987; Hauber *et al.*, 1988; Lochmüller *et al.*, 1989).

While Ty1, Ty2, and Ty4 integrate in a region-specific manner, Ty3 integration occurs distance-specifically (16–18 bp) upstream of tRNA genes (DelRey *et al.*, 1983; Brodeur *et al.*, 1983). In many cases, complex patterns of Ty elements and remnants thereof were encountered, representing the footsteps of multiple transposition and excision events at certain 'hot-spots of transposition' (Warmington *et al.*, 1986, 1987; Hauber *et al.*, 1988; Lochmüller *et al.*, 1989). Experimental proof for the tRNA gene regions in yeast to be preferred target sites for Ty1 transposition has been provided by an analysis using the entire chromosome III as a target (Ji *et al.*, 1993), and elevated transpositions into the upstream sequences of Pol III transcribed genes have been shown by using plasmid-based reporter target (Devine and Boeke, 1996). Ty3 is invariably targeted 1–4 bp from the initiation site of Pol III transcription (Chalker and Sandmeyer, 1992; Kirchner *et al.*, 1995).

A strict balance is needed between the level of transposition to maintain a population of transposition-competent elements and the amount of activity that can be tolerated by the organism. Obviously, in yeast such a compromise has been reached by a mechanism that directs integration of the elements to target sites that are the least hazardous for the yeast cell, i.e. the upstream flanking regions of tRNA genes. Whole-genome analysis reveals that these regions normally are extended and devoid of protein-coding genes. While the target site specificities in transposition of all types of Ty elements have been clearly established, mechanisms that dictate integration specificity are less well understood. In this regard, the best-studied example is that of Ty3 for which target preference has been shown to be dependent upon functional promoter elements within the Pol III transcribed gene (Chalker and Sandmeyer, 1992, 1993). Recent findings have detailed that Pol III transcription factors TFIIIB and TFIIIC in combination with the target sequence that contains binding sites for these factors, were sufficient for the integration of Ty3 elements, even in the absence of Pol III (Kirchner *et al.*, 1995). These authors' observations suggest that Ty3 and its integrase directly interact with a component or components of the Pol III transcription factors. Recent findings point to the possibility that retroelement integration generally is also affected by chromatin structure and by interactions ('tethering') between the integrase and chromatin-associated proteins (review: Curcio and Morse, 1996).

At a first glance, the genetic entities specifying tRNAs and Ty elements seem to have little in common. However, recent work has clearly established that tRNA genes and tRNAs influence several key steps in transposition (Voytas and Boeke, 1993; Curcio and Morse, 1996). Like many retroviruses, the elements Ty1 through Ty4 employ translational frameshifting to regulate the expression of their *gag* (TYA) and *pol* (TYB) gene products. Contrary to the overlapping reading frames separated by a -1 frameshift in many retroviruses, the Ty elements have a $+1$ frameshift. Translational regulation during frameshifting is mediated by the availability of particular rare tRNAs (Voytas and Boeke, 1993). Furthermore, like most retroelements, Tys use a specific tRNA as a primer for reverse transcription. These interactions between the Ty elements and the tRNAs again reflect the intricate ways in which these transposable elements and their host must have coevolved.

Architecture of the yeast genome
Gene density and gene organization

It is now well established that the gene density in all yeast chromosomes is similar: ORFs occupy on average 70 per cent of the sequences (excluding the ORFs contributed by the Ty elements). This leaves only limited space for the intergenic regions which can be thought to harbour the major regulatory elements involved in chromosome maintenance, DNA replication and transcription. Regarding transcription of protein-encoding genes, a variety of elements have been identified and characterized that are operative in transcriptional initiation, regulation and termination. Not all of the yeast genes are preceded by a canonical TATA box, and it remains open as to which type of AT-rich sequences or other elements can act as transcriptional initiation sites (Struhl, 1987). In some cases, terminator sequences have been defined, but no general consensus sequences can be deduced. The same holds true for polyadenylation sites and polyadenylation signal sequences. Where experimentally determined, it appears that there is a much larger variability to these sequences than in mammalian systems (Proudfoot, 1991). As in mammalian or plant systems, a number of regulatory *cis*-acting elements (upstream-acting sequences; UAS) and the corresponding *trans*-activating factors have been experimentally characterized in yeast (overview: Svetlov and Cooper, 1995). Also negative regulatory elements (upstream-repressing sequences; URS) have been shown to control the expression of some genes. However, in only a few instances, precise ideas on the intimate interplay of the various regulatory components mediating gene expression are beginning to evolve. The knowledge of the entire genome sequence combined with the powerful genetic tools available for yeast should now foster research along these lines.

Generally, ORFs appear to be rather evenly distributed among the two strands of the single chromosomes. In some chromosomes (I, II, VIII), there is a slight excess of coding capacity on one of the strands, the significance of which is not known. Figure 3.5 presents a scheme of how the single transcriptional units are organized along yeast chromosomes. Statistics indicates that there is no prevalence for one or the other type of gene arrangement, although arrays longer than eight genes that are transcriptionally oriented into the same direction can be found on several chromosomes. The extreme seems to be a region from chromosome VIII, where 17 in a run of 18 ORFs are located on the 'top' strand.

In the 'head-to-tail' arrangements, the intergenic regions between two consecutive ORFs are sometimes extremely short, raising the question whether they are maintained as separate units or coupled for transcription and translation. There are cases in which different functions have been combined in one genetic unit, but to the best of our knowledge, polycistronic messages have not been observed in yeast to date.

On a first view, the intervals between divergently transcribed genes might be interpreted to mean that their expression is regulated in a concerted fashion involving the common promoter region. This, however, seems not to hold for the majority of the genes and might be a principle reserved for a few cases, in which these genes belong to the same regulatory pathway (e.g. *GAL1/GAL10*; Bram *et al.*, 1986). By contrast, many examples are known, in which a constitutively expressed gene shares its upstream sequences with those of a highly controlled gene. Regarding the fact that most of the intergenic regions are relatively short (cf. legend to Figure 3.5), an intriguing question becomes apparent: are regulatory elements confined to these sequences or could they also be present in coding sequences of neighbouring genes located upstream? For example, experimental data obtained for several genes involved in meiosis point to this possibility (Smith *et al.*, 1990). This would imply that two different kinds of constraint were superimposed on sequences during evolution, one for

100

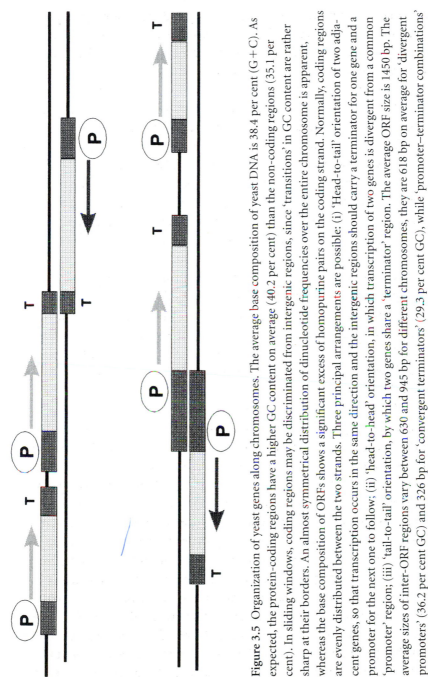

Figure 3.5 Organization of yeast genes along chromosomes. The average base composition of yeast DNA is 38.4 per cent (G+C). As expected, the protein-coding regions have a higher GC content on average (40.2 per cent) than the non-coding regions (35.1 per cent). In sliding windows, coding regions may be discriminated from intergenic regions, since 'transitions' in GC content are rather sharp at their borders. An almost symmetrical distribution of dinucleotide frequencies over the entire chromosome is apparent, whereas the base composition of ORFs shows a significant excess of homopurine pairs on the coding strand. Normally, coding regions are evenly distributed between the two strands. Three principal arrangements are possible: (i) 'Head-to-tail' orientation of two adjacent genes, so that transcription occurs in the same direction and the intergenic regions should carry a terminator for one gene and a promoter for the next one to follow; (ii) 'head-to-head' orientation, in which transcription of two genes is divergent from a common 'promoter' region; (iii) 'tail-to-tail' orientation, by which two genes share a 'terminator' region. The average ORF size is 1450 bp. The average sizes of inter-ORF regions vary between 630 and 945 bp for different chromosomes, they are 618 bp on average for 'divergent promoters' (36.2 per cent GC) and 326 bp for 'convergent terminators' (29.3 per cent GC), while 'promoter–terminator combinations' (34.2 per cent GC) are 517 bp in length on average.

maintaining function of coding sequences and the other for preserving regulatory sequences. By employing catalogues of consensus sequences of the known regulatory elements (Svetlov and Cooper, 1995; H. Feldmann, unpublished), one can detect many sites within the intergenic regions that could be thought to be functional, but at the same time, such sites are found scattered throughout the coding regions as well. Although the functional significance of these sequences might strictly depend on the regional context, it is difficult at the moment to discriminate between functional and non-functional elements merely by inspection of the sequence, even by developing more sophisticated computer algorithms. Finally, only experimental approaches will answer this problem.

Base composition and gene density

Average base composition has been found to be symmetrical over all chromosomes (the symmetry being even more apparent with dinucleotide frequencies), but this only reflects the almost equal numbers of ORFs encoded on each DNA strand of most of the yeast chromosomes. The base composition of ORFs themselves show a significant excess of homopurine pairs on the coding strand.

Regional variations of base composition with similar amplitudes were noted along chromosome III (Sharp and Lloyd, 1993) and XI (Dujon *et al.*, 1994), with major GC-rich peaks in each arm. The analysis of chromosome XI revealed an almost regular periodicity of the GC content, with a succession of GC-rich and GC-poor segments of ~50 kb each, a further interesting observation being that the compositional periodicity correlated with local gene density. Profiles obtained from similar analyses of chromosomes II (Feldmann *et al.*, 1994) and VIII (Johnston *et al.*, 1994) again showed these phenomena, albeit with less pronounced regularity. Other chromosomes also show compositional variation of similar range along their arms, with pericentromeric and subtelomeric regions being AT-rich, though spacing between GC-rich peaks is not always regular. In most cases, however, there is a broad correlation between high GC content and high gene density (Dujon, 1996).

Figure 3.6 exemplifies the results for chromosome II. GC-poor peaks coinciding with relatively low gene densities are located at the centromere (around coordinate 230) and at both sides of the centromere,

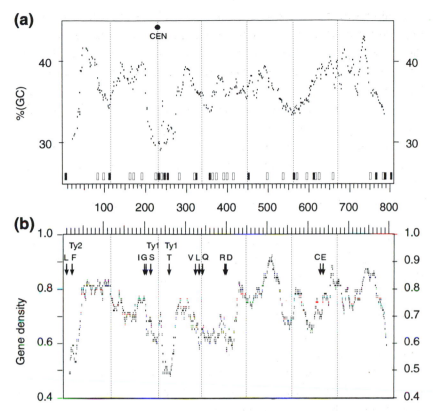

Figure 3.6a,b Periodic variation of GC content and gene density along chromosome II. (a) Compositional variation was calculated as described in Feldmann *et al.* (1994). Each point represents the average GC composition calculated from the silent positions only of the codons of 15 consecutive ORFs. Similar slopes were obtained when the GC composition was calculated from the entire ORFs or from the inter-ORF regions, or when the averages of 13–30 elements were plotted. The location of perfect ARS consensus sequences is indicated by the rectangles, whereby filled boxes represent those ARS fulfilling the criteria attributed to functional replication origins (Marahrens and Stillman, 1992; Bell and Stillman, 1992). (b) Gene density was expressed as the fraction of nucleotides within ORFs versus the total number of nucleotides in sliding windows of 30 kb. The arrows indicate the location of tRNA genes and intact Ty elements, respectively.

with a periodicity of *c.* 110 kb. These minima are more pronounced around coordinates 120, 340, and 560, while they are less so at coordinates 450 and 670. Remarkably, most of the tRNA genes reside in GC-poor 'valleys' and the Ty elements became eventually integrated into these regions. When analysing chromosome II for the occurrence of simple repeats, putative regulatory signals, and potential ARS elements, we noticed that the last mentioned were not found to be randomly distributed. In Figure 3.6a, we have listed the location of 36 ARS elements which completely conform to the 11 bp degenerate consensus sequence (Newlon, 1988; Van Houten and Newlon, 1990). Several of these were found associated at their 3' extensions with imperfect (1–2 mismatches) parallel and/or antiparallel ARS sequences or putative binding sites for ARS-binding factor 1, reminiscent of the elements reported to be critical for replication origins (Marahrens and Stillman, 1992; Bell and Stillman, 1992). Remarkably, these patterns are found within the GC-poor valleys, suggesting that functional replication origins might preferably be located in AT-rich regions. This phenomenon was also apparent from an analysis of chromosome XI and, more convincingly, when the distributions of functional replication origins mapped in chromosome VI (Shirahige *et al.*, 1993) or in 200 kb of chromosome III (Dershowitz and Newlon, 1993) were compared to the GC profiles of these chromosomes. Functional ARS elements have yet to be defined for the remainder of chromosome III and the other yeast chromosomes. In this context, it would be interesting to see whether the origins of replication reveal a regular spacing (Fangman and Brewer, 1992) and whether these and the chromosomal centromeres might maintain specific interactions with the yeast nuclear scaffold (Amati and Gasser, 1988). In all yeast chromosomes analysed thus far, ARS elements located in the subtelomeric regions are closely associated with binding sites for origin binding factors (Eisenberg *et al.*, 1988; Estes *et al.*, 1992).

As observed first in chromosome XI (Dujon *et al.*, 1994), not only the compositional periodicity is similar in many other yeast chromosomes but correlates with local gene density (Dujon, 1996), as is the case in more complex genomes in which isochores of composition are, however, much larger (Bernardi, 1993). Although the fairly periodic variation of base composition is now evident for all yeast chromosomes, its significance remains unclear. Several explanations for the compositional distribution and the location-dependent organization of individual genes have been offered. For example, the compositional periodicity of a yeast chromosome could reflect the evolutionary history of the chromosome, together with the folding of that chromosome, its attachment to the nuclear matrix or structural elements involved in chromosome segregation, or in the 'homology search' that precedes synapsis in the early meiotic prophase. Other possibilities could be tested experimentally. For example, transcription mapping of whole chromosomes could give a clue as to whether such rules may influence the expression of genes. Furthermore, long-range determination of DNaseI-sensitive sites may be used to find a possible correlation between compositional periodicity and chromatin structure along yeast chromosomes.

Organization of yeast telomeres

The organization of the yeast telomeres (Figure 3.7) has become clear from the work of E. Louis and his collaborators in conjunction with the chromosome sequences. All yeast chromosomes share characteristic telomeric and subtelomeric structures (Louis and Haber, 1992).

Telomeric ($C_{1-3}A$) repeats, some 300 nucleotides in length, are found at all telomere ends. 31/32 of the chromosome ends contain the X core subtelomeric elements (400 bp), and 21/32 of the chromosome ends carry an additional Y' element. There are two Y' classes, 5.2 kb and 6.7 kb in length both of which include an ORF for a putative RNA helicase of yet unknown function. Y's show a high degree of conservation but vary between different strains (Louis and Haber, 1992). Experiments with the *est1* (ever shortening telomeres) mutants, in which telomeric repeats are progressively lost, have shown that the senescence of these mutants can be rescued by a dramatic proliferation of Y' elements (Lundblad and Blackburn, 1993). Several additional functions have been suggested for these elements (review: Palladino and Gasser, 1994), such as extension of telomere-induced heterochromatin or protection of nearby unique sequences from its effects; a role in the positioning of chromosomes within the nucleus.

Comparisons between the chromosome termini revealed that they also share extended similarities in their subtelomeric regions by virtue of

105

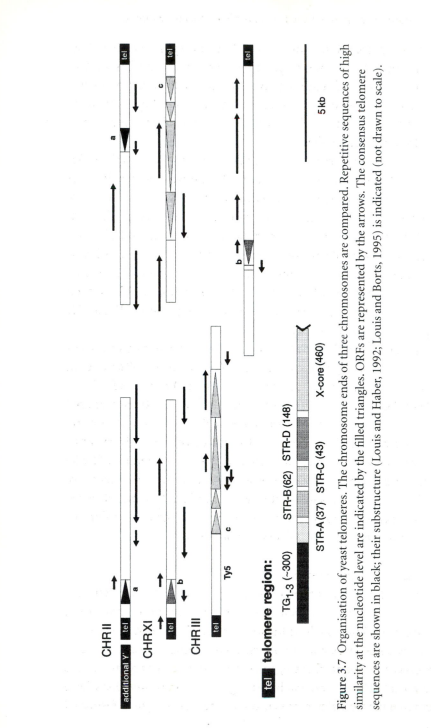

Figure 3.7 Organisation of yeast telomeres. The chromosome ends of three chromosomes are compared. Repetitive sequences of high similarity at the nucleotide level are indicated by the filled triangles. ORFs are represented by the arrows. The consensus telomere sequences are shown in black; their substructure (Louis and Haber, 1992; Louis and Borts, 1995) is indicated (not drawn to scale).

the occurrence of repetitious sequences of different types. Genetic redundancy is the rule at the ends of yeast chromosomes. The 'duplicated' regions contain copies of genes of known or predictable function as well as several ORFs the putative products of which exhibit high similarity; but their functions remain unclear as no homologues of known function have been found in the databases.

Figure 3.7 exemplifies this situation for chromosomes II, III and XI. The duplicated regions contain several ORFs the putative products of which exhibit high similarity. Also in chromosome VIII, extensive duplications being present on other chromosomes have been observed: 30 kb near the right telomere is more than 90 per cent identical to the similar region on the right arm of chromosome I. A smaller portion of this is also duplicated on the left arm of chromosome I. Shorter duplicated segments in the subtelomeric region of the left arm of chromosome VIII have been encountered in the subtelomeric regions of chromosomes III and XI (showing 54–94 per cent identity).

The unique constellation of repetitious sequences at the two ends of chromosome I has been pointed out already. Approximately 30 kb in each subtelomeric region carry similar (but non-essential) genes and a 15 kb repeat. These features are consistent with the idea that these terminal regions represent the yeast equivalent to heterochromatin and the occurrence of this type of DNA suggests that its presence gives this chromosome the critical length required for proper stability and function. The 30 kb region can be removed from each end without affecting vegetative growth, although chromosome stability is considerably reduced. Most likely, these repeated regions contribute to chromosome I size polymorphisms that have been observed (Bussey *et al.*, 1995).

Simple and complex repeats

Overall, the yeast genome is remarkably poor in repeated sequences. Complex repeats occurring in the yeast genome have been mentioned earlier. These include the telomeric and subtelomeric repeats at the chromosome ends and some extended internal repeats. Subtelomeric repeats or remnants thereof, and even short stretches of the simple sequence repeat $(TG_{1-3})_n$ normally sealing the chromosome ends have been encountered internal to some chromosomes. These types of internal

Table 3.1. *Simple repeat sequences on chromosome II corresponding to amino acid homopolymer stretches*

ORF	Gene	AA repeat	Nucleotide repeat
YBL084c	CDC27	Asn (>24)	TTA
YBL081w		Asn	CAA
		Ser	complex
YBL061c	SKT5	Arg	complex
YBL029w		Asn	TAA
		Ser & Pro	complex
YBL011w		Glu	GACGAA
YBL007c	SLA1	QQPQMMN	GT-ch
YBR016w		Gln	TACAAT
YBR040w		FLLI	CAAA
YBR067c	TIP1	Ser	complex
YBR112c	SSN6	Gln	GTT
YBR150c		Asn	TTA
		Asp (>21)	CAT
YBR289w	SNF5	Gln	CAACAG

repeats are probably relics of events during breakage and healing of chromosomes. Besides the Ty elements, it is the rDNA on chromosome XII that most significantly contributes to repetitiveness. A cluster of some 15 tandem repeats (2 kb each) containing the *CUP1* gene and contributing to polymorphic variation is found on chromosome VIII (Johnston *et al.*, 1994). Repeated stretches of short oligonucleotides exist. These include poly(A) or poly(T) tracts, alternating poly(AT) or poly(TG) tracts, and direct or inverted long repeats.

By applying the program PYTHIA (Milosavljevic and Jurka, 1993) to search for simple repeats, we detected at least 12 sets of regularly repeated di- or trinucleotides along chromosome II (Feldmann *et al.*, 1994) (Table 3.1). Concurrent examination of the chromosome II ORFs revealed that these represent repetitious codons for particular amino acids, such as asparagine, glutamine, arginine, aspartic acid, glutamic acid, proline and serine, thus forming homopeptide stretches. In some cases, the repetitious patterns are more complex, hence even more complex amino acid patterns result. Searches in the databases show that there are numerous proteins containing homopeptides built from the above amino acids, sometimes of considerable size, in yeast and other organisms. Although the role of such homopeptides is not well defined, it appears that they constitute specific domains enabling the respective proteins to fulfil specific functions.

A systematic study on the distribution and variability of trinucleotide repeats in the yeast genome revealed perfect and imperfect repeats ranging from 4 to 130 triplets, and the repartition of different triplet combinations was found to differ between ORFs and intergenic regions (Richard and Dujon, 1996). Examination of various laboratory strains revealed polymorphic size variations for all perfect repeats, compared with an absence of variation for the imperfect ones. These findings are particularly interesting regarding the fact that several human genetic disorders are caused by trinucleotide expansion. The yeast system may now provide an experimental approach to study the mechanisms of their expansion.

Genome organization and evolutionary aspects
Genetic redundancy in yeast

A survey of previous sequence data and sequences obtained in the yeast sequencing project suggest that there is a considerable degree of internal genetic redundancy in the yeast genome, which on the protein level has been estimated to be 30–40 per cent. Whereas an estimate of sequence similarity (both at the nucleotide and the amino acid level) becomes predictive at this stage, it still remains difficult to correlate these values to functional redundancy, because even in yeast only in a limited number of cases gene functions have been precisely defined.

In many instances, the duplicated sequences are confined to nearly the entire coding region of these genes and do not extend into the intergenic regions. Thus, the corresponding gene products share high similarity in terms of amino acid sequence or sometimes are even identical and, therefore, may be functionally redundant. However, as suggested by sequence differences within the promoter regions, gene expression should vary according to the nature of the regulatory elements or other (regulatory) constraints. This has been demonstrated experimentally in numerous examples (Table 3.2). It may well be that one gene copy is highly expressed while another one is poorly expressed. Turning on or off

Table 3.2. *Examples of duplicated genes in the yeast genome and their regulation*

Similarity	Function	Effect(s) of multiplicity	Functional exchange possible?	Examples of type of genes	Copies
Identical	Same	Gene dosage		Major tRNAs	Up to 12
Identical or very high	Same	Probably gene dosage		Ribosomal proteins	Mostly 2
Identical	Same	Gene dosage	No	Histones H2A, H2B, H3, H4	2
Identical or very high	Same	Probably gene dosage		Invertase and maltose metabolism (*SUC/MAL*)	Several
High (identical)	Same	Regulation different		ADP/ATP carrier	3
High (identical)	Same	Regulation different	No	Acid phosphatase (*PHO3/PHO5; PHO10; PHO11*)	4
Extended	Same	Regulation different	Yes	Pyruvate carboxylase (*PYC1/PYC2*)	2
Extended	Same	Differentiated expression	No	Chitin synthetase (*CHS1/CHS2/CHS3*)	3
Homologues		Not known	Not known	Mannosyltransferase (*KTR* genes)	5
Homologues		Not known	Not known	Several 'glycolytic' enzymes	Up to 5
Homologues	Similar	Not known	Not known	Pleiotropic drug resistance (*MDR* genes)	Many
Homologues	Similar	Not known	Not known	Various amino acid transporters	Several
Homologues	Different?	Not known	Probably no	Various transcription factors	Several
Homologues	Different?	Not known	Not known	Kinases of various types	Many
Extended	Different	Not known	No	Gene family of novel ATPases (AAA family)	Single

expression of a particular copy within a gene family may depend on the differentiated status of the cell (such as mating type, sporulation, etc.). Biochemical studies also revealed that in particular cases redundant proteins can substitute each other, thus accounting for the fact that a large portion of single gene disruptions in yeast do not impair growth or cause 'abnormal' phenotypes. This does not imply, however, that these redundant genes were *a priori* dispensable. Rather they may have arisen through the need to help adapt yeast cells to particular environmental conditions. These notions are of practical importance when carrying out and interpreting gene disruption experiments.

Duplicated genes in subtelomeric regions   Classical examples of duplicated genes in yeast are the *MEL*, *SUC*, *MGL* and *MAL* genes, which are involved in sugar metabolism and have been previously found as subtelomeric repeats in several yeast strains. In fact, yeast strains differ by the presence or absence of particular sets of these genes. For example, three genes mapped on chromosome II of wild-type strains, *MEL1*, *SUC3*, and *MGL2*, are absent from the strain αS288C. A comparison at the molecular level of αS288C with brewer's yeast strain C836 clearly shows that the *SUC* genes are present on chromosome II of the latter strain (Stucka, 1992). Regarding the genes involved in carbohydrate metabolism, the presence of multiple gene copies could be attributed to selective pressure induced by human domestication, as it appears that they are largely dispensable in laboratory strains (such as αS288C) which are no longer used in fermentation processes. Non-homologous recombination processes may account for the duplication of these and other genes residing in subtelomeric regions (e.g. Michels *et al.*, 1992), reflecting the dynamic structure of yeast telomeres in general. We have already mentioned the fact that the telomeres and subtelomeric regions of several yeast chromosomes share highly conserved segments, in some instances up to 30 kb, which carry duplicated genes the functions of which are largely unknown. One prominent example is the '*PAU* family' comprising more than 10 genes each encoding a highly conserved protein of 120–125 amino acids in length (Viswanathan *et al.*, 1994).

Duplicated genes internal to chromosomes   Additionally, there is a great variety of genes internal to chromosomes that appear to have

111

arisen from duplications, as suggested by the analyses of individual chromosomes. Before complete chromosome sequences became available, a multitude of genes have been known to occur in two or more identical or nearly identical copies located to different chromosomes, such as the histone genes; ribosomal protein genes; genes for ATP/ADP carriers; genes for amino acid and sugar transporters; genes for enzymes of the glycolytic pathway; and many others. Numerous examples can now be added, when the completed chromosomes are searched for similarity at the nucleotide as well as at the protein level.

Duplicated genes in clusters and cluster homology regions (CHRs) Remarkably, duplicated genes have also been found in clusters. There are at least three examples of this kind in chromosome II (Feldmann *et al.*, 1994). Another case is a cluster of three hexose transporter genes on chromosome VIII (Johnston *et al.*, 1994), which appear to be the result of a less recent gene duplication. Uncommon cases of gene duplications are represented by the aforementioned large clustered (tandem) gene family of membrane proteins on chromosome I, and a large cluster on chromosome VIII near *CUP1*. The *CUP1* gene encoding copper metallothionein, is contained in a 2 kb repeat that also includes an ORF of unknown function. The repeated region has been estimated to span 30 kb in strain αS288C, which could encompass 15 repeats, but the number of repeats varies among yeast strains.

An even more surprising phenomenon became apparent when the sequences of complete chromosomes were compared to each other, revealing that there are large chromosome segments in which homologous genes are arranged in the same order, with the same relative transcriptional orientations, on two or more chromosomes. Obviously, the genome has continued to evolve since this duplication occurred: genes have been inserted or deleted, Ty elements and introns have been lost and gained between the two sets of sequences. The occurrence of some 50 such cluster homology regions (CHRs) is now manifest for the yeast genome. To analyse the extent and pattern of redundancy in the yeast genome, a potent data structure, the HPT, has been developed at MIPS, allowing an all-against-all comparison of fixed-size blocks of nucleotides the results of which can be visualized by a graphical interface

(GSG) showing similarities both at the nucleotide and the protein level (Heumann *et al.*, 1996).

Chromosomes II and IV share the longest CHR, comprising a pair of pericentric regions of 170 and 120 kb, respectively, which share 18 pairs of homologous genes (equivalent to 15 per cent of the total gene content in these regions). It is most interesting that among these are 13 ORFs and five specific tRNA genes. In all, at least 10 CHRs (shared with chromosomes II, V, VIII, XII and XIII) can be recognized on chromosome IV (Figure 3.8). Remarkably, the entire chromosome XIV can be subdivided into several segments that are found duplicated on other chromosomes (Philippsen *et al.*, 1997). Using a minimum cluster size of 25 kb, shorter clusters can be identified if they are very compact and larger clusters can be constructed from multiple, collinear clusters. If optimized for maximum coverage, up to 40 per cent of the yeast genome is found to be duplicated in clusters, not including Ty elements and subtelomeric regions.

Timescale and mechanisms of gene duplications   The clustering of duplicated genes and the occurrence of extended regions of similarity compel us to consider the idea that entire genomic regions were duplicated, followed by rearrangements. These duplication events would appear to be ancient because the DNA sequence has clearly diverged outside the coding regions; moreover, such clusters even share a number of tRNA genes both in the same location and orientation.

Wolfe and Shields (1997), who finally defined 55 blocks of extended homology, suggested a convincing model that explains the gene duplications seen in today's yeast by (allo- or auto)-tetraploidization between two ancient strains of *S. cerevisiae* followed by reciprocal translocations and 85 per cent deletions of duplicated genes.

Additionally, other mechanisms have to be implicated to explain the occurrence of single copies of duplicated genes, preferably those found in the subtelomeric regions or as orphans outside the cluster regions. One can imagine, for example, that these could represent processed genes that were inserted into the genome relatively recently, a view which is consistent with the conservation of sequence only in the coding regions. However, all of these duplications would appear to have been

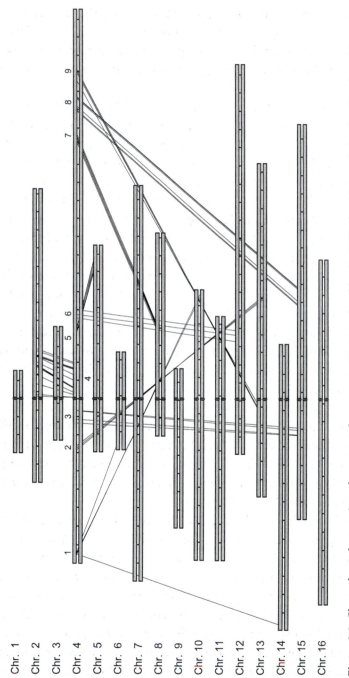

Figure 3.8 Cluster homology regions from yeast chromosome IV. As analysed by the HTP algorithm and visualized by the GSG (Heumann *et al.*, 1996), extended regions of protein similarity are detected in chromosome IV to occur in other yeast chromosomes.

created by integration of full-length complementary DNAs, because none appear to be pseudogenes, and this is unexpected in this model. In addition, some of the homologous gene pairs include introns in both genes, which suggest that at least these genes were not duplicated by this mechanism.

A recent analysis of Slonimski and his collaborators (P. P. Slonimski, pers. comm.) arrives at a different view stating that probably no ancient duplication of an ancestral genome has occurred. This study has revealed that half of the yeast genes (3000) occur in some 800 duplicated families whereas the rest does not. Of the duplicated genes 900 are found in 450 families of only two, while the residual genes form families with up to 100 members. Homology scores reveal that in evolutionary terms older and newer families exist and that more than 300 genes are evolving very fast. Most interestingly, more families exist that contain genes of unknown function than those containing genes of known function. Remarkably, the families with 10–110 members contain up to 70 per cent genes of unknown function. Among the genes of known function, no major duplications of metabolic proteins appear to have occurred, while major expansions are observed for genes encoding membrane proteins, factors involved in protein shaping, and in DNA or RNA wielding.

Whatever the relative timescale and mechanisms of duplications, these events followed by mutations affecting functional properties give a chance of improved environmental fitness. On the other hand, the high gene density in yeast indicates a strong tendency to maintain a compact genome, therefore compensatory mechanisms must exist to remove non-functional or superfluous gene copies.

An interesting problem intimately related to evolution is the origin of the organizational pattern of genes seen to date. Could we, for example, find any criteria, be they structural or functional in nature, that govern the regional arrangement of particular genes? In other words, is there an 'ordered grouping' of genes along the yeast chromosomes or are we left with a random succession of most of the genes? Presently, there are few clues to help answer these questions. In some respects, however, we do have indications for a particular location or grouping of genes, and some examples have already been mentioned above. In several cases, highly expressed genes are found associated with ARS elements, so that one could speculate that replication and efficient transcription are intimately coupled. In chromosome XI, it appears that highly expressed genes occur in 'clusters' within preferred regions (B. Dujon, pers. comm.). Clearly, the *MAL* and *SUC* loci, and the *GAL1/10–GAL7* locus represent examples in which functionally related genes involved in a particular metabolic pathway are closely associated to each other.

Sequence variation among yeast strains

The question as to what extent yeast strains differ with respect to their genetic content has implicitly been touched upon already. We have discussed a number of features that contribute to polymorphisms between different yeast strains: (i) variable number of gene copies from repeated gene families, (ii) individual patterns caused by the presence or absence of particular Ty elements, (iii) plasticity of the chromosome ends. Additionally, excisions or inversions of particular gene regions have been observed to give rise to polymorphisms. In all these instances, polymorphisms also become manifest through length differences between corresponding chromosomes. Chromosome breakage has been found to occur in yeast, resulting in karyotypes deviating from the 'normal' picture. However, sequence variations within the coding regions of individual genes seem to be rare and do not appear to contribute to polymorphism significantly, as far as we can tell from comparisons of homologous sequences obtained from different strains.

Detailed information on strain differences resulting from Ty insertions and/or deletions are available for chromosome II, where we have compared the Ty patterns from strains αS288C and C836, and local patterns from two other strains, YNN13 and M1417-c (Stucka, 1992), which showed remarkable differences. However, the sequences around the elements are well conserved among all these strains. Many more examples of this kind can be found in the literature. Altogether, this reveals a substantial plasticity of the yeast genome around tRNA gene loci, which appear to be the preferred target sites for Ty transpositions. As discussed above, the Ty integration machinery seems to detect regions of the genome that may represent 'safe havens' for insertion thus guaranteeing both survival of the host and the retroelement.

116

Genetic and physical maps compared

Comparing the physical map of the chromosomes from a particular yeast strain (in our case αS288C) to 'the' genetic map of yeast, one is immediately faced with the problem that the latter has been derived by using different yeast strains showing the aforementioned sequence variations. Predominantly, it has to be considered that some strains used in establishing the linkages have inversions or translocations that then might contribute to discrepancies between physical and genetic maps. Similarly, strain polymorphisms caused by the extended repetitious sequences or subtelomeric duplicated genes may lead to imprecisions of the genetic maps. Finally, the accuracy of genetic mapping will depend on the experimental approaches used.

Nevertheless, the genetic map of *S. cerevisiae* (Mortimer *et al.*, 1992) has been of considerable value to yeast molecular biologists before physical maps became available. In fact, we and others have used DNA probes from some known genes mapped to particular chromosomes for chromosomal walking. Finally, however, physical maps of all chromosomes of αS288C have been constructed without reference to the genetic maps.

For αS288C, generally, the comparison of the physical and genetic maps of the chromosomes reveals that most of the linkages have been established to give the correct gene order but that in many cases the relative distances derived from genetic mapping are imprecise. As an example, Figure 3.9 shows a comparison of the genetic with the physical map of chromosome II. Remarkably, the most critical deviation between the genetic and the physical maps was initially observed with chromosome XI (Dujon *et al.*, 1994) but could be corrected by repeating the genetic mapping of a segment located next to the left telomere (Simchen *et al.*, 1994). With the completion of the yeast genome sequence, co-linear genetic and physical maps of each chromosome have been compiled (Cherry *et al.*, 1997).

Altogether, the experience gained from the yeast project shows that genetic maps provide valuable information but that in some cases they may be misleading. Therefore, independent physical mapping and determination of the complete sequences were needed to unambiguously delineate all genes along chromosomes. At the same time, the differences

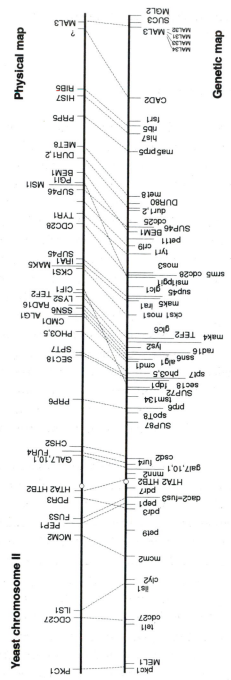

Figure 3.9 Comparison of genetic and physical maps from chromosome II. The genetic map with 71 mapped genes or markers was redrawn from Mortimer et al. (1992); the unmapped genes are listed below. The physical map was deduced from the complete sequence. Forty-two of the mapped genes and 11 of the unmapped genes could be unambiguously assigned to an ORF or an RNA gene from the sequence, on the basis of previous partial sequence data, use of probes or gene function; the assignment of four genes remained tentative. Thus, a total of 35 genes or markers remained unassigned on the physical map of chromosome II at the time the sequence had been determined.

found between various yeast strains demonstrate the need to use one particular strain as a reference system.

Relation of yeast to other genome projects
'Model' organisms

The impact of the study of model organisms on the human and other genome projects has been discussed recently (Miklos and Rubin, 1996). The systematic sequencing of one large model genome, the nematode *Caenorhabditis elegans* (105 Mb), has been tackled in a cooperation between the Sanger Centre in Cambridge and the Genome Sequencing Center at Washington University School of Medicine, St Louis (Hodgkin *et al.*, 1994). Sequence for the majority of genes (perhaps as much as 80 per cent) is presently available, so that the goal of finishing the entire sequence of the worm in 1998 seems realistic. Despite the fact that the genome size of *C. elegans* is eight times greater than that of yeast, the current estimates predict only 14 000 genes (twice the order of that of yeast): gene density is much lower in *C. elegans*, there are more and larger introns and longer intergenic regions than in yeast. The sequencing of other large model genomes such as the fruit fly *Drosophila melanogaster* (Spradling *et al.*, 1995) and the dicotyledonous plant *Arabidopsis thaliana* (Goodman *et al.*, 1995) has recently begun, but presently only defined segments of particular chromosomes are under study.

Prior to the release of the complete yeast genome sequence, two complete bacterial genomes had been made public: the 1.8 Mb sequence of *Haemophilus influenzae* (Fleischmann *et al.*, 1995) and the 0.6 Mb sequence of *Mycoplasma genitalium* (Fraser *et al.*, 1995). The sequence of the first archeon, *Methanococcus janaschii* (1.7 Mb) was released subsequently (Bult *et al.*, 1996). The small sizes of these genomes permitted the sequences to be solved by shotgun approaches eliminating the need for initial mapping efforts. Applying the same strategy, the sequence of at least two dozen other small microbial genomes is under way (Koonin *et al.*, 1996). Major attention is being given to several extremophile archea (e.g. *Haloferax volcanii*, *Methanopyrus kandleri*, *Pyrococcus furiosus*, *Sulfolobus solfataricus*), pathogenic bacteria (e.g. *Mycobacterium tuberculosis*), cyanobacteria and purple bacteria. Meanwhile, the entire sequences of *Mycoplasma pneumoniae* (Himmelreich *et al.*, 1996), *Synechocystis* sp. (Kaneko *et al.*, 1996) and *Helicobacter pylori* (Tomb *et al.*, 1997) have appeared. Genome sequencing of *E. coli* (Regalado, 1995) and *B. subtilis* (Devin, 1995) has been finalized in collaboration by a number of laboratories in Europe, Japan, and the USA, and pertinent data can be found in the databanks.

Other fungi

Now that the entire sequence of a laboratory strain of *S. cerevisiae* has been obtained, the complete sequences of other yeasts of industrial or medical importance is within our reach. Such knowledge would considerably accelerate the development of productive strains needed in other areas (e.g. *Kluyveromyces, Yarrowia*) or the search for novel antifungal drugs. It may even be unnecessary to finish the entire genome whenever a yeast or fungal genome displays considerable synteny with that of *S. cerevisiae*. In fact, recent studies with a filamentous fungus (*Ashbya gossypii*) have shown that many of its ORFs demonstrate considerable similarity to those of *S. cerevisiae* and that at least a quarter of the clones in an *A. gossypii* gene bank contain groups of genes in the same order and relative position as their *S. cerevisiae* counterparts (Altmann-Jöhl and Philippsen, 1996). On the other hand, there is probably no synteny between the *S. cerevisiae* and *Schizosaccharomyces pombe* genomes because of their evolutionary distance of some 1000 million years.

The human–yeast connection

The availability of the complete yeast genome sequence not only provides further insight into genome organization and evolution in yeast but extends the catalogue of novel genes detected in this organism. Many of these may be of particular value to yeast molecular biologists only, but of general interest may be those that are homologues to genes that perform differentiated functions in multicellular organisms or that might be of relevance to malignancy. Although the role of these genes has still to be clarified, yeast may offer a useful experimental system to identify their function. On the other hand, the wealth of information to be expected

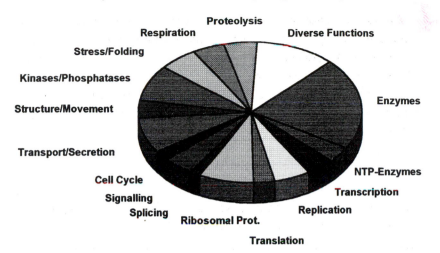

Figure 3.10 Human genes of known function with significant similarity to yeast genes. More than 22 per cent of the ORFs on the yeast chromosomes have significant scores in similarity searches against human genes in the present data libraries. The approximate functional classification of the proteins is shown.

clearly demands that new routes are explored to investigate the functions of novel genes.

Comparing the catalogue of human sequences available in the databases with the ORFs on the completed yeast chromosomes at the amino acid level reveals that more than 30 per cent of the yeast genes have homologues among the human genes of known function. Figure 3.10 shows an attempt to classify these genes according to (gross) function. As expected, most of the genes categorized in this way represent basic functions in both organisms.

Undoubtedly, the most compelling protagonists among these homologues are yeast genes that bear substantial similarity to human 'disease genes'. Recently, comparative studies along these lines have been published (Tugendreich *et al.*, 1994; Bassett *et al.*, 1996). Table 3.3 summarizes and extends these findings. It will be most interesting to continue this approach; the specific tools for searches of homologous genes in various organisms have been developed and are publicly available through a homepage on the Web (see Appendix).

Table 3.3. *Human disease genes with similarity to yeast genes. The positionally cloned genes are listed in order of decreasing statistical significances of the best match in the data banks*

Human disease	Human gene	Yeast gene	Yeast protein description
Hereditary non-polyposis colon cancer	MSH2	MSH2	DNA mismatch repair enzyme
Hereditary non-polyposis colon cancer	MLH1	MLH1	DNA mismatch repair enzyme
Cystic fibrosis	CFTR	YCF1	Cadmium resistance protein
Glycerol kinase deficiency	GK	GUT1	Glycerol kinase
Bloom syndrome	BLM	SGS1	Mismatch repair enzyme
Adrenoleukodystrophy, X-linked	ALD	PAL1	Phenyl ammonia lyase
Ataxia telangiectasia	ATM	TEL1	Telomere associated gene, chr. II
Pleitrophic lateral sclerosis	SOD1	SOD1	Superoxide dismutase
Myotonic dystrophy	DM	YPK1	cAMP-dependent protein kinase
Lowe syndrome	??		
Neurofibromatosis, type 1	NF1	IRA2	Inhibitory regulator of *ras*-cAMP, chr. II
Choriodermia	CHM	GDI1	GDP dissociation inhibitor
Diastrophic dysplasia	DTD	SUL1	Sulfate transport protein
Lissencephaly	LIS1	MET30	Methionine pathway factor
Thomsen disease	CLC1	GEF1	Chloride channel protein
Wilms tumour	WT1	FZF1	Sulfite resistance protein
Achondroplasia	FGFR3	IPL1	Protein kinase
Menkes disease	MNK	PCA1	Copper-transporting ATPase, chr II
Multiple endocrine neoplasia 2A	RET	CDC15	Cell division control protein 15
Duchenne muscular dystrophy	DMD	MLP1	Myosin-like protein
Aniridia	PAX6	PHO2	Regulator in phosphate metabolism
Gonadal dysgenesis	SRY	ROX1	Hypoxic function transcription repressor
Breast cancer, early onset	BCRA1	RAD18	DNA repair protein
Chronic granulomatous disease	NCF1	AGA1	α-agglutinin core protein
Epidermolytic palmoplantar keratoderma	KRT9	MLP1	Myosin-like protein
Wardenburg syndrome	PAX3	RPB1	RNA polymerase, subunit 10
Familial polyposis coli	APC	AMYH	Adenylate cyclase
Neurofibromatosis, Type 2	NF2	YNL161	Putative protein kinase
Retinoblastoma	RB1	CYP1	Regulator of O$_2$-dependent genes
Wiskott–Aldrich syndrome	WASP	CLA4	Protein kinase
Xeroderma pigmentosum	RAD27	YKL113	Nucleotide excision repair enzyme

The next goal: functional analysis in the yeast genome

From the beginning, it was evident to anyone engaged in the project that the determination of the entire sequence of the yeast genome should only be considered a prerequisite for functional studies of the many novel genes to be detected (Goffeau, 1994). Following these considerations, a European project has been initiated that will be devoted to a systematic functional analysis of – at least – those chromosomes which have been sequenced within the EU sequencing project (Oliver, 1996). Similar activities have been started in the international collaborating laboratories for the other chromosomes (Johnston, 1996). In the European programme called EUROFAN (*European Functional Analysis Network*), a first goal will be to investigate systematically the phenotypes resulting from disruption (and possibly overexpression) of the yeast genes of unknown function. A special set of yeast strains has been constructed for this purpose using a PCR-mediated gene replacement technique for the deletion of individual genes (Wach *et al.*, 1994).

Concurrently, it is intended to systematically construct complete transcriptional maps of entire chromosomes. One approach that may be of major importance in functional analysis is two-dimensional gel electrophoresis (Blomberg *et al.*, 1995; Wilkins *et al.*, 1996). The technique is extremely sensitive in that it is capable of analysing thousands of proteins from a single sample. The nature of each single spot can be identified through amino acid analysis, preferably with the help of mass spectrometry (Wilm *et al.*, 1996). Moreover, the technique can reveal whether proteins become modified as a consequence of 'environmental' changes. Differences in expression of the genetic information can be monitored by computer analysis of the resulting gel images. Along with this endeavour, the development of refined *in silicio* analysis methods will improve prediction of function (e.g. Casari *et al.*, 1996). These data will then be used as a basis for intensified functional analyses: relevant genes or groups of genes that are suggested to be involved in particular functions will be attributed to consortia of specialized laboratories for further exploitation.

The wealth of fresh and biologically relevant information collected from the yeast sequences and the functional analyses have an impact on other large scale sequencing projects. Clearly, of outstanding importance will remain those genes that are homologues to genes that perform differentiated functions in multicellular organisms or that are of relevance to malignancy. Given the high evolutionary conservation of a multitude of basal functions from yeast to man and the experimental advantages of the yeast system, it will be of great benefit to combine these potentials to assist the human genome project.

Acknowledgements

I wish to thank all colleagues who have actively contributed to the success of the yeast genome project, particularly those who participated in sequencing chromosome II (Feldmann *et al.*, 1994). I gratefully acknowledge help with information and computer analyses by B. Dujon, W. Mewes, K. Kleine, A. Zollner and J. Hani. The work carried out in this laboratory was supported by the EU under the BAP, BRIDGE and BIOTECH II Programmes, the Bundesminister für Forschung und Technologie, and the Fonds der Chemischen Industrie.

References

Altman-Jöhl, R. and Philippsen, P. (1996). AgTHR4, a new selection marker for transformation of the filamentous fungus *Ashbya gossypii*, maps in a four-gene cluster that is conserved between *A. gossypii* and *Saccharomyces cerevisiae. Molec. Gen. Genet.* 250:69–80.

Altschul, S. F., Gish, W., Miller, W., Myers, E. W. and Lipman, D. J. (1990). Basic local alignment search tool. *J. Mol. Biol.* 215:403–410.

Amati, B. B. and Gasser, S. M. (1988). Chromosomal ARS and CEN elements bind specifically to the yeast nuclear scaffold. *Cell* 54:967–978.

Bairoch, A., Bucher, P. and Hofmann, K. (1995). The PROSITE database: its status in 1995. *Nucleic Acids Res.* 24:189–196.

Bassett, D. E., Boguski, M. S. and Hieter, P. (1996). Yeast genes and human disease. *Nature* 379:589–590; http://www.ncbi.nlm.gov/XREFdb/

Bell, S. P. and Stillman, B. (1992). ATP-dependent recognition of eukaryotic origins of DNA replication by a multiprotein complex. *Nature* 357:128–134.

Bernardi, G. (1993). The isochore organization of the human genome and its evolutionary history: a review. *Gene* 135:57–66.

124

Blomberg, A. *et al.* (1995). Interlaboratory reproducibility of yeast protein patterns analyzed by immobilized pH gradient two-dimensional gel electrophoresis. *Electrophoresis* 16:1935–1945.

Bowman, S., Churcher, C., Badcock, K. *et al.* (1997). The nucleotide sequence of *Saccharomyces cerevisiae* chromosome XIII. *Nature* 387 (Suppl.):90–93.

Bram, R., Lue, N. F. and Kornberg, R. D. (1986). A GAL family of upstream activating sequences in yeast: roles in both induction and repression of transcription. *EMBO J.* 5:603–608.

Broach, J. R. Pringle, J. R. and Jones, E. W. (1981). *The Molecular and Cellular Biology of the Yeast Saccharomyces.* Cold Spring Harbor Laboratory Press, Cold Spring Harbor, New York.

Brodeur, G. M., Sandmeyer, S. B. and Olson, M. V. (1983). Consistent association between sigma elements and tRNA genes in yeast. *Proc. Natl. Acad. Sci. USA* 80:3292–3296.

Bult, C., White, O., Olsen, G. J. *et al.* (1996). Complete genome sequence of the methanogenic archaeon, *Methanococcus jannaschii. Science* 273:1058–1073.

Burke, D. T., Carle, G. F. and Olson, M. V. (1987). Cloning of large segments of exogenous DNA into yeast by means of artificial chromosome vectors. *Science* 236:806–812.

Bussey, H., Kaback, D. B., Zhong, W. *et al.* (1995). The nucleotide sequence of chromosome I from *Saccharomyces cerevisiae. Proc. Natl. Acad. Sci. USA* 92:3809–3813.

Bussey, H., Storms, R. K., Ahmed, A. *et al.* (1997). The nucleotide sequence of *Saccharomyces cerevisiae* chromosome XVI. *Nature* 387 (Suppl.):102–105.

Califano, A. and Rigoutsos, I. (1993). FLASH: A Fast Look-up Algorithm for String Homology. In: Proceedings of the First International Conference on Intelligent Systems for Molecular Biology, Bethesda, MD., pp. 56–64.

Carle, G. F. and Olson, M. V. (1985). An electrophoretic karyotype for yeast. *Proc. Natl. Acad. Sci. USA* 82:3756.

Casari, G., de Daruvar, A., Sander, C. and Schneider, R. (1996). Bioinformatics and the discovery of gene function. *Trends Genet.* 12:244–245.

Chalker, D. L. and Sandmeyer, S. B. (1992). Ty3 integrates within the region of RNA polymerase III transcription initiation. *Genes Dev.* 6:117–128.

Chalker, D. L. and Sandmeyer, S. B. (1993). Sites of RNA polymerase III transcription initiation and Ty3 integration at the U6 gene are positioned by the TATA box. *Proc. Natl. Acad. Sci. USA* 90:4927–4931.

Cherry, J. M., Ball, C., Weng, S. *et al.* (1997). Genetic and physical maps of *Saccharomyces cerevisiae. Nature* (Suppl.) 387:67–74.

Churcher, C., Bowman, S., Badcock, K. *et al.* (1997). The nucleotide sequence of *Saccharomyces cerevisiae* chromosome IX. *Nature* 387 (Suppl.):84–87.

Clarke, L. and Carbon, J. (1976). A colony bank containing synthetic ColE hybrid plasmids representative of the entire *E.coli* genome. *Cell* 9:91–99.

Clayton, R. A., White, O., Ketchum, K. A. and Venter, J. C. (1997). The first genome from the third domain of life. *Nature* 387:459–462.

Curcio, M. J. and Morse, R. H. (1996). Tying together integration and chromatin. *Trends Genet.* 12:436–438.

Dear, S. and Staden, R. (1991). A sequence assembly and editing program for efficient management of large sequencing projects. *Nucleic Acids Res.* 19:3907–3911.

Decottignies, A. and Goffeau, A. (1997). Complete inventory of the yeast ABC proteins. *Nature Genet.* 15:137–145.

Del Rey, F. J., Donahue, T. F. and Fink, G. R. (1983). The histidine tRNA genes of yeast. *J. Biol. Chem.* 258:8175–8182.

Dershowitz, A. and Newlon, C. S. (1993). The effect on chromosome stability of deleting replication origins. *Mol. Cell. Biol.* 13:391–398.

Devereux, J., Haeberli, P. and Smithies, O. (1984). A comprehensive set of sequences and analysis programs for the VAX. *Nucleic Acids Res.* 12:387–395.

Devin, K. M. (1995). The Bacillus subutilis genome project: aims and progress. *Trends Biotechnol.* 13:210–216.

Devine, S. E. and Boeke, J. D. (1996). Integration of the yeast retrotransposon Ty1 is targeted to regions upstream of genes transcribed by RNA polymerase III. *Genes Dev.* 10:620–630.

Dietrich, F. S., Mulligan, J., Hennessy, K. *et al.* (1997). The nucleotide sequence of *Saccharomyces cerevisiae* chromosome V. *Nature* 387 (Suppl.):78–81.

Dujon, B. (1996) The yeast genome project: what did we learn? *Trends Genet.* 12:263–270.

Dujon, B., Alexandraki, D., Andre, B. *et al.* (1994). Complete nucleotide sequence of yeast chromosome XI. *Nature* 369:371–378.

Dujon, B., Albermann, K., Aldea, M. *et al.* (1997). The nucleotide sequence of *Saccharomyces cerevisiae* chromosome XV. *Nature* 387 (Suppl.):98–102.

Eigel, A. and Feldmann, H. (1982). Ty1 and delta elements occur adjacent to several tRNA genes in yeast. *EMBO J.* 1:1245–1250.

Eisenberg, S., Civalier, C. and Tye, B. K. (1988). Specific interaction between a *Saccharomyces cerevisiae* protein and a DNA element associated with certain autonomously replicating sequences. *Proc. Natl. Acad. Sci. USA* 85:743–746.

126

El-Mabrouk, N. and Lisacek, F. (1996). Very fast identification of RNA motifs in genomic DNA. Application to tRNA search in the yeast genome. *J. Mol. Biol.* 264:46–55.

Estes, H. G., Robinson, B. S. and Eisenberg, S. (1992). At least three distinct proteins are necessary for the reconstitution of a specific multiprotein complex at a eukaryotic chromosomal origin of replication. *Proc. Natl. Acad. Sci. USA* 89:11156–11160.

Fangman, W. L. and Brewer, B. J. (1992). A question of time: replication origins of 58 eukaryotic chromosomes. *Cell* 71:363–366.

Fairhead, C. and Dujon, B. (1994). Transcript map of two regions of chromosome XI of *Saccharomyces cerevisiae* for interpretation of sympatic sequencing results. *Yeast* 10:1403–1413.

Feldmann, H., Aigle, M., Aljinovic, G. *et al.* (1994). Complete DNA sequence of yeast chromosome II. *EMBO J.* 13:5795–5809.

Fields, S. and Song, O. K. (1989). A novel genetic system to detect protein–protein interactions. *Nature* 340:245–246.

Fink, G. R., Boeke, J. D. and Garfinkel, D. J. (1986). The mechanism and consequences of retrotransposition. *Trends Genet.* 2:118–123.

Fishant, G. A. and Burks, C. (1991). tRNAscan: identifying potential tRNA genes in genomic DNA sequences. *J. Mol. Biol.* 220:659–671.

Fleischmann, R. D., Adams, M. D., White, O. *et al.* (1995). Whole-genome random sequencing and assembly of *Haemophilus influenzae* Rd. *Science* 269:496–512.

Fraser, C. M., Gocayne, J. D., White, O. *et al.* (1995). The minimal gene complement of *Mycoplasma genitalium*. *Science* 270:397–403.

Galibert, F., Alexandraki, D., Baur, A. *et al.* (1996). Complete nucleotide sequence of *Saccharomyces cerevisiae* chromosome X. *EMBO J.* 15:2031–2049.

Garrells, J. I. (1996). YPD: a database for the proteins of *Saccharomyces cerevisiae*. *Nucleic Acids Res.* 24:46–49; http://quest7.proteome.com/YPDhome.html

Goffeau, A. (1994). Yeast genes in search of functions. *Nature* 369:101–102.

Goffeau, A., Nakai, K., Slonimski, P. P. and Risler, J. L. (1993). The membrane proteins encoded by yeast chromosome III genes. *FEBS Lett.* 325:112–117.

Goffeau, A., Barrell, B. G., Bussey, H. *et al.* (1996). Life with 6000 genes. *Science* 274:546–567.

Goffeau, A. *et al.* (1997a) The yeast genome directory. *Nature* 387 (Suppl.):1–105.

Goffeau, A., Park J., Paulsen, I. T., Jonniaux, J. L., Dinh, T., Mordant, P. and Sailer, M. H. Jr (1997b). Multi-drug-resistant transport proteins in yeast: complete inventory and phylogenetic characterization of yeast open reading frames with the major facilitator superfamily. *Yeast* 13:43–54.

127

Goodman, H. M., Eckers, J. R. and Dean, C. (1995). The genome of *Arabidopsis thaliana. Proc. Natl. Acad. Sci. USA* 92:10831–10835.

Guthrie, C. and Fink, G. R. (1991). Guide to yeast genetics and molecular biology. *Methods Enzymol.* 169.

Hartl, F. U., Pfanner, N., Nicholson, D. W. and Neupert, W. (1989). Mitochondrial protein import. *Biochim. Biophys. Acta* 988:1–45.

Hauber, J., Stucka, R., Krieg, R. and Feldmann, H. (1988). Analysis of yeast chromosomal regions carrying members of the glutamate tRNA gene family: various transposable elements are associated with them. *Nucl. Acids Res.* 16:10623–10634.

Heumann, K., Harris, C. and Mewes, H. W. (1996). Exhaustive analysis of genetic redundancy in *S. cerevisiae*. Pp. 98–108 in *Proceedings of the Fourth International Conference on Intelligent Systems for Molecular Biology*. AAAI Press, Menlo Park.

Himmelreich, R., Hilbert, H., Plagens, H., Pirkl, E. and Herrmann, R. (1996). Complete sequence analysis of the genome of the bacterium *Mycoplasma pneumoniae. Nucleic Acids Res.* 24:4420–4449.

Hodgkin, J., Plasterk, R. H. A. and Waterston, R. H. (1994). The nematode *Caenorhabditis elegans* and its genome. *Science* 270:410–440.

Huang, M. E., Chuat, J. C., Thierry, A., Dujon, B. and Galibert, F. (1994). Construction of a cosmid contig and of an *Eco*RI restriction map of yeast chromosome X. *DNA Seq.* 4:293–300.

Ito, H., Fukuda, Y., Murata, K. and Kimura, A. (1983). Transformation of intact yeast cells treated with alkali cations. *J. Bacteriol.* 153:163–168.

Jacq, C., Alt-Mörbe, B., Arnold, W. *et al.* (1997). The nucleotide sequence of *Saccharomyces cerevisiae* chromosome IV. *Nature* 387 (Suppl.): 75–78.

Ji, H., Moore, D. P., Blomberg, M. A. *et al.* (1993). Hot spots for unselected Ty1 transposition events on yeast chromosome III are near tRNA genes and LTR sequences. *Cell* 73:1007–1013.

Johnston, J. R. (1994). *Molecular Genetics of Yeast: a Practical Approach*. Oxford University Press, Oxford.

Johnston, M. (1996). Towards a complete understanding of how a simple eukaryotic cell works. *Trends Genet.* 12:242–243.

Johnston, M., Andrews, S., Brinkman, R. *et al.* (1994). Complete nucleotide sequence of *Saccharomyces cerevisiae* chromosome VIII. *Science* 265:2077–2082.

Johnston, M., Hillier, L., Riles, L. *et al.* (1997). The nucleotide sequence of *Saccharomyces cerevisiae* chromosome IX. *Nature* 387 (Suppl.): 87–90.

128

Kalegoropoulos, A. (1995). Automatic intron detection in nuclear DNA sequences of *Saccharomyces cerevisiae*. *Yeast* 11:555–565.

Kaneko, T., Sato, S., Kotani, H. *et al.* (1996) Sequence analysis of the genome of the unicellular cyanobacterium *Synechocystis* sp. strain PCC6803. II. Sequence determination of the entire genome and assignment of potential protein-coding regions. *DNA Res.* 3:109–136.

Kirchner, J., Connolly, C. M. and Sandmeyer, S. B. (1995). Requirement of RNA polymerase III transcription factors for in vitro position-specific integration of a retrovirus-like element. *Science* 267:1488–1491.

Klein, P., Kanehisa, M. and Delisi, C. (1985). The detection and classification of membrane spanning proteins. *Biochim. Biophys. Acta* 815:468–476.

Koonin, E. V., Mushegian, A. R. and Rudd, K. E. (1996). Bacterial genome sequences. *Curr. Biol.* 6:404–416.

Levy, J. (1994). Sequencing the yeast genome: an international achievement. *Yeast* 10:1689–1706.

Lindegren, C. C. (1949). *The Yeast Cell, Its Genetics and Cytology*. Educational Publishers, St Louis.

Lloyd, A. T. and Sharp, P. M. (1992a). Codons: a microcomputer program for codon usage analysis. *J. Hered.* 83:239–240.

—— (1992b). Evolution of codon usage patterns. *Nucleic Acids Res.* 20:5289–5295.

Lochmüller, H., Stucka, R. and Feldmann, H. (1989) A hot-spot for transposition of various Ty elements on chromosome V in *Saccharomyces cerevisiae*. *Curr. Genet.* 16:247–252.

Louis, E. J. (1994). Corrected sequence for the right telomere of *S. cerevisiae* chromosome III. *Yeast* 10:271–274.

Louis, E. J. and Borts, R. H. (1995). A complete set of marked telomeres in *Saccharomyces cerevisiae* for physical mapping and cloning. *Genetics* 139:125–136.

Louis, E. J. and Haber, J. E. (1992). The structure and evolution of subtelomeric Y' repeats in *Saccharomyces cerevisiae*. *Genetics* 119:303–315

Lowe, T. M. and Eddy, S. R. (1997). tRNAscan-SE: a program for improved detection of transfer RNA genes in genomic sequences. *Nucleic Acids Res.* 25:955–964.

Lundblad, V. and Blackburn, E. H. (1993). An alternative pathway for yeast telomere maintenance rescues *est1* sequences. *Cell* 73:347–360.

Marahrens, Y. and Stillman, B. (1992). A yeast chromosomal origin of DNA replication defined by multiple functional elements. *Science* 255:817–823.

Mewes, H. W. and Heumann, K. (1995). Genome analysis: pattern search in biological macromolecules. *Proc.6th Ann. Symp. on Combinatorial Pattern Matching, LNCS* 937:261–285.

Mewes, H. W., Albermann, K., Bähr, M. *et al.* (1997). Dictionary of the yeast genome. *Nature* 387 (Suppl.): 9–32.

Michels, C. A., Read, E., Nat, K. and Charron, M. J. (1992). The telomere-associated *MAL3* locus of *S. cerevisiae* is a tandem array of repeated genes. *Yeast* 8:655–665.

Miklos, G. L. G. and Rubin, G. M. (1996) The role of the genome project in determining gene function: insights from model organisms. *Cell* 86:521–529.

Milosavljevic, A. and Jurka, J. (1993). Discovering simple DNA sequences by the algorithmic significance method. *CABIOS* 9:(4): 407–411.

Mortimer, R. K. and Schild, D. (1991). Pp. 11–26 in J. R. Broach, J. R. Pringle and E. W. Jones, eds. *The Molecular and Cellular Biology of the Yeast Saccharomyces.* Cold Spring Harbor Press, Cold Spring Harbor, New York.

Mortimer, R. K., Contopoulou, R. and King, J. S. (1992). Genetic and physical maps of *Saccharomyces cerevisiæ*, Edition 11. *Yeast* 8:817–902.

Murakami, Y., Naitou, M., Hagiwara, H. *et al.* (1995). Analysis of the nucleotide sequence of chromosome VI from *Saccharomyces cerevisiae. Nature Genet.* 10:261–268.

Newlon, C. S. (1988). Yeast chromosome replication and segregation. *Microbiol. Rev.* 52:568–601.

Newman, A. (1994). RNA splicing. Activity in the spliceosome. *Curr. Biol.* 4:462–464.

Niedenthal, R. K., Riles, L., Johnston, M. and Hegemann, H. J. (1996). Green fluorescent protein as a marker for gene expression and subcellular localization in budding yeast. *Yeast* 12:773–786.

Oliver, S. G. (1996). A network approach to the systematic analysis of yeast gene function. *Trends Genet.* 12:241–242.

Oliver, S. G., van der Aart, Q. J., Agostoni-Carbone, M. L. *et al.* (1992). The complete nucleotide sequence of yeast chromosome III. *Nature* 357:38–46.

Olsen, G. J. and Woese, C. R. (1996). Lessons from an Archaeal genome: what are we learning from *Methanococcus janaschii? Trends Genet.* 12:377–379.

Olson, M. V. (1993). The human genome project. *Proc. Natl. Acad. Sci. USA* 90:4338–4344.

Olson, M. V., Dutchik, J. E., Graham, M. Y. *et al.* (1986). Random-clone strategy for genomic restriction mapping in yeast. *Proc. Natl. Acad. Sci. USA* 83:7826–7830.

Saccharomyces cerevisiae: the first complete genome sequence

Palladino, F. and Gasser, S. M. (1994). Telomere maintenance and gene repression, a common end? *Curr. Opin. Cell Biol.* 6:373–379.

Pavesi, A., Conterio, F., Bolchi, A., Dieci, G. and Ottonello, S. (1994). Identification of new eucaryotic tRNA genes in genomic DNA databases by a multistep weight matrix analysis of transcriptional conttrol regions. *Nucleic Acids Res.* 22:1247–1256.

Pearson, W. R. and Lipman, D. J. (1988). Improved tools for biological sequence comparison. *Proc. Natl. Acad. Sci. USA,* 85:2444–2448.

Percudani, R., Pavesi, A. and Ottonello, S. (1997). Transfer RNA gene redundancy and translational selection in *Saccharomyces cerevisiae. J. Mol. Biol.* 268:322–330.

Philippsen, P., Kleine, K., Pöhlmann, R. *et al.* (1997). The nucleotide sequence of *Saccharomyces cerevisiae* chromosome XIV and its implications. *Nature* 387 (Suppl.): 93–98.

Proudfoot, N. (1991). Poly(A) signals. *Cell* 64:671–674.

Regalado, A. (1995) Commotion over *E. coli* project. *Science* 267:1899–1900.

Richard, G.-F. and Dujon, B. (1996). Distribution and variability of trinucleotide repeats in the genome of the yeast *Saccharomyces cerevisiae. Gene* 174:165–174

Riles, L., Dutchik, J. E., Baktha, A. *et al.* (1993). Physical maps of the smallest chromosomes of *Saccharomyces cerevisiae* at a resolution of 2.6 kilobase pairs. *Genetics* 134:81– 150.

Roman, H. (1981) Development of yeast as an experimental organism. In: J. N. Strathern, E. W. Jones and J. R. Broach, eds. *The Molecular and Cellular Biology of the Yeast Saccharomyces.* Cold Spring Harbor Laboratory Press, Cold Spring Harbor, New York.

Sandmeyer, S. B. (1992). Yeast retrotransposons. *Curr. Opin. Genet. Dev.* 2:705–711.

Scholler, P., Schwarz, S. and Hoheisel, J. D. (1995). High-resolution cosmid mapping of the left arm of *Saccharomyces cerevisiae* chromosome XII; a first step towards an ordered sequencing approach. *Yeast* 11:659–666.

Sharp, P. M. and Li, W. H. (1987). The codon adaptation index: a measure of directional synonymous codon usage bias, and its potential applications. *Nucleic Acids Res.* 15:1281–1295.

Sharp, P. M. and Lloyd, A. T. (1993). Regional base composition variation along yeast chromosome III: evolution of chromosome primary structure. *Nucleic Acids Res.* 21:179–183.

Shirahige, K., Iwasaki, T., Rashid, M. B., Ogasawa, N. and Yoshikawa, H. (1993). Localisation and characterization of autonomously replicating sequences from chromosome VI of *Saccharomyces cerevisiae*. *Mol. Cell. Biol.* 13:5043–5056.

Simchen, G., Chapman, K. B., Caputo, E. *et al.* (1994). Correction of the genetic map from chromosome XI of *Saccharomyces cerevisiae*. *Genetics* 138:283–287.

Smith, H. E., Yu, S. S. Y., Neigeborn, L., Driswell, S. E. and Mitchell, A. P. (1990) Role of *IME1* expression in regulation of meiosis in *Saccharomyces cerevisiae*. *Mol. Cell. Biol.* 10:6103–6113.

Spradling, A. C. *et al.* (1995). Gene disruptions using P transposable elements: an integral component of the *Drosophila* genome project. *Proc. Natl. Acad. Sci. USA* 92:10824–10830.

Sprinzl, M., Stegborn, C., Hübel, F. and Steinberg, S. (1996). Compilation of tRNA sequences and sequences of tRNA genes. *Nucleic Acids Res.* 24:68–72.

Strathern, J. N., Jones, E. W. and Broach, J. R. (1981). *The Molecular and Cellular Biology of the Yeast Saccharomyces*. Cold Spring Harbor Laboratory Press, Cold Spring Harbor, New York.

Struhl, K. (1987). Promoters, activator proteins, and the mechanism of transcriptional initiation in yeast. *Cell* 49:295–297.

Stucka, R. (1992). Thesis. University of Munich.

Stucka, R. and Feldmann, H. (1994). Cosmid cloning of Yeast DNA. Pp. 49–64 in J. Johnston, ed. *Molecular Genetics of Yeast: a Practical Approach*. Oxford University Press, Oxford.

Svetlov, V. V. and Cooper, T. G. (1995). Review: compilation and characteristics of dedicated transcription factors in *Saccharomyces cerevisiae*. *Yeast* 11:1439–1484.

Tettelin, H., Agostini Carbone, M. L., Albermann, K. *et al.* (1997). The nucleotide sequence of *Saccharomyces cerevisiae* chromosome VII. *Nature* 387 (Suppl.): 81–84.

Thierry, A. and Dujon, B. (1992). Nested chromosomal fragmentation in yeast using the meganuclease *I-Sce* I: a new method for physical mapping of eukaryotic genomes. *Nucleic Acids Res.* 20:5625–5631.

Thierry, A., Gaillon, L., Galibert, F. and Dujon, B. (1995). Construction of a complete genomic library of *Saccharomyces cerevisiae* and physical mapping of chromosome XI at 3.7 kb resolution. *Yeast* 11:121–135.

Thierry-Mieg, J. and Durbin, R. (1992). ACeDb, *C. elegans* database. *Cahiers IMA BIO* 5:15–24.

Tomb, J.-F., White, O., Kervalage, A. R. *et al.* (1997). The complete genome sequence of the gastric pathogen *Helicobacter pylori*. *Nature* 388:539–547.

132

Tugendreich, S., Bassett Jr, D. E., McKusick, V. A., Boguski, M. S. and Hieter, P. (1994). Genes conserved in yeast and humans. *Human Genome Res.* 3:1509–1517.

Van Houten, J. V. and Newlon, C. S. (1990). Mutational analysis of the consensus sequence of a replication origin from yeast chromosome III. *Mol. Cell. Biol.* 10:3917–3925.

Vassarotti, A. and Goffeau, A. (1992). Sequencing the yeast genome: the European effort. *Trends in Biotechnology* 10:15–18.

Vassarotti, A., Dujon, B., Mordant, P., Feldmann, H., Mewes, H. W. and Goffeau, A. (1995). Structure and organization of the European Yeast Genome Sequencing Network. *J. Biotechnol.* 41:131–137.

Viswanathan, M., Muthukumar, G., Cong, Y. S. and Lenard, J. (1994). Seripauperins of *Saccharomyces cerevisiae*: a new multigene family encoding serine-poor relatives of serine-rich proteins. *Gene* 148:149–153.

Voytas, D. F. and Boeke, J. D. (1993). Yeast retrotransposons and tRNAs. *Trends Genet.* 9:421–427.

Wach, A., Brachat, A., Pohlmann, R. and Philippsen, P. (1994). New heterologous modules for classical or PCR-based gene disruptions in *Saccharomyces cerevisiae*. *Yeast* 10:1793–1808.

Warmington, J. R., Anwar, R., Newlon, C. S. *et al.* (1986). A 'hot-spot' for transposition on the left arm of yeast chromosome III. *Nucleic Acids Res.* 14:3475–3485.

Warmington, J. R., Green, R. P., Newlon, C. S. and Oliver, S. G. (1987). Hot-spots for Ty transposition. *Nucleic Acids Res.* 15: 8963–8982.

Wilkins, M. R. *et al.* (1996). From proteins to proteome: large scale protein identification by two dimensional electrophoresis and amino acid analysis. *Biotechnology* 14:61–65.

Wilm, M., Shevchenko, A., Honthaeve, T. *et al.* (1996). Femtomole sequencing of proteins from polyacrylamide gels by nano- electroscopy mass spectrometry. *Nature* 379:466–469.

Wolfe, K. and Shields, D. C., (1997). Molecular evidence for an ancient duplication of the entire yeast genome. *Nature* 387:708–713.

Yoshikawa, A. and Isono, K. (1990). Chromosome III of *Saccharomyces cerevisiae*: an ordered clone bank, a detailed restriction map and analysis of transcripts suggest the presence of 160 genes. *Yeast* 6:383–401.

Zou, S., Ke, N., Kim, S. J. and Voytas, D. F. (1996). The *Saccharomyces* retrotransposon Ty5 integrates preferentially into regions of silent chromatin at the telomeres and mating loci. *Genes Dev.* 10:634–645.

133

Appendix

A number of information resources on yeast are available on the Internet [S. Walsh and B. Barrell (1996) *Trends Genet.* 12:276–277]. Valuable information coupled with data libraries, routines for searches and data retrieval can be found in several home pages on the World Wide Web. Sequence data can also be retrieved from various databases by FTP or e-mail. Likewise, these resources offer information and special services via e-mail addresses.

Information on the yeast genome and related projects, search and query facilities

http://www.mips.biochem.mpg.de/ yeast/
http://www.embl-ebi.ac.uk
http://www.sanger.ac.uk/yeast/home.html
http://genome-www.stanford.edu
http://www.ncbi.nlm.nih.gov/
http://www.ncbi.nlm.nih.gov/XREFdb
http://quest7.proteome.com/YPDhome.html
http://expasy/hcuge.ch/cgi.bin/list?yeast.txt

e-mail addresses

mewes@mips.biochem.mpg.de [information on the European yeast project]
barrell@sanger.ac.uk [information on the yeast project at Cambridge]
yeast-curator@genome.stanford.edu [information on the American yeast project]
NetServ@ebi.ac.uk [general information on databases]
DataLib@ebi.ac.uk [general information on databases]

Data retrieval by FTP

ftp://mips.embnet.org/yeast/
ftp://ftp.ebi.ec.uk/pub/databases/yeast
ftp://ftp.ebi.ec.uk/pub/databases/lista
ftp://genome-ftp.stanford.edu/yeast/genome_seq

4 Interactions between pathway-specific and global genetic regulation and the control of pathway flux

A. R. HAWKINS, K. A. WHEELER, L. J. LEVETT,
G. H. NEWTON AND H. K. LAMB

Introduction

It is evident that living organisms regulate their metabolism in order to balance the conflicting demands of providing building blocks for synthesis and growth whilst at the same time providing energy derived from catabolism. These metabolic pathways are regulated by well-characterized enzymatic methods such as feedback inhibition and, once appropriate levels of enzymes have been provided, can be efficiently regulated at the level of the control of metabolic flux by the myriad of interactions between the enzymes and metabolites. Many essential enzymes and proteins are produced constitutively, but others are regulated at the level of gene transcription and/or translation. It is likely that in the most primitive cells, metabolic pathways evolved before transcription regulatory mechanisms. Transcriptional control may have evolved, therefore, to act as a damping mechanism, smoothing out the effects of more rapid changes caused by the build-up and dissemination of pathway metabolites. Or such regulation may be required to target the production of enzymes to particular tissues or organelles, or to allow the organism to use preferred carbon or nitrogen sources. But one of the driving forces for the cell may have been energy conservation: ensuring that enzymes are only supplied when they are needed, and stopping unnecessary enzyme synthesis. In this possible latter role, the cell has had to develop metabolite recognition mechanisms and signal transduction pathways that are able to sense particular cellular metabolites and couple this to transcription control mechanisms. Transcriptional control may be of particular importance when the organism is required to direct appropriate levels of

135

flux through potentially competing anabolic and catabolic pathways that share common metabolites.

Types of regulation

Regulation of the genes and enzymes of the fungal cell can be classified into several levels of control. Transcription mechanisms that regulate metabolic pathways can be divided into two broad categories: pathway-specific and global control. *Pathway-specific control* is typical of catabolic systems required for the utilization of alternative carbon, nitrogen, sulfur and phosphorus sources, and is characterized by the presence of signal transduction pathways that are able to recognize the presence of pathway metabolites and to stimulate transcription of appropriate genes. This control is mediated at the level of transcription initiation, but can be overridden by global control mechanisms. The paradigm for pathway-specific control is the galactose utilization pathway in *Saccharomyces cerevisiae.*

Global control mechanisms apparently function to allow the organism both to use a wide array of carbon, nitrogen, sulfur and phosphorus sources and to allow some discrimination so that when, for example, several carbon or nitrogen sources are available as a mixture, only preferred sources are used. Global control systems are, typically, able to recognize small organic effector molecules and to determine the transcription activation status of pathway-specific control mechanisms. The mechanisms by which this global control is exerted include the control of transcription initiation and post-translational control. Good examples of global control are afforded by carbon catabolite repression and nitrogen metabolite repression in filamentous fungi, and general amino acid control in yeast.

Pathway-specific transcription regulation and control of metabolic flux

The quinate and shikimate pathways of *Neurospora crassa* and *Aspergillus nidulans*

The quinate utilization (*qut*) pathway of *Neurospora crassa*, *Aspergillus nidulans* and other microbial eukaryotes is a dispensable alternative carbon utilization pathway that exploits quinate, which constitutes around 10 per cent of leaf litter. Production of the three enzymes necessary to convert quinate to protocatechuate is induced by the presence of quinate (Figure 4.1). This induction is subject to carbon catabolite repression: if quinic acid is provided as a mixed carbon source with glucose, then the induction of the three quinate pathway enzymes is substantially reduced. Two of the intermediates in the catabolism of quinate, dehydroquinate (DHQ) and dehydroshikimate (DHS), are also metabolites in the prechorismate section of the shikimate pathway (see Figure 4.1) where they are interconverted by two structurally unrelated isoenzymes (types I and II dehydroquinases), which have different reaction mechanisms. The dehydroquinase isoenzymes therefore provide a good example of convergent evolution. One isoenzyme (type I) comprises one domain of the pentadomain AROM protein which is active in the shikimate pathway and the other (type II) enzyme is active in the quinate pathway. In the shikimate pathway, the AROM protein is produced constitutively at a low level.

The original dissection of the quinate and shikimate pathways started in the 1960s using *Neurospora crassa*. Initial genetic evidence identified four linked genes, one for each of the three *qut* pathway enzymes (designated *qa* in *N. crassa*), and one interpreted to encode a control protein that could be split into two distinct regulatory domains. The two types of allele at this regulatory locus were characterized by either a fast (*qa-1F*) or slow (*qa-1S*) complementation response when combined in heterokaryons with certain mutant alleles encoding nonfunctional quinate pathway enzymes. Not surprisingly, these early genetic data were interpreted within a framework provided by prokaryotic operon model systems (such as that of Jacob and Monod), an interpretation strengthened by parallel studies on the prechorismate section of the shikimate pathway. Later molecular data showed that the *qa-1* gene was in fact two closely linked genes, with the *qa-1F* mutants identifying an activator protein and the *qa-1S* mutants identifying a repressor protein.

Mutants singly lacking the enzymes necessary for the synthesis of chorismate from 3-deoxy-D-arabino-heptulosonic acid-7-phosphate were identified; one class, however, was missing. Mutants in dehydroquinase were not recovered because the equivalent *qut* pathway isoenzyme

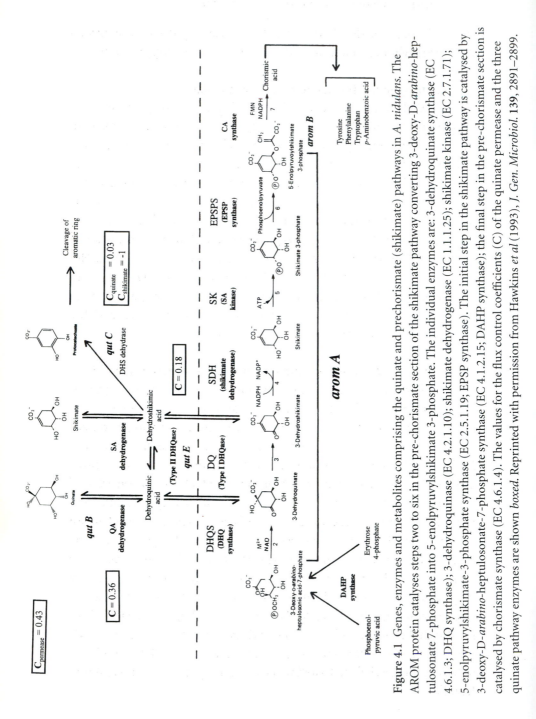

Figure 4.1 Genes, enzymes and metabolites comprising the quinate and prechorismate (shikimate) pathways in *A. nidulans*. The AROM protein catalyses steps two to six in the pre-chorismate section of the shikimate pathway converting 3-deoxy-D-*arabino*-heptulosonate 7-phosphate into 5-enolpyruvylshikimate 3-phosphate. The individual enzymes are: 3-dehydroquinate synthase (EC 4.6.1.3; DHQ synthase); 3-dehydroquinase (EC 4.2.1.10); shikimate dehydrogenase (EC 1.1.1.25); shikimate kinase (EC 2.7.1.71); 5-enolpyruvylshikimate-3-phosphate synthase (EC 2.5.1.19; EPSP synthase). The initial step in the shikimate pathway is catalysed by 3-deoxy-D-*arabino*-heptulosonate-7-phosphate synthase (EC 4.1.2.15; DAHP synthase); the final step in the pre-chorismate section is catalysed by chorismate synthase (EC 4.6.1.4). The values for the flux control coefficients (C) of the quinate permease and the three quinate pathway enzymes are shown *boxed*. Reprinted with permission from Hawkins *et al* (1993). *J. Gen. Microbiol.* **139**, 2891–2899.

could substitute in the shikimate pathway. causing the corresponding shikimate pathway mutants to elude the mutant screening procedure.

Cross pathway flux of common intermediates

When the mutation causing the loss of the shikimate pathway dehydroquinase was placed in a genetic background that was wild-type with respect to the quinate pathway, it proved to have interesting properties. Such strains were auxotrophic for the aromatic amino acids, but produced the quinate pathway enzymes constitutively, even in the absence of exogenously supplied quinate. This constitutive production of the quinate pathway enzymes was attributed to the accumulation of DHQ generated by the shikimate pathway leading to induction of the quinate pathway and provided the first *in vivo* evidence that the pools of DHQ and DHS produced by the quinate and shikimate pathways were not kept entirely separate. The presence of metabolites other than quinate could therefore lead to the induction of the quinate pathway enzymes, with the most likely candidates being DHQ, shikimate and DHS.

Does the AROM protein channel the common intermediates ?

Genetic analysis *of N. crassa* shikimate pathway mutants (designated *arom*) showed that the mutants lacking steps two to six in the shikimate pathway (see Figure 4.1) were tightly linked forming a 'cluster gene'. In wild-type strains, the presence of shikimate pathway-derived DHQ and DHS does not lead to internal induction of the *qut* pathway enzymes. Taken in conjunction with the finding that AROM consisted of two identical pentafunctional polypeptides, a channelling hypothesis was proposed: that the AROM served as an efficient channelling mechanism to keep the shikimate pathway-derived DHQ and DHS separate from the quinate pathway. That a mutant AROM, lacking the dehydroquinase activity leaks DHQ, and that all five enzyme activities could be separately assayed *in vitro* using purified protein suggested that any channelling function could not be complete. The channelling hypothesis was subsequently challenged by others who could find no *in vivo* or *in vitro* evidence for such a function and, as described below, later work on metabolic flux has helped to define the common pool of intermediates in the cell.

Two control genes and a permease in the quinate pathway of *A. nidulans*

At this point, work on the equivalent quinate utilization (*qut*) pathway in *A. nidulans* began to make a contribution to the question of the regulation of enzyme production and control of metabolic flux. Genetic analysis of the ability to utilize quinate as a carbon source identified three classes of mutant corresponding to different complementation groups that pleiotropically affected the production of all three of the *qut* pathway enzymes (Table 4.1). One class (*qutD* in *A. nidulans*; *qa-Y* in *N. crassa*) encoded a permease necessary to transport quinate into the mycelium at neutral pH. A second class mainly gave rise to recessive mutations that had a non-inducible phenotype but also rarely to a dominant non-inducible phenotype. These mutations (*qutA* in *A. nidulans*; *qa-1F* in *N. crassa*) were interpreted as identifying a positively acting regulatory protein. The third class were identified in *N. crassa* (designated *qa-1S*) in a colorimetric plate test designed to identify mutants that constitutively produced the quinate pathway enzymes in the absence of quinate. Such mutants were found to be recessive. In *A. nidulans*, the equivalent gene (*qutR*) was identified as a second site suppressor of *qutD⁻* mutations that, when outcrossed, conferred a recessive phenotype of constitutive production of the quinate pathway enzymes in the absence of quinate. Dominant mutations conferring the phenotype of quinate non-utilization and apparently mapping at the *qa-1S* and *qutR* loci were identified in *N. crassa* and *A. nidulans* and found to revert at a high frequency to a mutant phenotype of constitutive production of the quinate pathway enzymes in the absence of quinate. This third class of mutant was interpreted as identifying a negatively acting control gene.

Transcription activator binding sites, zinc binuclear clusters and autoregulation

In both *N. crassa* and *A. nidulans*, the genes involved in the utilization of quinate and its wild-type regulation have been shown to map in a cluster. Both clusters have been isolated and the genes encoding the quinate pathway enzymes shown to be regulated at the level of transcriptional control. Short, approximately palindromic sequences have been identified in the promoters of the *qa* and *qut* genes as the potential

Table 4.1. *The genes comprising the* qut *cluster in* Aspergillus nidulans

Locus	Mutant phenotype	Function	Comments
qutA	Recessive loss of induction of all three enzymes	Encodes a transcription activator	The QUTA protein has a zinc binuclear cluster motif and is related to the two amino-terminal domains of the pentadomain AROM protein
qutR	Recessive constitutive production of all three enzymes; dominant or semi-dominant loss of induction for all three enzymes	Encodes a transcription repressor	The QUTR amino acid sequence is related to the three carboxy-terminal domains of the pentadomain AROM protein
qutB	Recessive loss of quinate dehydrogenase	Encodes quinate dehydrogenase	
qutC	Recessive loss of dehydroshikimate dehydratase	Encodes dehydroshikimate dehydratase	
qutE	Recessive loss of the type II	Encodes a type II dehydroquinase	
qutD	Recessive loss of induction for all three enzymes	Encodes a quinate permease	The QUTD protein has several membrane-spanning hydrophobic motifs, the mutant phenotype is pH repairable, and the qutD gene has two putative introns
qutG	Not determined	Unknown, possibly encodes a phosphatase	The QUTG amino acid sequence is highly similar to bovine myo inositol monophosphatase, the qutG gene has four putative introns
qutH	Not determined	Unknown	
aromA	Unable to synthesis the aromatic amino acids	Encodes a pentadomain protein that catalyses steps two to six in the prechorismate section of the shikimate pathway	The AROMA protein is homologous with the QUTA and QUTR proteins

binding sites for the QA-1F and QUTA proteins. The binding of the *N. crassa* QA-1F protein to these sequences has been confirmed. The *A. nidulans* QUTA activator protein controls the production of its own mRNA and that of the *qutR* repressor gene in an autoregulatory circuit. Further, the QUTA and QA-1F proteins contain zinc binuclear cluster motifs of the 6 Cys type; a peptide of 129 amino acids from the QUTA protein containing this motif binds zinc *in vitro*.

The activator protein has two domains

Genetic analysis of 23 non-inducible *qutA⁻* mutants revealed that they all map within the amino-terminal half of the encoded QUTA protein. One dominant mutation, *qutA382*, introduces a stop codon such that the encoded protein is approximately half the size of the native protein. This truncated protein can bind DNA but cannot activate transcription. A second dominant mutation, *qutA214*, is missense, changing 457E→K in a region of localized high negative charge that is just distal to the *qutA382* mutation and potentially identifies a transcription activation domain. The positions and phenotypes of these mutations suggests that the QUTA protein is composed of at least two domains and that the amino-terminal domain can fold and function in the absence of the carboxy-terminal 50 per cent of QUTA (Figure 4.2).

The repressor protein works by binding to the activator protein

The phenotypes of the strains with mutant control genes do not give any direct information on how the activator and repressor proteins interact to control transcription. Indirect information from *A. nidulans*, however, does suggest that the QUTR protein mediates its repressing effect by binding directly to the QUTA protein. When *A. nidulans* is transformed with plasmids that contain the wild-type *qutA* gene, transformants that constitutively produce the quinate pathway enzymes can be isolated. The constitutive phenotype of these transformants is associated with an increased copy number of the transforming *qutA* gene and elevated *qutA* mRNA levels. Conversely, when *A. nidulans* is transformed with plasmids that contain the *qutR* gene under the control of the constitutive *pgk* promoter, transformants with a super-repressed phenotype (unable to use quinate as a carbon source) were isolated. This super-repressed phenotype

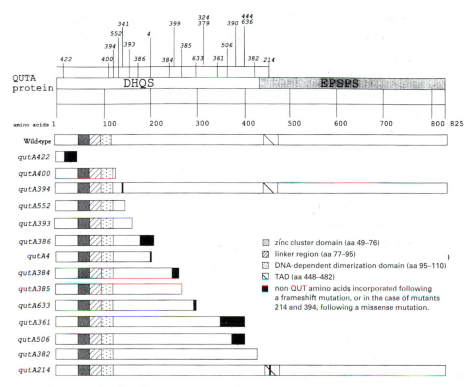

Figure 4.2 Predicted proteins produced by translation of the *qutA** gene and *qutA* mutants. DHQS, the region of the QUTA protein homologous with the AROM protein DHQ synthase domain; EPSP, the region of the QUTA protein homologous with the AROM protein EPSP synthase domain. Reprinted with permission from Levesley *et al.* (1996) *Microbiology*, 142, 87–98.

is associated with an increased copy number of the transforming *qutR* gene and elevated *qutR* mRNA levels. Such copy number-dependent phenotypes argue that the levels of the QUTA and QUTR proteins were elevated in the high copy number transformants.

When diploid strains were formed by combining haploid strains that contained high copy numbers of either the *qutA* gene (constitutive phenotype) or the *qutR* gene (super-repressing; non-inducible phenotype), the resulting diploid phenotype is one of quinate-inducible production of the quinate pathway enzymes, in a manner similar to

143

wild-type. The simplest interpretation of these observations is that the QUTR repressor protein mediates its repressing activity through a direct interaction with the QUTA activator protein.

There are significant comparisons to be made between these experiments and the genetic experiments on the control of galactose utilization in *Saccharomyces cerevisiae*. GAL4 is a positively acting transcription-activating protein that switches on the transcription of the genes necessary for the utilization of galactose. GAL80 is a negatively acting transcription regulating protein that mediates its action on galactose utilization by a post-translational interaction with the GAL4 protein. Certain dominant alleles encoding a mutant GAL4 protein that produced the galactose pathway enzymes constitutively in the absence of galactose were shown to complement some dominant mutant alleles encoding a super-repressing GAL80 protein. The resulting diploid strains had the wild-type inducible phenotype and proved to be the first evidence that the GAL4 and GAL80 proteins directly interacted with one another. The current generalized model for the molecular control of transcription that we use as a guide for further experimentation is shown in Figure 4.3.

Control of metabolic flux in the quinate and shikimate pathways

Are the common quinate and shikimate pathway intermediates kept separate? If not, why do the intermediates produced constitutively by the shikimate pathway not induce the transcription of the *qut* genes in the absence of quinate ? These questions have led previously to controversy over whether or not the complex penta-enzymatic AROM protein acts to channel these intermediates in wild-type strains, but more modern analysis has aimed to determine the control of quinate and shikimate pathway flux. This research makes use of metabolic control analysis (MCA) which is used increasingly in research into the regulation of the cell.

An overview of metabolic control analysis

As the knowledge of chemical transformations in metabolism has developed and the enzymes catalysing these reactions have been characterized, the need to explain how metabolic homeostasis at the physiological level is conducted at the molecular level has become increasingly important. A variety of mechanisms including feedback inhibition, cooperativity, covalent modification of enzymes and the control of enzyme synthesis and degradation all have the potential to alter the observed flux through a pathway (Fell, 1992). Controversies have often arisen about which enzymes, coenzymes and metabolites control metabolism. These controversies have resulted from the qualitative nature of the questions being addressed to the individual steps of a system. Many analyses have started from the viewpoint that there is a single 'rate-limiting step' which has complete control over flux through the metabolic pathway. It is unlikely, however, that such a simple qualitative measure will adequately explain flux through any pathway. In certain instances the overproduction of proposed 'rate-limiting' enzymes has been shown to have negligible effects on the resultant pathway flux. For example, phosphofructokinase (PFK-1,6) of yeast has been cited as the rate-limiting step in steady-state glycolytic flux for many years, but recent observations that neither a reduction nor an increase in the concentration of PFK-1,6 has any apparent effect on glycolytic flux suggests that this enzyme is not rate limiting. Therefore, a more sophisticated approach is required to study the control exerted over metabolic flux.

Theoretical analysis of the potential of different molecular mechanisms to control flux has been attempted and three variants of the analysis arose: metabolic control analysis (MCA), biochemical systems theory and flux-oriented theory. All of the theories include a form of sensitivity analysis in which the magnitude of the effect of a small change in a parameter on a metabolic system property is mathematically related to the properties of the components of the system. The difference between the approaches of the different theories is controversial but the underlying mathematics is similar. The one main difference is the choice of the type of parameter which is changed for the determination of the sensitivities.

MCA has become used most frequently, which may be due to the ease with which it can be applied to biological systems; important parameters that can be analysed in MCA are enzyme concentration or activity. MCA was developed in the 1970s but only recently has its potential importance to the understanding of flux control in metabolic systems been widely recognized. A discussion of some of the basic tenets that underlie MCA is presented in Box 4.1 (adapted from Fell, 1992).

145

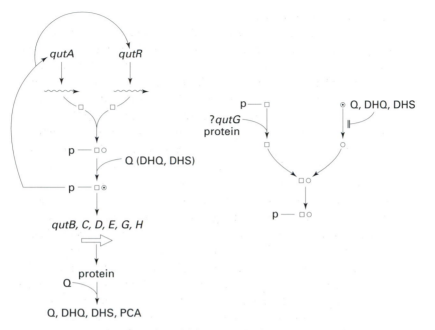

Figure 4.3 The general model for control of transcription of the quinate gene cluster in *A. nidulans*. (1) The QUTA (*square*) and QUTR (*circle*) proteins are transcribed from low-level constitutive mRNAs from the *qutA* and *qutR* genes. (2) The QUTR protein is proposed to bind to the QUTA protein occluding the surface acidic residues in its transcription activation domain thereby negating its ability to stimulate the transcription of the other *qut* genes. At this stage the QUTR protein is also proposed to potentiate the QUTA protein by phosphorylation to produce a form of the protein with enhanced potential transcription activation ability once it is dissociated from the QUTR protein. This proposal is based upon the clear similarity of the amino-terminal amino acid sequence of the QUTR protein with metabolic kinase enzymes. In this regard, it is significant that the C-terminal half of the QUTA activator protein is homologous to the AROM protein EPSP synthase domain, an enzyme that recognizes and binds shikimate 3-phosphate. The equivalent domain in the QUTA activator protein may have retained the ability to bind a metabolite that is a substrate for the putative metabolic kinase activity associated with the QUTR repressor protein. The proposal is further strengthened by the observation that in *S. cerevisiae* the GAL 4 transcription activator protein has three isoforms, two of which (the most transcriptionally active) are derived by phosphorylation of the native protein. (3) Some QUTA protein, native or

Caption to figure 4.3 (*cont.*)

phosphorylated, is always unbound and stimulates a very low level transcription of the other *qut* genes, thus ensuring that the quinate permease is
always present in the membrane at a low level. (4) When quinate is present
in the growth medium, it is taken up by the quinate permease. Sufficient
quinate is then available to bind to the QUTR protein leading to its dissociation from the QUTA protein and thereby to free a pool of QUTA protein
which then initiates an autoregulatory circuit ensuring the high level transcription of all the *qut* genes. (5) Under laboratory conditions a steady state
emerges as the rate of quinate uptake reaches its maximum value, and
quinate, DHQ and DHS are thereby able to bind to the pool of QUTR
protein, ensuring maximum flux of intermediates through the quinate
pathway. (6) As the quinate in the growth medium becomes exhausted, the
rate of uptake into the mycelium drops, leading to a large reduction in the
steady-state concentrations of quinate, DHQ and DHS. This drop in
metabolite concentrations frees a pool of QUTR protein that is able to bind
to and inactivate the QUTA protein, thereby leading to a rapid reduction in
the transcription of the *qut* gene cluster. We speculate that the QUTG
protein may have a role at this point by acting to dephosphorylate a
metabolite bound to the QUTA protein leading to a less transcriptionally
active form thereby helping to rapidly shut down transcription of the *qut*
gene cluster. Reprinted with permission from Hawkins *et al.* (1994) pp.
195–220 in *Aspergillus: 50 years on.* Progress in Industrial Microbiology, vol
29, pp 195–220, eds S. D. Martinelli and J. R. Kinghorn.

Box 4.1 Metabolic control analysis (MCA)

In its simplest form, MCA depends upon certain assumptions :

1   the system is a single unit in which all reactions are connected *via*
common metabolites

2   the system is at steady state, so that the rate of formation of every
metabolite is equal to its rate of degradation – all concentrations remain
constant with time. For a steady state to occur while there is a flow of
metabolites through the system, a source must exist to provide a 'pool' of
substrate(s) and a 'sink' into which the product(s) of the pathway flows.
However, the steady state of a system is approached asymptotically and so
no biological system is likely to be in a true steady state

Box 4.1 (*cont.*)

3   the metabolites are distributed homogeneously among the enzymes that act upon them

4   the rate of enzyme action is directly proportional to the enzyme concentration

5   enzymes are parameters in the analysis, not variables. A parameter is a property of the experimental system set by the experimenter, such as temperature, or by nature, such as the K_m of an enzyme, which remain constant during the experiment. A variable however, only attains a constant value during the steady state; the steady state magnitude of any variable is set by all of the parameters

6   all metabolites are in the free form; the amounts of enzyme-bound metabolites are presumed to not affect the properties of the steady state in most cases.

MCA attempts to relate the overall properties of a metabolic system to the properties of its components, in particular, the enzymes.

There are two approaches to studying the control of metabolism in intact systems. The individual components can be studied *in vitro*, or the system can be studied as a whole *in vivo*. Both approaches are required for a full understanding of metabolic control.

The elasticity coefficient, ϵ:
Consider a segment of metabolic pathway:

$$S_0 \xrightarrow[V_1]{E_1} S_1 \xrightarrow[V_2]{E_2} S_2 \xrightarrow[V_3]{E_3} S_3$$

then S_0 (source), S_1, S_2 and S_3 (sink) are the metabolites and each one of the enzymes, E_1, E_2 and E_3 catalyses a local reaction velocity V_1, V_2 and V_3 respectively. If a change δS_1, in the concentration of the substrate S_1 is made, keeping the concentration of all other metabolites constant, then a response δV_2, will be observed in the corresponding local reaction velocity V_2. The reaction rates are termed 'local' as they are catalysed by individual enzymes within a system.

The ratio of the response $\delta V_2/V_2$ to the change in the metabolite concentration, $\delta S_1/S_1$ is defined as the elasticity coefficient, ϵ, of the enzyme, E_2 with respect to the metabolite S_1 :

$$\epsilon \frac{S_1}{E_2} = \frac{\delta V_2/V_2}{\delta S_1/S_1}$$

Therefore, the elasticity coefficient of an enzyme for a particular metabolite can be determined from the response curve of the local reaction velocity, V, to changes in the concentration of the metabolite, S. The elasticity coefficient for any given metabolite concentration is then estimated from the slope of the tangent to the curve, $\delta V/\delta S$, at the desired metabolite concentration, [S], multiplied by the scaling factor [S]/V, where V is equal to the velocity at the desired substrate concentration.

This coefficient has no units as only fractional changes are being considered. The elasticity coefficient describes the quantitative response of a local reaction to a small change in the concentration of any one metabolite which interacts with the enzyme so as to modify its net activity. Therefore, there are as many elasticity coefficients for each enzyme as there are metabolites which associate with the enzyme in a concentration-dependent manner, (substrates, co-substrates, products and allosteric effectors). As enzyme concentration is a factor in any local reaction rate, each enzyme has its own elasticity coefficient. In almost all cases, this elasticity coefficient has a value of 1; a change in local reaction rate is proportional to any change in enzyme concentration. Elasticity coefficients can be either positive or negative; metabolic products have negative values whereas substrates generally have positive values. However, at higher substrate concentrations, negative elasticity coefficients can occur. The absolute value of an elasticity coefficient is an enzymatic parameter and as such, its value is fixed for a prescribed set of conditions but will vary as conditions vary.

The control coefficients

In a metabolic system, the individual steps are not isolated: each step is coupled to adjacent reactions and so any changes in either a metabolite or enzyme concentration will be transmitted throughout the entire system. The intermediary metabolite concentrations will adjust to a new steady state so that, in the case of a stationary steady state, the reaction velocities of the pathway are identical and equal to the flux, J, of the pathway.

The flux control coefficient, C

The response of a pathway flux to a small change in the activity of any one enzyme will be a consequence of the change in the concentration of all the metabolites throughout the pathway. This response is described by the flux control coefficient, C, of the enzyme, E_1. The flux control coefficient, C, is defined as the fractional change in pathway flux, δJ, as a function of the fractional change in enzyme concentration, δE_1 and is described mathematically as:

Box 4.1 (*cont.*)

$$C_{E1}^{J} = \frac{\delta J/J}{\delta E_1/E_1}$$

The flux control coefficient, C, of an enzyme can be determined for a particular enzyme activity, typically the wild-type activity, from the response curve of the *in vivo* flux, J, to changes in the activity of the enzyme under consideration. The flux control coefficient is estimated from the tangent to the curve, $\delta J/\delta E$, at the specified enzyme concentration, E, multiplied by a scaling factor, E/J, where the value of J is equal to the flux at the specified enzyme concentration.

There are as many flux control coefficients for a pathway as there are enzymes or carrier molecules in the system. The control of a global characteristic, such as flux, is shared quantitatively among all of the enzymes and carriers of the system. This coefficient has no units as only fractional changes are considered. Individual flux control coefficients have different values which can be either positive or negative. The greater the value of the coefficient, the greater the contribution of the corresponding enzyme to the overall flux through the system. Flux control coefficients must be determined experimentally whilst the enzymes are still in the intact system, as the magnitude of one flux control coefficient is not a property of the enzyme *per se* but depends on the activities of all the other enzymes and carriers and hence the magnitude of the other flux control coefficients.

The summation theorem

By definition, the sum of all the flux control coefficients that relate to a particular pathway must be equal to some limiting value. Rigorous and logical analysis showed this summation to be 1, unity. This is termed the summation theory and is described mathematically as:

$$\sum_{All\ E} C_E^J = 1$$

If there are n enzymes in a pathway, then the average value for each coefficient will be $1/n$. In a large pathway, most of the enzymes will have a small coefficient and, therefore, contribute little to the overall control of flux.

The connectivity theorem

The response of the whole system in terms of altered flux and metabolite concentration to any changes in a parameter, such as enzyme concentration, is generated by the joint action of all the local responses. These changes in metabolite concentration and each of the fluxes will depend on the elasticity coefficients of the system. A connectivity theorem was established which demonstrated that the global responses of the system (the flux control coefficients) are generated solely by the local responses (the elasticity coefficients) at all steps in the system. The connectivity theorem states that the ratio of the flux control coefficients of adjacent steps in a pathway is equal to the ratio of the two elasticity coefficients with respect to the shared metabolite. This is shown mathematically as:

$$\frac{C_{E_1}}{C_{E_2}} = -\frac{\varepsilon_{E_2}}{\varepsilon_{E_1}}$$

This connectivity theorem is widely regarded as the most important part of MCA as it provides an understanding of how the kinetics of the enzymes, represented by the elasticity coefficients, affect the values of the flux control coefficients. Large elasticity coefficients with values < -1 or >1, can arise through allosteric interactions but the largest feasible values are likely to be exhibited by near-equilibrium reactions when both the substrate and product elasticity coefficients for the enzyme of interest will be large. There is a tendency for these large elasticity coefficients to be associated with small flux control coefficients. The basic concepts of MCA have been modified and extended such that it can now be applied to branched, channelled and group transfer pathways and even single large perturbations in a parameter such as enzyme activity. These extensions to MCA often result in a more complicated mathematical analysis and largely described in terms of model theoretical systems.

Control of shikimate pathway flux: channelling or mass action effects?

As the *qut* pathway and the shikimate pathways share the metabolites DHQ and DHS, with their interconversion being catalysed by two separate dehydroquinase isoenzymes, the pentafunctional AROM protein was, on the basis of genetic data, proposed to have an efficient channelling effect keeping the shikimate pathway-derived DHQ and DHS separate from the enzymes of *qut* pathway. Although not originally intended as such, early *in vivo* experiments altering the concentrations of AROM and *qut* pathway enzymes in *A. nidulans* have brought some of the methods of MCA to bear on this question of channelling by AROM. Overproduction of the AROM protein up to fivefold over the wild-type level allows *qut* pathway mutants lacking the *qut* dehydroquinase to use quinate as sole carbon source, albeit very poorly. This effect is presumably due to DHQ from the *qut* pathway entering the AROM protein and leaving after conversion to DHS whereupon it is converted to PCA by dehydroshikimate dehydratase.

In contrast, increasing the *in vivo* concentration of the AROM protein above this level leads to an inability to grow in the presence of quinate. This is presumed to be due to the increased flux in the shikimate pathway leading to the accumulation of a toxic level of a shikimate pathway metabolite(s) because growth is not restored by the provision of an additional carbon source. The dehydroshikimate dehydratase enzyme catalyses the reaction at the bifurcation point in the utilization of DHQ and DHS in the *qut* and shikimate pathways (see Figure 4.1). Constitutive amplification of this enzyme in absence of the quinate has shown that shikimate pathway-derived DHS can be fluxed into the *qut* pathway with a concomitant partial auxotrophic requirement for the aromatic amino acids. Although these data are open to different interpretations, they demonstrate that AROM is leaky *in vivo* and that the observations discussed can be explained in terms of a mass-action effect, without the need to impute a channelling function for the AROM protein.

Control of quinate pathway flux

At the usual *in vitro* growth pH of 6.5, quinate enters the mycelium by means of a specific permease, and is converted to PCA. The extent of control on metabolic flux exerted by the permease and the three enzymes was investigated by applying the techniques of MCA. The flux control coefficients for each of the three quinate pathway enzymes were determined empirically by varying the *in vivo* concentrations of the three quinate pathway enzymes and measuring the effect on overall pathway flux. The values (C) were: quinate permease ($C = 0.43$); quinate dehydrogenase ($C = 0.36$); dehydroquinase ($C = 0.18$) and DHS dehydratase ($C = 0.03$). Attempts were made to partially decouple the quinate permease from the control over flux by measuring flux at pH 3.5 (when a significant percentage of the soluble quinate is protonated and able to enter the mycelium without the aid of a permease), leading to an increase of approximately 50 per cent in the flux control coefficient for the dehydroquinase. These experiments were consistent with the view that the quinate permease exerts a high degree of control over pathway at pH 6.5.

All three *qut* pathway enzymes were purified and their elasticity coefficients were derived. The relative ratios of the flux control coefficients for the various quinate pathway enzymes, and how this control shifts between them, was determined over a range of possible metabolite concentrations. These calculations were able to successfully predict the hierarchy of control observed under the standard laboratory growth conditions. The calculations imply that the control exerted by the quinate pathway enzymes is stable and relatively insensitive to changing metabolite concentrations in the range most likely to correspond to the *in vivo* values.

Prospects for pathway engineering

One of the goals of biotechnology is to attempt to increase flux through commercially important pathways. An ever-increasing number of genes encoding equivalent enzymes from different species is becoming available giving the potential to substitute functionally equivalent enzymes that have different elasticity coefficients in chosen pathways. This strategy therefore offers the possibility of pathway engineering by substituting enzymes with elasticity coefficients 'tailored' to help circumvent a particular problem.

The effect of substituting the type I 3-dehydroquinases from *Salmonella typhi* and the *A. nidulans* AROM protein, and the *Mycobacterium tuberculosis* type II 3-dehydroquinase in the quinate pathway, however, provides a cautionary tale. The genes encoding the type I dehydroquinases from *S. typhi* and *A. nidulans*, and the type II dehydroquinase from *M. tuberculosis* have been expressed in mutant strains of *A. nidulans* where they substituted for the usual (*qutE* encoded) quinate pathway type II dehydroquinase. Production of these heterologous dehydroquinases led to significant negative effects on quinate pathway flux. Overproduction of the enzymes from *S. typhi* and *A. nidulans* led to a diminution of pathway flux caused by a lowering of *in vivo* quinate dehydrogenase levels. Conversely however, production of the *M. tuberculosis* enzyme led to an increase in the level of quinate dehydrogenase levels above those of wild-type. It was speculated that these changes in quinate pathway enzyme activity may be due to changes in the internal pool sizes of quinate and DHQ leading to either stabilization or destabilization of quinate dehydrogenase activity. Effects on enzyme stability by changing metabolite pool sizes may turn out to be a significant consideration for the rational design of pathway flux manipulation.

Themes arising from these studies

It is clear that the common intermediates DHQ and DHS are not kept separate in the quinate and shikimate pathways, and that the relative pathway flux of the two pathways is radically different. Flux through the quinate pathway is 30-fold higher than the shikimate pathway, and the quinate pathway enzyme dehydroshikimate dehydratase has a shikimate pathway flux control coefficient of minus 1. Therefore in this particular instance, if the quinate pathway enzymes were constitutively produced in the absence of quinate there would be a chronic drain on the shikimate pathway. In wild-type strains, this effect of cross-pathway fluxing of metabolites by quinate pathway enzymes will only occur when the quinate pathway enzymes have exhausted the supply of exogenously supplied quinate and the quinate pathway is in the process of being shut down. Filamentous fungi that can utilize quinate are soil organisms and in order for the carbon source to be utilized efficiently, it has to be in solution. The water for this solution will come from intermittent rainfall or dewfall, thereby causing quinate to be available in pulses; under such conditions of substrate pulsing dehydroshikimate dehydratase has the potential to drain shikimate pathway intermediates into the quinate pathway causing starvation for the aromatic amino acids. When this cross-pathway fluxing by dehydroshikimate dehydratase is induced artificially, the organism responds by increasing the concentration of the AROM enzyme thereby relieving the drain on shikimate pathway intermediates by a simple mass action effect. It is clear therefore that the organisms have developed strategies to deal with fluctuating metabolite concentrations and their effect on pathway flux, and that in wild-type strains there is a subtle interplay between transcriptional and flux control.

154

One final question is still left unanswered: if the shikimate pathway is constitutively expressed and the AROM enzyme is leaky, why do the metabolites DHQ and DHS not lead to induction of the quinate pathway in the absence of exogenously supplied quinate? The answer to this question almost certainly lies in the observation that the shikimate pathway flux is low relative to the quinate pathway flux. The answer to the question is that in wild-type strains the rate of leakage and pool size of the metabolites is probably too low to cause internal induction of the quinate pathway; it is only in strains containing a non-functional AROM 3-dehydroquinase that the rate of substrate leakage and subsequent pool size is sufficient to cause internal induction of the quinate pathway.

Evolution of transcription-regulating proteins by enzyme recruitment

The problem of multiple recognition events

The molecular model for the control of the quinate pathway in *A. nidulans* (which is also applicable to *N. crassa*) shown in Figure 4.3 proposes that three primary recognition events are required to engage the autoregulatory circuit. These recognition events are: first, the recognition of cognate DNA-binding sites by the activator proteins (QUTA in *A. nidulans*, QA-1F in *N. crassa*); second, the recognition of the activator proteins by the repressor proteins (QUTR in *A. nidulans* and QA-1S in *N. crassa*) proteins; and third, the recognition and binding of quinate pathway metabolites by the repressor proteins as a precursor to the presumed allosteric changes which mediate transcriptional regulation of the quinate utilization pathway. Such a mechanism would place strong constraints upon the molecules involved, in that the activator proteins must evolve the ability to recognize and bind particular nucleotide sequences and that the repressor proteins must evolve the ability to bind pathway-specific metabolites; additionally, however, the activator and repressor proteins have to recognize and bind to one another. This problem of the underlying mechanism for the selection of the ability of a particular protein to recognize different classes of macromolecule is a fundamental problem in the evolution of transcriptional regulatory systems that are sensitive to modulation by pathway-specific metabolites.

Can enzymes provide mechanisms for metabolite recognition ?

Elements (enzymes) containing one of the functions (metabolite recognition) necessary for transcription regulation are already present within the cell. An attractive idea, therefore, is that the ability to recognize specific substrates which is obviously inherent in enzymes can be exploited and incorporated into transcription-regulating proteins. It is possible that selection for the modification of an enzyme so that it acquires a regulatory function would place constraints on the protein such that it could perform neither function well. This problem could be resolved by selection acting on duplicated copies of the gene encoding the enzyme leading to two homologous structures, one an enzyme and one a regulatory protein that may or may not have retained the ability to recognize the substrate for the enzyme homologue. It has been predicted that regulatory proteins evolving by such a route may still contain in their amino acid sequences 'fossil' evidence of their evolutionary origin.

A good example of enzymes changing their function is provided by the discovery that lens crystallins are multifunctional. In order to focus light a lens must have a higher refractive index than the surrounding medium and this is achieved within the eye by having very high concentrations of soluble structural proteins called crystallins. Many of these crystallins are not specialized structural proteins, are not even lens specific and remarkably some taxon-specific crystallins turn out to be metabolic enzymes recruited to a new role by changes in gene expression without gene duplication or loss of enzyme activity. This selection for disparate activities (light refraction versus enzyme activity) within the same molecule has resulted in some modifications to the structure of the protein. Still other crystallins turn out to be related to metabolic enzymes but are enzymatically inactive; these crystallins have evolved divergently from duplicated copies of the gene encoding the original enzyme.

The quinate pathway repressor is related to the three carboxy-terminal domains of AROM

All the known genes involved in quinate utilization have been isolated and their nucleotide sequences determined, and an analysis of these sequences has uncovered an ancient and complex evolutionary relationship with the enzymes of the prechorismate section of the shikimate

156

pathway. Computer analysis of predicted amino acid sequences has demonstrated the QUTR and QA-1S repressor proteins are related to the three carboxy-terminal domains of the pentadomain AROM protein. These amino acid sequence alignments suggest a simple molecular model for how the cell is able to sense the presence of *qut* pathway inducers and initiate production of the three enzymes necessary for quinate metabolism. These metabolites, with the exception of quinate, are recognized as substrates by the three C-terminal domains of the AROM protein, therefore it has been proposed that the *qut* pathway repressor protein has retained the ability to recognize and bind these metabolites but has lost catalytic activity towards them. Binding of the quinate pathway metabolites by the repressor proteins is thought to initiate allosteric changes in the protein necessary for the initiation of transcription. It is thought that these allosteric changes affect the ability of the QUTR protein to recognize QUTA thereby altering the position of the equilibrium favouring dissociation of the two proteins.

The quinate pathway activator proteins are related to the two amino-terminal domains of AROM

Although the above sequence alignments suggest a plausible molecular mechanism for the proposed recognition of the pathway inducer molecules, they raise the larger problem of how a protein that is homologous to three shikimate pathway enzymes is able to recognize and bind to a transcription activator protein. Clues for a possible solution to this conundrum come in the form of two observations: first, QUTR and QA-1S correspond, in size at least, not only to the three carboxy-terminal domains of AROM, but also to the carboxy-terminal one-fifth of the EPSP synthase domain (see Figure 4.3); and second, the size of the QUTA and QA-1F activator proteins corresponds to the remainder of the two amino-terminal domains of AROM.

A detailed comparison of the amino acid sequences of these activator proteins and AROM has led to the proposal that the QUTA and QA-1F proteins are homologous to the complete amino-terminal domain and four-fifths of the EPSP synthase domain of AROM. On the basis of these sequence alignments, the quinate pathway activator and repressor proteins have been proposed to have their origins in the splitting of a duplicated copy of the *arom* gene (encoding the AROM protein) or its

157

precursor, with the subsequent recruitment of a zinc binuclear cluster into the amino-terminus of the amino-terminal DHQ synthase-like domain. This sequence alignment shows that the level of overall sequence identity is very low at around 11 per cent with 18–25 per cent overall sequence similarity. These values are below those typically accepted as statistically significant and fall within the 'background noise' of random similarity that can be generated because the proteins being compared fortuitously may have similar amino acid compositions. This problem of low-level sequence identity is compounded by the use of gaps that, if used liberally, can generate this apparent level of identity and similarity between amino acid sequences that are known to be unrelated. In the absence of structural data, how then is it possible use these low-level amino acid sequence alignments to test the hypothesis of common evolutionary origin?

Other evidence for the proposed evolutionary origins QUTA and QUTR ?

The key feature of these sequence alignments is that they make specific predictions about the positions of domain boundaries and the functions of the domains identified in this manner. Analysis of the nucleotide positions of the *qutA* mutations discussed above reveals that the dominant mutation *qutA382* (conferring a non-inducible phenotype) is precisely at the predicted domain boundary between the DHQ synthase and EPSP synthase domain. This mutant therefore will produce a truncated version of the QUTA protein which corresponds to the stable DHQ synthase domain of the AROM protein, thereby supporting the bi-domain structure of the QUTA protein. The mutation *qutA214* (conferring a non-inducible phenotype) localizes a putative transcription activation domain in the extreme amino-terminus of the second domain which is proposed to be homologous with EPSP synthase. The phenotype of these two mutants is exactly that predicted by the bi-domain model for the activator protein and is consistent with the known biophysical characteristics of the AROM domains.

The 3-dehydroquinase domain of the AROM protein shows the greatest degree of amino acid sequence identity with the quinate pathway repressor proteins, and the equivalent domain from the QUTR protein has been overproduced in and purified in bulk from *Escherichia coli.*

158

Biophysical analysis of the AROM 3-dehydroquinase and the QUTR 3-dehyroquinase-like domains has shown that they have similar physical characteristics that derive from a highly conserved tertiary structure.

It is evident therefore that by taking account of the total biology of a set of potentially distantly related proteins that have changed function it is possible to design genetic and biophysical experiments that can critically test the hypothesis of common evolutionary origin.

The re-use of a successful modular design: evolution of a dispensable pathway

The enzymes and regulatory proteins of the *qut* pathway appear to be an extreme example whereby a small group of protein shapes (modules) have been coopted to modified and/or new function by a complex series of gene duplications and subsequent fusion and fission events. In addition to the homologies discussed so far, comparative amino acid sequence alignments have shown that the quinate permease is apparently related to a group of sugar transporters, and that the translation products of the *qutG* and *qa-X* genes of *A. nidulans* and *N. crassa* respectively are related to inositol monophosphatase. The proposed modular nature of the *qut* and prechorismate (shikimate) pathways is summarized in Figure 4.4. The *qut* pathway is an entirely dispensable alternative carbon utilization pathway; therefore the observation that it apparently has been assembled mainly by coopting modules or domains from the ancient shikimate pathway demonstrates a basic principle, that of re-using and modifying a successful basic modular design.

Global control and the evolution of transcription-regulating proteins

The GCN2 protein and general amino acid control in
S. cerevisiae

In yeast, the genes comprising the amino acid biosynthetic pathways are expressed at a low basal level; however, when cells are starved of any one of several amino acids, transcription of all these genes is induced 2- to10-fold. This induction is positively mediated by the GCN4 protein

159

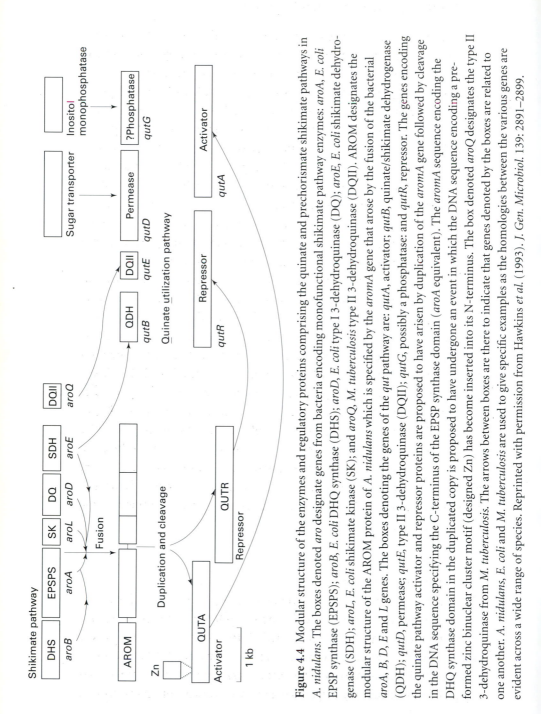

Figure 4.4 Modular structure of the enzymes and regulatory proteins comprising the quinate and prechorismate shikimate pathways in *A. nidulans*. The boxes denoted *aro* designate genes from bacteria encoding monofunctional shikimate pathway enzymes: *aroA*, *E. coli* EPSP synthase (EPSPS); *aroB*, *E. coli* DHQ synthase (DHS); *aroD*, *E. coli* type I 3-dehydroquinase (DQ); *aroE*, *E. coli* shikimate dehydrogenase (SDH); *aroL*, *E. coli* shikimate kinase (SK); and *aroQ*, *M. tuberculosis* type II 3-dehydroquinase (DQII). AROM designates the modular structure of the AROM protein of *A. nidulans* which is specified by the *aromA* gene that arose by the fusion of the bacterial *aroA*, *B*, *D*, *E* and *L* genes. The boxes denoting the genes of the *qut* pathway are: *qutA*, activator; *qutB*, quinate/shikimate dehydrogenase (QDH); *qutD*, permease; *qutE*, type II 3-dehydroquinase (DQII); *qutG*, possibly a phosphatase; and *qutR*, repressor. The genes encoding the quinate pathway activator and repressor proteins are proposed to have arisen by duplication of the *aromA* gene followed by cleavage in the DNA sequence specifying the C-terminus of the EPSP synthase domain (*aroA* equivalent). The *aromA* sequence encoding the DHQ synthase domain in the duplicated copy is proposed to have undergone an event in which the DNA sequence encoding a pre-formed zinc binuclear cluster motif (designed Zn) has become inserted into its N-terminus. The box denoted *aroQ* designates the type II 3-dehydroquinase from *M. tuberculosis*. The arrows between boxes are there to indicate that genes denoted by the boxes are related to one another. *A. nidulans*, *E. coli* and *M. tuberculosis* are used to give specific examples as the homologies between the various genes are evident across a wide range of species. Reprinted with permission from Hawkins *et al.* (1993). *J. Gen. Microbiol.* 139: 2891–2899.

whose production is regulated by amino acid availability, but at the level of translation initiation. The GCN2 protein functions by de-repressing the GCN4 protein and possible mechanisms for how this de-repression is linked to amino acid availability have been provided by an analysis of GCN2 domain structure and function. GCN2 has 1590 amino acids and contains at least two recognizable domains. The amino acid sequence 530–910 has significant sequence identity with the catalytic domain of protein kinases. The carboxy terminal region of GCN2, amino acids 920–1450, is related (22 per cent identity) to the histidyl-tRNA synthetases. Substitution of a highly conserved lysine residue in the presumed ATP-binding site of GCN2 impairs the de-repression of histidine biosynthetic genes under GCN4 control, strongly suggesting that protein kinase activity is required for the GCN2 positive regulatory function.

These results suggest that amino acid starvation could be linked to GCN4 function by the domain functions of the GCN2 protein. It may be that the GCN2 histidyl-tRNA synthetase-like domain responds to amino acid starvation by detecting the presence of uncharged tRNA, thereby activating the adjacent protein kinase domain. GCN4 activity is then de-repressed by increasing translation of the preformed GCN4 mRNA. Subsequently, the GCN2 protein was shown to specifically phosphorylate the alpha subunit of rabbit or yeast eukaryotic translation initiation factor 2 (elF-2) and to associate with the ribosomes. The mammalian elF2 alpha factor has extensive homology with GCN2 and when produced *in vivo* in yeast shown to substitute functionally for the GCN2 protein. Taken together these observations show that the recruitment of enzymes to a transcription regulatory role is not just a curiosity confined to the quinate pathway.

Concluding remarks

The use of fungi in research on genetic regulation can address fundamental questions of how transcriptional regulation of the genes encoding metabolic enzymes is integrated into mechanisms that allow discrimination between preferred metabolite sources, whilst at the same time maintaining the overall physiological homeostasis. What has emerged from an initially bewildering array of control mechanisms and recognition events provided by the examples above are a few relatively simple principles that underpin the molecular mechanisms employed. First, the transcription-regulating proteins studied are characterized by a multidomain nature and appear to have solved the problem of specific metabolite recognition/interaction by coopting domains provided by enzymes to a new function as part of a biosensing mechanism. Second, a successful modular design can be re-used. The quinate pathway appears to be an extreme example of this phenomenon, with the proteins – enzymes, permease and transcription-regulating proteins – all apparently having their origins in duplicated copies of genes encoding enzymes or permeases that are active in metabolic pathways of ancient lineage. Possibly the best illustration of this principle is provided by the module, which in prokaryotic lineage is a shikimate dehydrogenase, whilst in microbial eukaryotes the same module has evolved a second activity becoming a quinate/shikimate dehydrogenase. Additionally however, this same module in the microbial eukaryotic lineage has evolved a possible biosensing function as the C-terminal domain of the tri-domain quinate pathway transcription repressor protein. This variant of the module thereby regulates the transcription of the gene encoding the bifunctional quinate/dehydrogenase version of the module demonstrating the utility and flexibility of re-using a successful modular design.

Acknowledgements

Work in the authors' laboratory was supported by the UK Research Councils and the Wellcome Trust.

Recommended reading

Bentley, R. (1990). The shikimate pathway – a metabolic tree with many branches. *Crit. Rev. Biochem. Mol. Biol.* 25:307–384.

Coggins, J. R. and Boocock, M. R. (1986). Pp. 259–281 in *Multidomain Proteins: Structure and Evolution* (Hardie, D. G. and Coggins, J. R. eds.). Elsevier, Amsterdam.

Fell, D. A. (1992). Metabolic control analysis: a survey of its theoretical and experimental development. *Biochem. J.* 286:313–330.

162

Giles, N. H., Case, M. E., Baum, J., Geever, R., Huiet, L., Patel, V. and Tyler, B. (1985) Gene organization and regulation in the *qa* (quinic acid) gene cluster of *Neurospora crassa*. *Microbiol. Rev.* 49:338–358.

Hawkins, A. R. and Lamb, H. K. (1995). The molecular biology of multidomain proteins: selected examples. *Eur. J Biochem.* 232:7–18.

Johnston, M. (1987). A model fungal gene regulatory mechanism: the *GAL* genes of *Saccharomyces cerevisiae*. *Microb. Rev.* 51:458–476.

Martinelli, S. D. and Kinghorn, J. R. (eds) (1994). *Aspergillus* 50 years on. *Progress in Industrial Microbiology*, 29. Elsevier, Amsterdam.

Wistow, G. (1993). Lens crystallins: gene recruitment and evolutionary dynamism. *TIBS* 18:301–306.

Further reading

Anton, I. A., Duncan, K. and Coggins, J. R. (1987). A eukaryotic repressor protein, the *qa-1S* gene product of *Neurospora crassa*, is homologous to part of the AROM multifunctional enzyme. *J. Mol. Biol.* 197:367–371.

Carpenter, E. P., Hawkins, A. R., Frost, J. W. and Brown, K. A. (1998). Structure of dehydroquinate synthase reveals an active site capable of multistep catalysis. *Nature* 394:299–302.

Kell, D. B., van Dam, K., Westerhoff, H. V. (1989). Control analysis of microbial growth and productivity. Pp. 61–93 in *Society of General Microbiology Symposium* 44 (Baumberg, S., Hunter, I., Rhodes, M. eds.). Cambridge University Press, Cambridge.

Rolfes, R. J. and Hinnebusch, A. G. (1993). Translation of the yeast transcriptional activator GCN4 is stimulated by purine limitation: implications for activation of the protein kinase GCN2. *Mol. Cell Biol.* 13:5099–5111.

Wheeler, K. W., Lamb, H. K. and Hawkins, A. R. (1996). Control of metabolic flux through the quinate pathway in *Aspergillus nidulans*. *Biochem J.* 315:195–205.

Lamb, H. K., Moore, J. D., Lakey, J. H., Levett, L. J., Wheeler, K. W., Lago, H., Coggins, J. R. and Hawkins, A. R. (1996). Comparative analysis of the QUTR transcription repressor protein and the three C-terminal domains of the pentafunctional AROM enzyme. *Biochem. J.* 313:941–950.

5 Hyphal cell biology

G. ROBSON

The hyphal mode of growth

The dominance of filamentous fungi within the ecosystem is attributed to their common mode of growth, extending as branched filaments (hyphae) which can rapidly spread across uncolonized substrates. The success of this growth habit for exploiting the natural environment can be judged on a number of factors: the extraordinary diversity of fungal species (estimated at three million, second only to the insects), their distribution in virtually every habitat on the planet and the parallel evolution of a similar growth habit by another important class of soil microorganisms, the prokaryotic streptomycetes. Clearly the ability of a microbe to rapidly colonize new substrates by concentrating growth at its apex, is well suited for life as a heterotroph in a heterogenous environment.

Spore dormancy and germination

Spores are products of both sexual and asexual reproduction and act as units of dispersal in fungi. The majority of spores germinate to produce one or more germ tubes and a new fungal mycelium when the spore settles on an appropriate substrate under favourable environmental conditions. When a spore is faced with unfavourable conditions such as lack of nutrients, low temperature, an unfavourable pH or the presence of an inhibitor (e.g. on a plant surface), the spore remains dormant. Spores under these conditions are *exogenously dormant* and will only germinate when the environmental conditions become favourable. Some fungal species produce spores that fail to germinate immediately, even under favourable conditions because of factors within the spore such as nutrient impermeability or the presence of endogenous inhibitors. Spores exhibiting these characteristics are termed *endogenously dormant*. Dormancy within these spores is usually broken by ageing, when nutrients can begin to enter or the endogenous inhibitors leach out, reducing the concentration within the spore.

Prior to the emergence of a germ tube, fungal spores undergo a process of swelling during which spores increase in diameter up to four-fold due the uptake of water. During this phase, the metabolic activity increases greatly and protein, DNA and RNA production all increase. This is followed by the emergence of one or more germ tubes that extend outwards from the spore in a polarized manner (Figure 5.1).

Colony formation

Following germination, the extension rate of the germ tube increases exponentially towards a maximum rate at which point the hypha attains a linear extension rate. The maximum rate of hyphal extension varies greatly between different fungal species and is also dependent on environmental conditions such as temperature, pH and nutrient availability. Before the maximum rate of extension is attained, a lateral branch is formed to produce a new growing hypha which also accelerates towards a maximum rate. As the germling continues to grow, new lateral branches are formed at an exponential rate. Although individual hyphae in the developing mycelium eventually attain a maximum linear rate, the overall growth of the mycelium is therefore exponential (Figure 5.2). During early growth, nutrients surrounding the young mycelium are in excess and the mycelium is *undifferentiated*. During undifferentiated growth, the mean rate of hyphal extension is dependent on the specific growth rate of the organism (the rate of growth per unit time) and the degree of branching. This is quantified as follows;

$$E = \mu G$$

Where E = mean hyphal extension rate (μm h^{-1}), μ = specific growth rate (h − 1) and G = hyphal growth unit (μm, the mean length of hypha associated with each branch).

As the mycelium develops further, nutrients in the centre are increasingly utilized and a zone of nutrient depletion begins to form

165

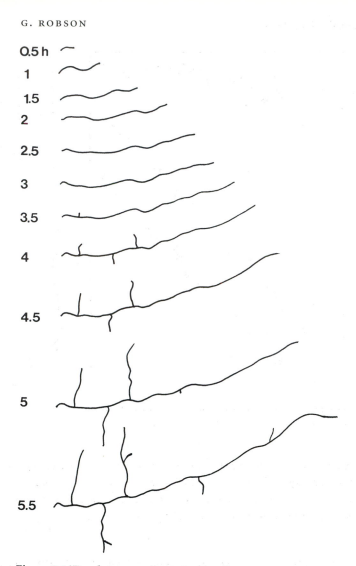

Figure 5.1 Development of a young germling over time.

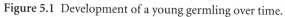

beneath the colony. At this point growth at the centre of the young colony begins to decrease and as the colony expands further, ceases altogether. Therefore as the colony develops, growth becomes restricted to its periphery where nutrients are still available and the mycelium is *differentiated*. This peripheral ring, which is growing at the maximum rate, is the *peripheral growth zone*. The maximum rate of extension of

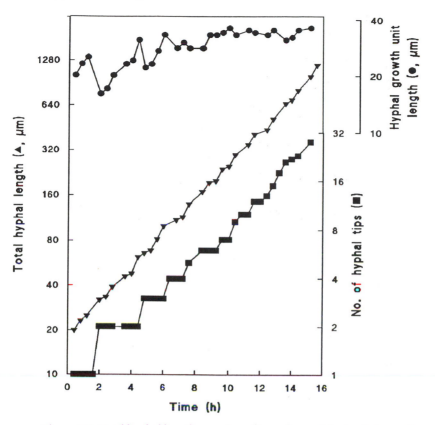

Figure 5.2 Total hyphal length, number of branches and the hyphal growth unit of a developing fungal mycelium. Notice that initially the hyphal growth unit fluctuates as new branches are formed but quickly becomes constant after more than five branches have been formed.

hyphae at the peripheral growth zone can be measured by the rate at which the colony expands. This rate or *colony radial growth rate* is dependent on the specific growth rate and the width of the peripheral growth zone and is defined as:

$$Kr = w\mu$$

where Kr is the colony radial growth rate (μm h^{-1}), w is the width of the peripheral growth zone (μm) and μ is the specific growth rate (h^{-1}). The fungal colony therefore grows outward radially at a linear rate, continually

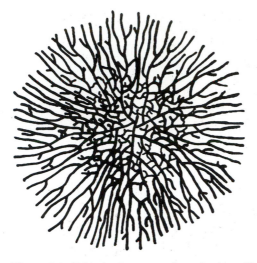

Figure 5.3 A developing young fungal colony. Notice how the growing hyphae are orientated outward into uncolonized regions whilst the production of branches ensures the mycelium efficiently colonizes available substrate.

growing into unexploited substratum. As it does so, the production of new branches ensures the efficient colonisation and utilization of the substratum (Figure 5.3). The hyphae of many fungi can alter their direction of growth to avoid growing into each other and move into uncolonized regions of the substratum. The avoidance mechanism or *autotropism* (Figure 5.4) is particularly evident at low hyphal densities, such as the margin of the growing colony. The ability of hyphae to sense the presence of another hypha is thought to be due either to a localized depletion of oxygen around the hypha, a higher concentration of carbon dioxide or the presence of a secreted metabolite.

As nutrients become depleted at the centre of the colony and metabolic products accumulate, spore production is often initiated. Therefore, different parts of the colony are at different physiological ages, with the youngest actively extending hyphae at the edge of the fungal colony and the oldest, non-growing, sporulating mycelium at the centre (Figure 5.5).

Unlike colonies formed by unicellular bacteria and yeast, where colony expansion is the result of the production of daughter cells and

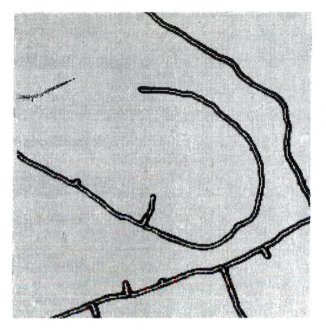

Figure 5.4 A fungal hypha has detected the presence of an adjacent hypha and reorientated its growth in an autotropic response.

Physiological Age

Figure 5.5 Schematic representation of a growing fungal colony illustrating the gradient in physiological age from the youngest growth at the margin to the oldest mycelium at the colony centre. As the colony continues to grow, the oldest mycelium at the colony centre may produce staling products and secondary metabolites which diffuse into the agar medium and differentiate to form spore-bearing structures.

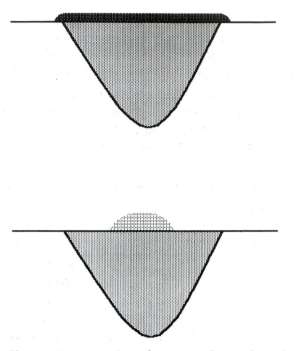

Figure 5.6 A comparison of nutrient utilization beneath a bacterial and a fungal colony. In the upper picture, the edge of the growing fungal colony advances beyond the nutrient-depleted region beneath the colony. By contrast, in the lower panel, the slower rate of expansion of a bacterial colony results in the nutrient-limited diffusion of the colony which limits the size it can attain.

occurs only slowly, the ability of filamentous fungi to concentrate all their growth capacity at the hyphal apex allows the colony to expand far more rapidly. The fungal colony therefore expands at a rate which exceeds the rate of diffusion of nutrients from the surrounding substratum. As a consequence, nutrients under the colony are rapidly exhausted whereas the hyphae at the edge of the colony have only a minor effect on the substrate concentration. By contrast, the rate of expansion of bacterial and yeast colonies is extremely slow and less than the rate of diffusion (Figure 5.6). Colonies of unicellular organisms quickly become diffusion limited and therefore, unlike fungal colonies, can only attain a finite size.

Hyphal growth

Polarized growth of fungal hyphae is achieved by restricting growth to the hyphal apex. The cell wall at the hyphal tip has viscoelastic properties and yields to the internal turgor pressure within the hypha. Further behind the tip, the wall is rigid and resistant to the turgor forces. Turgor pressure generated within the hypha therefore acts as the driving force for hyphal extension. Hyphal growth is supported by the continuous flow of vesicles generated within the cytoplasm behind the tip. The vesicles are derived from the endoplasmic reticulum and processed through Golgi bodies before migrating towards the extending apex. The vesicles are readily seen in longitudinal sections of fixed hyphae under electron microscopy and consist of two main size classes, the smallest ranging from 20 to 80 nm and the largest from 80 to 200 nm. The smaller group, which has been isolated from hyphae and shown to contain an inactive form of the major wall biosynthetic enzyme chitin synthase, are known as *chitosomes*. Fusion of the vesicles with the membrane at the hyphal apex releases the biosynthetic machinery for wall assembly and also adds new membrane to the growing hypha. Synthesis of the hyphal wall and plasma membrane are therefore coordinately regulated. The development of such a highly polarized vesicular pathway not only supports rapid hyphal extension, but also acts as a transport mechanism for extracellular enzymes. Extracellular enzymes are secreted into the environment at the hyphal tip catalysing the degradation of complex polymers such as lignocellulose. It is possible to estimate the number of vesicles that must fuse with the extending tip to support tip extension. For *Neurospora crassa*, the maximum hyphal extension rate of 25 μm min^{-1} requires 38 000 vesicles to fuse with the apical membrane per minute! Despite the presence of such a highly polarized secretory system in filamentous fungi, remarkably little is known about the mechanisms by which vesicles flow in a polarized fashion to the tip, or their means of transportation. Recently, detailed studies of growing hyphae at high magnification by video-enhanced microscopy have revealed that extension is not strictly constant, but occurs as a series of cyclic pulses. The duration of the pulses is dependent upon the fungal species, varying from 2 to 15 sec (Figure 5.7).

In many, but not all filamentous fungi, an electron dense body is present in the cytoplasm just distal to the growing apex and appears to be

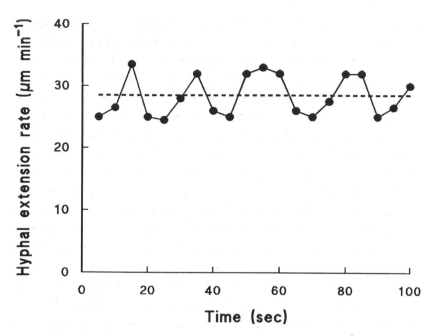

Figure 5.7 The extension rate of fungal hyphae does not attain a maximum linear rate as previously thought, but oscillates regularly around a mean giving rise to 'pulsed' growth.

intimately associated with hyphal extension and growth. This body, or *Spitzenkörper* is composed of a complex of vesicles when observed in longitudinal sections of the hypha under electron microscopy. Recent evidence has shown that the position of the *Spitzenkörper* controls both the direction of growth of the growing hyphae and the rate of extension. Satellite *Spitzenkörpers*, which appear to separate from the main body and migrate to the lateral cell wall, have been shown to precede the initiation of new branches. This has led to the concept of the *Spitzenkörper* as a *vesicle supply centre*, which acts as a focal point for the migrating apical vesicles. In this model, vesicles migrate to the *Spitzenkörper* and then radiate outwards equally in all directions. Using a computer model based on this concept, a tube with a tapering tip analogous to a growing hypha could be generated. Moreover, the direction of growth could be altered by moving the vesicle supply centre upwards or downwards, mimicking the movement observed of the *Spitzenkörper* during changes in the direction of hyphal growth. The *Spitzenkörper* is therefore thought to play a critical role in the mechanism of polarized growth in these fungi, controlling not only hyphal extension rate and direction of growth, but also in generation of new lateral branches. Despite the evidence that the *Spitzenkörper* is involved in tip growth, it is clearly not the only way in which polarized growth can occur as lower fungi, for example *Saprolegnia ferax,* and other polarized systems, for example pollen tubes, lack a *Spitzenkörper* or equivalent body. To maintain its position behind the growing hypha, the *Spitzenkörper* must be anchored in some manner to the hyphal apex. The precise nature of this anchor is currently unknown but two classes of cytoskeletal proteins, the *microtubulins* and *actin* appear likely candidates. The microtubulins are heteropolymers of tubulin dimers containing peptides derived from different genes to form a tubulin protofilament. Thirteen protofilaments are connected laterally to form microtubule filaments, which are found in the hypha as single elements and as bundles. Microtubules are involved in nuclear division, forming the spindle pole bodies involved in chromosome separation during mitosis and meiosis and are also involved in flagellar movement of motile zoospores of fungi of the order Mastigomycotina. There is also evidence to support the role of microtubulins in the motility of organelles, including nuclei, mitochondria and vesicles within the hypha. Disruption of the microtubules with drugs such as colchicine and taxol, inhibits organelle motility and positioning and results in an abnormal distribution of microvesicles. Actin filaments are composed of an array of polymerized actin monomers that, in fungi, may be the product of a single gene. In hyphae, actin has been found located at growing tips as a fibrillar network radiating from the apex as slender cables and in localized areas as actin plaques within the cytoplasm (Figures 5.8 and 5.9). Disruption of actin also affects organelle motility and to an accumulation of microvesicles.

Recent evidence from other polarized tip growing systems such as pollen tubes, has revealed the presence of an apical gradient of *calcium* at the hyphal tip, with the concentration highest immediately below the tip. The maintenance of such a gradient is thought to be due to the presence of Ca^{2+} ion channels at the tip, allowing the flow of calcium ions down a concentration gradient from the outside to the inside of the cell. The

173

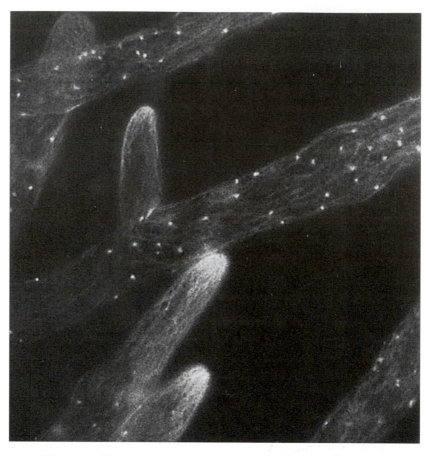

Figure 5.8 Fluorescence micrograph showing actin microfibrils concentrated at the hyphal tips and actin plaques in older regions. (From Heath, I. B. (1987) *Eur. J. Cell Biol.* 44:10–16.)

concentration of Ca^{2+} in eukaryotic cells is highly regulated and maintained at low levels in the cytoplasm by *calcium pumping ATPases*. Calcium-pumping ATPases are located on the plasma membrane, pumping calcium out of the cell and the vacuolar membranes allows sequestration and storage of the ion inside vacuoles within the hypha. Although the role and function of a calcium gradient in polarized cell growth has yet to be established, the presence of a tip high calcium gradient is likely to be an important factor in establishing cell polarity. Ca^{2+}

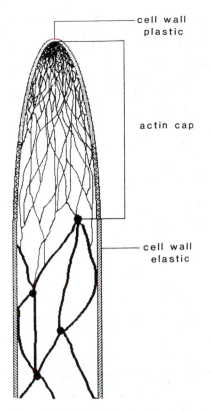

cell wall
plastic

actin cap

cell wall
elastic

Figure 5.9 Model of actin distribution in growing hyphae. The proliferation of actin microfibrils at the tip is thought to play an essential structural role at the growing apex.

ions are known to regulate the assembly of the cytoskeleton and to aid in vesicle fusion with membranes.

Yeast mycelial dimorphism

The majority of fungi grow either as round or spherical cells (yeasts) or as a polarized branching mycelium (filamentous). However, some fungi are capable of growing as either form, depending on the environmental conditions surrounding them. Such fungi are found throughout the fungal kingdom and are termed *dimorphic.* Dimorphism is common amongst several human and animal pathogens (e.g. *Candida albicans,*

Paracoccidiodes brasilensis, Histoplasma capsulatum and *Blastomyces dermatitidis*) as well as a number of plant pathogens (e.g. *Ustilago maydis, Ophiostoma ulmi* and *Rhodosporidium sphaerocarpum*). Whilst most dimorphic fungi are members of the Ascomycotina, dimorphism also occurs in other classes (e.g. *Mucor rouxii*, a member of the Zygomycotina). *Candida albicans* is an important opportunistic pathogen of humans and the majority of research has focused on understanding the mechanisms involved in dimorphism in this organism. Normally, infections are relatively superficial and restricted to mucosal membranes causing both oral and vaginal candidosis (thrush). However, for individuals who are immunocompromised, for example following immunosuppressive therapies or the advent of AIDS, candida infections often become systemic and invasive and have a high rate of morbidity. Invasive candidosis has been associated with the presence of the hyphal form of the organism, whereas superficial infections are generally associated with the yeast phase, so implicating the transition from yeast to hypha as an important event in the pathology of the organism. However, the human pathogens *Histoplasma capsulatum*, and *Paracoccidiodes brasiliensis* are pathogenic only in the yeast phase and a mutant strain of *C. albicans* was still able to cause infections and death in mice, though the extent and spread of the organism was reduced. A wide range of environmental parameters have been shown to induce a yeast to hypha transition in candida, including serum, *N*-acetylglucosamine, proline and a shift from an acidic to a more alkaline medium. This suggests that a number of independent signal transduction systems exist within candida for each factor, particularly as mutants which have lost the ability to undergo a yeast to mycelium transition by one stimulus are still capable of forming germ tubes when exposed to other stimuli.

The fungal cell wall

The fungal *wall* defines the shape of the fungal hypha and provides the mechanical strength to resist the internal *turgor pressure*. The hyphal wall of most filamentous fungi consists of an inner primary wall composed of *chitin* microfibrils (a polymer of *N*-acetylglucosamine) and an outer or secondary wall composed largely of β-1-3 and β-1-6 *glucans* (a polymer of glucose). The exceptions are members of the Oomycotina,

Figure 5.10 Scanning electron microscopy image of a chemically treated fungal hypha revealing the chitin microfibrillar network. (From Gooday, G. W. and Trinci, A. P. J. (1980). *The Eukaryotic Microbial Cell.* Society for General Microbiology Symposium 30. Cambridge University Press, Cambridge.)

where cellulose rather than chitin forms the inner microfibrillar component of the wall. β-1-3 glucans and mannoproteins are also present as components of the outer wall matrix. Removal of the outer wall with lytic enzymes has revealed the architecture of the inner chitin wall which is composed of microfibrils formed by the aggregation of the chitin polymers by hydrogen bonding (Figure 5.10). The chitin inner wall is cross-linked to the outer β-glucan components and forms a major component of the walls of most fungi. Synthesis of the cell wall occurs at the plasma membrane of the growing hyphal tip. *Chitin synthase* catalyses the formation of chitin from the precursor UDP-*N*-acetylglucosamine and appears as both the active enzyme and an inactive zymogen, requiring activation by cleavage of a peptide by an endogenous protease. Recently, three genes, chitin synthase I, II and III have been cloned from the yeast *Saccharomyces cerevisiae* and serve different functions in the cell. CHS1 acts as a repair enzyme and is involved in synthesizing chitin at the point where the daughter and mother cells separate, CHS2 is involved in septa formation and CHS3 is involved in chitin synthesis of the cell wall. In filamentous fungi the situation appears to be even more complex. At least six chitin synthase genes have been isolated from *Aspergillus fumigatus* and it seems likely that each may play a role in specific stages of fungal growth. However, the precise functions of the individual genes remains to be determined.

β-1-3 glucan, like chitin, is synthesized by a membrane-associated β-1-3 *glucan synthase* which utilizes UDP-glucose as the substrate, inserting glucose into the β-glucan chains. β-1-3 glucan synthase activity is found both in the membrane and cytosolic fractions of fungal mycelia and is stimulated by GTP. Although genes encoding a β-1-3 glucan synthase have been isolated from a number of fungi, it is not clear whether a gene family for β-1-3 glucan synthase exists in fungi as it does for chitin synthase.

As chitin and β-glucans are not found in mammalian cells (chitin is present in the exoskeleton of many insects and some molluscs) and so is a potentially powerful target for antifungal development, surprisingly few antifungal agents are used which are directed against wall biosynthesis. Two classes of nucleoside peptides, the *polyoxins* and the *nikkomycins* act as potent and specific inhibitors of chitin synthesis, competing as analogues of UDP-*N*-acetylglucosamine. However, their toxicity has prevented them from being exploited clinically as antifungal agents and their use as agrochemical fungicides hampered by the rapid emergence of resistant fungal strains. Another class of naturally occurring antibiotics, the *echinocandins*, are specific inhibitors of β-glucan biosynthesis. Initially, this class of compounds showed only a narrow spectrum of

178

activity toward fungi and was useful only when administered intravenously. More recently however, semisynthetic echinocandins have been synthesized with a broader antifungal spectrum, higher potency, and in a form which can be taken orally. This class of compounds is likely to become extremely important in the treatment of human fungal infections and has stimulated researchers to reappraise the cell wall as an important target for antifungal development.

Wall biosynthesis

The mature wall below the extension zone is rigid and resistant to internal turgor pressure, whereas the wall at the tip where growth occurs is plastic and malleable, yielding to the internal turgor pressure of the hypha and driving extension. Previously, the simultaneous delivery of both wall synthetic and lytic enzymes at the growing hyphal tip was proposed to account for the malleable nature of the hyphal tip, with the lysins continuously breaking the newly formed polymers at the tip and serving both to loosen the polymers at the tip and to create new sites for polymer extension. More recently, a new model has been formed based on experimental work which examined the fate of radiolabelled glucosamine and glucose incorporation into the wall at the growing tip. By exposing growing hyphae to radiolabelled glucosamine and glucose for short intervals followed by 'chasing' with unlabelled substrate, it was possible to determine the time taken for newly incorporated material at the tip to become incorporated into the wall. Initially, both glucosamine and glucose were readily extracted from the tip; however, as the radiolabelled components were displaced by the continuous supply of new precursors to the tip, they became progressively more difficult to extract from the hypha. These studies have revealed that wall precursors released at the tip, become progressively cross-linked and more rigidified as they are displaced down the hyphal tip (Figure 5.11). In this model, there is no requirement for the presence of lytic enzymes in order to produce a plastic, malleable hyphal tip.

The fungal membrane

The membranes of fungi, like those of other eukaryotic cells function to provide a barrier between the hypha and its environment, and are

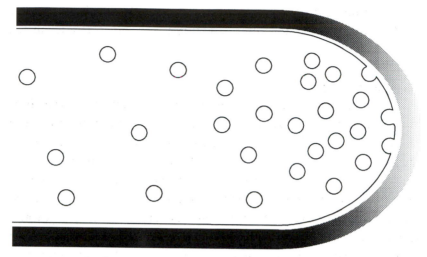

Figure 5.11 Schematic diagram of a growing fungal hypha illustrating the gradient in wall plasticity. Vesicles migrating to the tip fuse with the plasmalemma releasing wall precursors at the apex. Newly formed chitin becomes progressively cross-linked to the β-1-3 glucan component of the wall resulting in increasing wall rigidification away from the apex.

composed of a phospholipid bilayer into which are anchored various proteins. Sterols are also a critical component of the fungal membrane and serve to regulate membrane fluidity and the activity of membrane-associated enzymes and transport mechanisms. In mammalian cells, cholesterol is the chief sterol in the membrane, whilst in the majority of fungi it is *ergosterol,* the exceptions being the Chytridiomycotina and Oomycotina, where the dominant sterol is also cholesterol. This difference in the primary sterol component of fungal and mammalian cells has allowed the development of two classes of antifungal agents, the *polyenes* and the *azoles.*

The polyene antifungals, which includes nystatin and amphotericin, bind hydrophobically to the ergosterol component of the membrane forming pores leading to a loss of cell membrane integrity. The azole antifungals, which include the triazoles and imidazoles, act primarily by inhibiting 14-α-sterol demethylase which is involved in ergosterol biosynthesis, resulting in ergosterol depletion in the membrane and the accumulation of 14-α-methylsterols. The subsequent alteration in the sterol composition of the membrane results in changes in membrane fluidity, transport processes and wall biosynthesis and ultimately in cell death.

Nutrient uptake

In order for the fungi to grow, external nutrients must be assimilated across the plasma membrane. To successfully achieve the absorption of nutrients from the surrounding environment, the fungi possess a diverse range of specific transport proteins in the plasma membrane. Three main classes of nutrient transport occur in fungi, facilitated diffusion, active transport and ion channels. Fungi usually contain two transport mechanisms for the assimilation of solutes such as sugars and amino acids. The first is a constitutive *low affinity transport* system which allows the accumulation of solutes when present at a high concentration outside of the hypha. This process of facilitated diffusion is not energy dependent and does not allow accumulation of solutes against a concentration gradient. When the solute concentration is low (as is often the case in the environment), a second class of carrier proteins is induced that have a *higher affinity* for the solute and mediate the energy-dependent uptake of the solute against a concentration gradient at the expense of ATP. In order to assimilate nutrients at low concentrations, fungi create an electrochemical proton gradient by pumping out hydrogen ions from the hyphae at the expense of ATP via proton pumping *ATPases* in the plasma membrane. The proton gradient created provides the electrochemical gradient which drives nutrient uptake as hydrogen ions flow back down the gradient.

Thus fungi are capable of adapting their transport mechanisms according to the external solute concentration, ensuring continued uptake under different nutrient concentrations. *Ion channels* are highly regulated pores in the membrane which allow influx of specific ions into the cell when open. A number of ion channels have been described in fungi by patch clamping studies analogous to studies conducted on mammalian cells. Patch clamping involves measuring the current flowing across the cell membrane, which can be used to study the flow of various ions across the membrane. In fungi, the cell wall has first to be removed by incubating mycelium in an osmotic stabilizer and a mixture of lytic

181

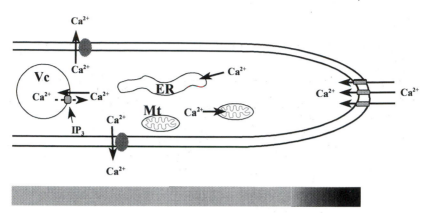

Gradient in intracellular calcium

Figure 5.12 Model of calcium homeostasis in the hyphal tip. The high calcium gradient is thought to be generated by stretch-activated calcium channels located at the apex. Further back, calcium levels are maintained at a lower, constant level by the concerted action of calcium-pumping ATPases which pump calcium out of the hypha and sequestration into organelles. In addition, the intracellular messenger inositol trisphosphate (IP_3) has been shown to mobilize calcium from the vacuole where the resulting transient rise in intracellular calcium levels may trigger a number of downstream events. Vc, vacuole; Mt, mitochondria; ER, endoplasmic reticulum.

enzymes which digest away the cell wall producing naked sphaeroplasts or protoplasts. Two Ca^{2+} stimulated K^+ channels have been identified in *Saprolegnia ferax* which carry an inward flux of K^+ ions and are thought to be involved in regulating the internal turgor pressure of the hypha. More recently, the presence of a mechanosensitive or stretch-activated Ca^{2+} channel has also been described in this organism. Stretch-activated channels are opened when the membrane is under mechanical stress and may play an important role in the generation of the tip high calcium gradient observed in this organism (Figure 5.12).

Nutrient sensing

When readily utilizable sugars, such as glucose or fructose are added to fungi which are derepressed (starved) for glucose or other sugars, there follows a number of rapid metabolic responses which are mediated by a

182

Glucose

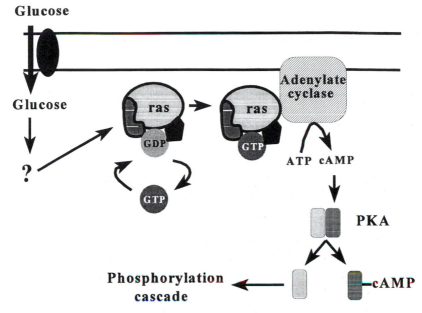

Figure 5.13 Schematic diagram showing the activation of the ras-cAMP pathway in fungi. The influx of glucose catalyses the exchange of GDP for GTP leading to a conformation change in the ras complex which then activates membrane-bound adenylate cyclase. The rise in cAMP activates cAMP-dependent protein kinase (PKA) by binding to the regulatory domain of the enzyme, which releases the active catalytic subunit. PKA activation triggers the phosphorylation and activation of a number of downstream targets.

transient rise in the levels of *cAMP*. These responses include the inhibition of gluconeogenesis and the activation of glycolysis and trehalase, which is involved in trehalose degradation, and are mediated by the activation of *cAMP-dependent protein kinase* (PKA) (Figure 5.13). The increase in cAMP is due to the activation of the membrane-bound enzyme adenylate cyclase as a result of the activation of a class of small GTP-binding proteins, the RAS proteins. Activation of RAS leads to the exchange of RAS-bound GDP for GTP, causing a conformational change leading to the stimulation of adenylate cyclase activity. The RAS protein complex includes an intrinsic GTPase which converts GTP-bound RAS to GDP-bound RAS, thus returning RAS to its resting state in the absence of an activator. Currently, the mechanism by which glucose stimulates the RAS complex is unknown but the available evidence strongly suggests that the RAS pathway forms part of a global mechanism for nutrient sensing.

Further reading

Gooday, G. W. (1993). Cell envelope diversity and dynamics in yeasts and filamentous fungi. *J. Appl. Bacteriol.* 74:12S–20S.

Gow, N. A. R. (1994). Growth and guidance of the fungal hypha. *Microbiology* 140:3193–3205.

Heath, I. B. (1990). The roles of actin in tip growth of fungi. *Int. Rev. Cytol.* 123:95–127.

Jackson, S. L. and Heath, I. B. (1993). Roles of calcium in hyphal tip growth. *Microbiol. Rev.* 57:367–382.

Jennings, D. H. (1986). Morphological plasticity in fungi. *Sym. Soc. Exp. Biol.* 40:329–346.

Peberdy, J. F. (1994). Protein secretion in filamentous fungi: trying to understand a highly productive black box. *Trends Biotechnol.* 12:50–57.

Prosser, J. I. and Tough, A. J. (1991). Growth mechanisms and growth kinetics of filamentous microorganisms. *Biotechnology* 10:253–274.

Wessels, J. G. H. (1993). Wall growth, protein excretion and morphogenesis in fungi. *New Phytol.* 123:397–413.

6 Asexual sporulation: conidiation

T. H. ADAMS AND J. K. WIESER

Introduction

Asexual sporulation is generally the most prolific reproductive mode for fungi. Asexual spores of higher fungi are called conidia, which are non-motile asexual propagules made from the side or tip of specialized sporogenous cells and do not form through progressive cleavage of the cytoplasm. The process of conidiation is complex and involves temporal and spatial regulation of gene expression, cell specialization and intercellular communication. However, the genetic mechanisms controlling fungal sporulation have only been addressed in detail in two well-studied ascomycetes, *Aspergillus nidulans* and *Neurospora crassa*. In this chapter we will describe the genetic regulation of development in *A. nidulans*. It is presumed that variations on this theme will apply in many cases to understanding conidiogenesis in other fungi as well.

The *A. nidulans* asexual reproductive cycle can be divided into three conceptual stages: (1) a growth phase that is required for cells to acquire competence to respond to induction signals; (2) initiation of the developmental pathway; (3) the events leading to sporulation.

Colony formation

Vegetative growth in *A. nidulans* begins with the germination of a spore. Spore germination leads to the formation of tubular structures, termed hyphae, that grow in a polar fashion by apical extension to form a network of interconnected hyphae known as a mycelium. The mycelium forms a radially symmetrical colony that expands indefinitely at a constant rate of about 0.5 mm h^{-1} (at 37 °C). The fungal mycelium appears to be an amorphous collection of equivalent vegetative cells. However, it is clear that the various cells of the mycelium must interact to form an ordered network with different hyphae having distinct roles in the acquisition of

185

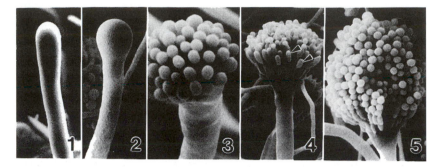

Figures 6.1–6.5 Conidiophore formation: scanning electron micrographs of the stages of conidiation. Figure 6.1 Early conidiophore stalk. Figure 6.2 Vesicle formation from tip of stalk. Figure 6.3 Developing metulae. Figure 6.4 Phialides developing. Figure 6.5 Mature conidiophores bearing chains of conidia. (Reproduced with permission from Mims *et al.*, 1988. *Protoplasma* 44:132–141.)

nutrients from the environment and in determining the precisely timed development of specialized reproductive structures.

Approximately 16 hours after spore germination, aerial hyphal branches are formed in the centre of the colony and some of these branches subsequently differentiate into conidiophores, the term applied to the asexual spore-bearing structures. It takes between 6 and 8 hours from the initiation of aerial growth to the formation of the first asexual spore or conidium, so the asexual cycle can be reinitiated within about 24 hours of the original spore germination. Later on, developmental initiation moves out towards the edge of the colony leaving the oldest conidiophores in the centre and the newly forming conidiophores at the margin of the growing colony. Typically, hyphae must be exposed to an air-aqueous interface to stimulate conidiophore production (but see below).

Conidiophore morphology

The formation of conidiophores in a colony is a complex process that can be divided into several morphologically distinct stages (Figures 6.1–6.5). This process begins with the growth of a conidiophore stalk that elongates by apical extension of an aerial branch (Figure 6.1). The conidiophore stalk differs from a vegetative hyphae in at least three easily recognizable ways. First, the stalk cell extends from a specialized thick-walled

186

cell, termed a footcell, that anchors the stalk to the growth substratum. In maturing conidiophores, the footcell can be distinguished from other mycelial cells by the presence of a two-layered wall in which the outer layer is continuous with the rest of the mycelium while the inner layer is unique to the footcell and the emerging conidiophore stalk. The second distinguishing feature of the conidiophore stalk is that it has a wider diameter than typical vegetative aerial hyphae. Finally, unlike vegetative aerial hyphae that grow indefinitely and are capable of branching, conidiophore stalks rarely branch and their length is relatively determinate. For *A. nidulans*, the conidiophore stalk attains a height of about 100 μm but this is a species-specific trait and varies among the different species of *Aspergillus*.

After apical extension of the conidiophore stalk ceases, the tip begins to swell and reaches a diameter of ~ 10 μm (Figure 6.2). This structure, known as the conidiophore vesicle is continuous with the stalk in that it is not separated by a septum. Multiple nuclei align around the outside of the vesicle and undergo a more or less synchronous division associated with the synchronous production of buds from the vesicle surface to form a layer of primary sterigmata termed metula. Each conidiophore head contains about 60 metula and each metular bud forms a cigar-shaped cell that is about 6 μm long and 2 μm wide and contains a single nucleus (Figure 6.3). Metulae bud twice to produce a layer of about 120 uninucleate sterigmata termed phialides. This metular budding is polar so that both phialides are produced from the end of the metula that is farthest away from the vesicle (Figure 6.4). Phialides subsequently give rise to chains of uninucleate spores called conidia (Figure 6.5). Because each phialide can make upwards of 100 spores, the total number of conidia from one conidiophore can exceed 10000.

Growth versus development during colony formation
Acquisition of developmental competence
Conidiation does not usually occur in *A. nidulans* until cells have gone through a defined period of vegetative growth. This was demonstrated by taking advantage of the fact that under normal media conditions, *A. nidulans* can be maintained indefinitely in the vegetative stage of its life cycle by growing submerged in liquid medium. Asexual sporulation

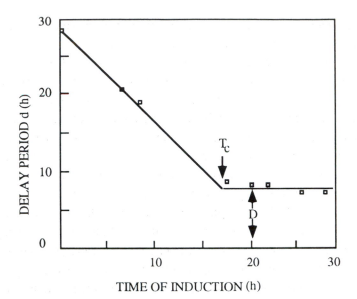

Figure 6.6 Experimental determination of competence (T_c). Spores were inoculated into liquid medium at $t = 0$ and allowed to germinate in sub-merged culture. At times indicated on the horizontal axis, mycelia were transferred to a solid substrate and observed to determine the earliest time of conidiophore formation. The delay time (d) is the interval between transfer to solid media and the first appearance of conidiophores. The time T_c is the time after which $d = D = $ constant. (Reproduced with permission from Champe and Simon 1992. *Morphogenesis: an Analysis of Development of Biological Form*. Marcel Dekker, New York.)

only rarely takes place and sexual sporulation never occurs unless vegetative hyphae are exposed to air (see below). This trait provides a means for synchronous developmental induction through harvesting hyphae grown in submerged culture and then exposing them to air. By varying the times hyphal cultures are maintained in submerged culture, it has been shown that there is a minimum growth period before cells respond to induction. If development is induced following at least 18 hours of vegetative growth, the time required to observe the initial developmental structures remains constant at about 5 hours (Figure 6.6). If the submerged growth period is shorter than 18 hours, the time required for development increases correspondingly. These results have been interpreted to mean that cells require

188

approximately 18 hours of growth before they are competent to respond to the inductive signal provided by exposure to air.

Although the acquisition of competence is the earliest described conidiation-specific event in the *A. nidulans* life cycle, it is also the least understood. Neither the concentration of a limiting nutrient such as glucose, nor continuous transfer to fresh medium alters the timing with which cells become competent. However, precocious conidiation mutants with a decreased time requirement for acquisition of competence have been isolated suggesting a genetic component to competence. Physiological changes ranging from changes in inducibility of enzyme systems to altered glucose uptake have been correlated with cultures obtaining developmental competence, but how these changes are regulated and which events occur first are not known. In this regard, it is interesting that these same physiological changes take place more quickly in precocious conidiation mutants than in wild-type strains. Taken together, these results have been interpreted to mean that this early aspect of *Aspergillus* conidiophore development occurs as an integral part of the life cycle rather than as a response to unfavourable environmental conditions.

Developmental induction

Developmentally competent liquid-grown hyphae from *A. nidulans* normally do not conidiate until they are exposed to air. The mechanism by which exposure of competent hyphae to air serves to induce conidiation is not understood. The signal may be cell surface changes induced by the air–water interface at the hyphal surface.

Conidiation in *A. nidulans* strains that have the wild-type allele of the velvet gene (*veA$^+$*) is promoted by red light. Fifteen to 30 minutes of red light activates the developmental program. *A. nidulans* strains (including most laboratory strains) that carry the *veA1* mutation, conidiate in the absence of light suggesting that the *veA* product has a negative regulatory role.

Starvation stress-induced development

The block in conidiation under water can be overcome when nutrients are limited or in response to other stresses (Figure 6.7). Although

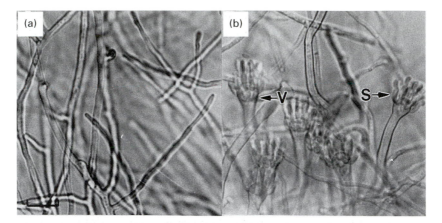

Figure 6.7a,b Starvation induced conidiation in submerged culture. A wild-type strain was grown in glucose minimal medium for 18 h and then shifted to fresh minimal medium (a) or minimal media lacking a carbon source (b). Conidiophores formed within 12 h of a shift to nitrogen starvation media (development also occurred upon shifting to a carbon starvation media), but no conidiophores were seen in the culture shifted to minimal containing a proper carbon or nitrogen source. (Reproduced with permission from Lee and Adams, 1996. *EMBO J.* 15:303.)

these conditions are apparently sufficient to cause development in submerged culture, it is not yet clear whether such nutritional stress has a direct role in developmental activation for hyphae grown on the surface of a plate. While nutritional stress may be sufficient to cause sporulation, it is apparently not required for development to take place. The nature of the signal to induce sporulation is not yet clear.

Genetics of sporulation
Rationale for the use of *A. nidulans* as a model for the study of sporulation

A. nidulans is a particularly useful organism in which to study sporulation. The asexual life cycle of *A. nidulans* includes production of complex, highly ordered, multicellular structures. Conidiation can be observed macroscopically due to conidial pigmentation. As conidiophores and conidia are dispensable, it is a simple task to obtain defective mutants. Conidiation mutants are easily manipulated to construct strains with desired genotypes for epistasis studies or for mapping mutations.

Many molecular genetic techniques are available in *A. nidulans* (see Chapter 4). The most important of these is a DNA-mediated transformation system. Genes of interest can be cloned by complementation of a mutant phenotype by transformation using genomic DNA libraries. Integrated cosmids can then be recovered from the DNA of the transformant. It is feasible to disrupt or delete a specific gene from the genome. Thus a combination of classical and molecular genetics can be used to isolate a gene by complementation of a mutant phenotype and show that the correct gene has been isolated.

Furthermore, transcriptional promoters can be analysed *in vivo* by fusing the promoter of interest to a reporter gene like the *Escherichia coli lacZ* gene and introducing the construct into the fungus. β-galactosidase activity can be determined at various times during development by isolating crude protein from cell extracts. *In situ* staining is also possible and is one way to determine the cellular location of a given protein. Strong inducible promoters can be used to drive the expression of a gene of interest for study of the developmental effects of misscheduled expression.

Sporulation-specific gene functions

Differentiation of the multiple cell types making an *A. nidulans* conidiophore is correlated with activation of several hundred genes. However, to date the functions of only a small number of these genes has been determined. Timberlake showed that there are approximately 1200 diverse mRNAs that accumulate to varying concentrations specifically during conidiation. By contrast, Martinelli and Clutterbuck used mutational analysis to estimate that only 45–100 genes are uniquely required for asexual sporulation. This large discrepancy in predicted gene number could be explained if some genes detected based only on expression patterns encode redundant or incremental functions that would not be detectable by simple visual examination of mutants. In keeping with this idea, a deletion of a 38 kb region from *A. nidulans* containing numerous spore-specific genes did not result in detectable changes in development. Furthermore, subtle defects in sporulation, like the spore wall defect resulting from deletion of the *rodA* gene (see below) might be difficult to detect in broad screens based on tedious visual examination of colonies.

Genes required in pigmentation

Wild-type *A. nidulans* spores have a characteristic dark-green pigment in their walls. It is straightforward to identify mutants with abnormal pigmentation. These mutants identified genes involved in pigment synthesis that serve as excellent reporters of developmentally regulated gene expression.

Though many loci contribute to the nuances of spore colour (see below) two genes, *wA* and *yA*, are central to the process of spore pigmentation. The *wA* gene is predicted to encode a polyketide synthase; mutations in *wA* result in white-spored strains that lack melanin and an electron-dense outer spore wall layer that contains α 1,3 glucan. *yA* is predicted to encode a p-diphenol oxidase known as conidial laccase; *yA* mutants are yellow spored. Genetic analysis suggests that *wA* catalyses the synthesis of a yellow-pigmented polyketide that is converted to the green form by the product of the *yA* gene. The transcripts of both *yA* and *wA* are developmentally regulated by the central regulatory pathway described below. *wA* is apparently directly regulated by *wetA*, while *yA* is activated by both *brlA* and *abaA*.

Many other genes are involved in the production of green-pigmented wild-type spores. Mutations at *ygA* or *yB* give rise to yellow-spored conidiophores. *drkA* and *drkB* are involved in production of the sac that encloses individual conidia. Mutations in *chA* (chartreuse), *fwA* (fawn), *bwA* (brown), or *dilA* and *B* (dilute conidia colour) result in modified spore colour.

Pigmentation of the conidiophore has been shown to require the activity of two genes *ivoA* and *ivoB* that when mutated give rise to colourless conidiophores. The products of these genes are responsible for synthesis of a melanin-like pigment from N-acetyl-6-hydroxytryptophan. Expression of *ivoA* and *ivoB* is probably controlled by the activity of *brlA*.

Spore wall proteins

Another class of genes encodes proteins that make up specific components of the spore wall. *rodA* and *dewA* encode proteins that contribute to the hydrophobicity of conidia, presumably facilitating their dispersal

by air. Both of these genes were shown to belong to a general class of proteins found in fungi termed hydrophobins. Disruption of the *rodA* gene
resulted in conidia that were very easily wettable compared with wild-
type and gave colonies with a dark centre due to the abnormal accumulation of liquid on the conidiophores. This phenotype arose from the loss of
the conidial rodlet layer. Disruption of *dewA* also resulted in spore wall
defects but the phenotype was much more subtle. *dewA* mutant conidia
were not wettable with water alone but were susceptible to wetting with
dilute detergent. The *dewA* protein was found to be located in the spore
wall. *rodA* and *dewA* double mutants are even less hydrophobic than
either single mutant indicating that *dewA* increases hydrophobicity independent of the rodlet *rodA*-encoded layer.

Fungal hydrophobins have been found in several species of
Basidiomycetes and Ascomycetes. They are structural components found
in the walls of spores or of aerial hyphae and contribute to the hydrophobicity of these structures. Fungal hydrophobins share little sequence identity at the amino acid level, but have certain structural features in
common. They are small proteins with from 96 to 187 amino acids. They
all contain eight cysteine residues arranged in a conserved pattern and
they have a similar arrangement of hydrophobic domains and thus
exhibit similar hydrophobicity plots.

Execution of development: The central regulatory pathway

Genetic and biochemical studies of *A. nidulans* identified two critical genes, *brlA* and *abaA*, that are specifically required for conidiation.
Together with a third gene called *wetA*, *brlA* and *abaA* have been proposed to define a central regulatory pathway controlling conidiation-
specific gene expression and determining the order of gene activation
during conidiophore development and spore maturation. Mutation in
any one of these three genes blocks asexual sporulation at a specific stage
in conidiophore morphogenesis and prevents expression of a broad class
of developmentally regulated mRNAs.

brlA encodes an early regulator of development

The best characterized developmental regulatory gene is *brlA*. The
phenotype of *brlA* null mutants has been described as 'bristle' because

Figure 6.8a,b Conidiophores of a wild-type strain (a) or a *brlA* mutant strain (b) are shown. In the wild-type strain, the stalks grow to a fairly uniform height and bear the other structures of the conidiophores, while the stalks of a *brlA* mutant grow somewhat indeterminately and fail to elaborate the other structures of the conidiophore. Arrows indicate stalks. (Reproduced with permission from Adams *et al.*, 1988. *Cell* 54: 354–362.)

these mutants are blocked early in development and fail to make the transition from polar growth of the conidiophore stalk to swelling of the conidiophore vesicle. Instead, *brlA* mutants differentiate conidiophore stalks that grow somewhat indeterminately, reaching heights 20–30 times taller than wild-type conidiophores and giving the colony a 'bristly' appearance (Figure 6.8). These *brlA* mutants also fail to accumulate mRNAs for a host of developmentally regulated genes including *abaA* and *wetA*. Thus, *brlA* is thought to activate development-specific gene expression beginning at the time of conidiophore vesicle formation. Two lines of evidence indicate that *brlA* activity continues to be required during spore formation.

194

First, numerous hypomorphic mutant alleles of *brlA* have been described that support more extensive conidiophore development than *brlA* null mutants, having varying degrees of vesicle and sterigmata development, but never spores. These hypomorphic mutants presumably maintain partial BrlA function and result in altered expression patterns for numerous genes encoding development-specific activities. Second, developmental induction experiments using a temperature-sensitive *brlA* allele, *brlA42^{ts}*, showed that development was arrested following a shift to the restrictive temperature for BrlA42 activity, regardless of the developmental stage.

The wild-type *brlA* gene encodes two overlapping transcription units, designated *brlAα* and *brlAβ*, which each accumulate to detectable levels early in development at about the time conidiophore vesicles first appear. *brlAβ* transcription initiates about 1 kb upstream of *brlAα* transcription which begins within *brlAβ* intronic sequences. The *brlAβ* transcript encodes two open reading frames (ORFs) that begin with AUGs, a short upstream ORF (μORF) and a downstream reading frame that encodes the same polypeptide as *brlAα* except that it includes 23 additional amino acids at the NH-terminus (Figure 6.9). Mutations that block expression of either transcript alone cause abnormal development. However, multiple copies of either *brlAα* or *brlAβ* can compensate for loss of the other gene. These results are consistent with the hypothesis that the *brlAα* and *brlAβ* transcription units are individually essential for normal development but the products of each gene have redundant functions. *brlAα* and *brlAβ* are controlled by different mechanisms suggesting that this complex locus has evolved to provide a mechanism to separate responses to the multiple regulatory inputs activating and maintaining *brlA* expression throughout development (see below).

The BrlA polypeptide contains two directly repeated 'zinc finger' DNA binding motifs, indicating that *brlA* encodes a nucleic acid-binding protein. In addition, *A. nidulans brlA* expression in *Saccharomyces cerevisiae* results in *brlA*-dependent activation of *Aspergillus* genes in *S. cerevisiae* providing that they contain so-called BREs (BrlA Response Elements; [C/A][G/A]AGGG[G/A]). Although attempts to demonstrate BrlA binding to the BRE element *in vitro* have been unsuccessful to date, several developmentally regulated genes including *abaA*, *wetA*, *rodA*, and

195

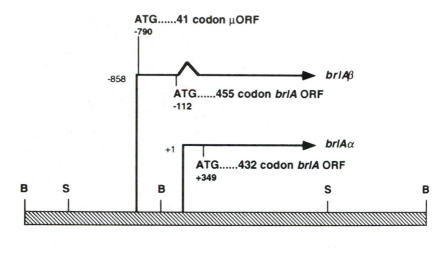

Figure 6.9 Organization of the *brlA* locus. The *brlAα* and *brlAβ* transcripts are indicated by arrows and the *brlAα* transcription start site is designated (+1). ATGs indicate the predicted translation start sites of the three open reading frames. Reproduced with permission from Han *et al.*, 1996. *EMBO J.* 12:2450.

yA contain multiple BREs upstream of their transcription start sites. This observation is consistent with the idea that *brlA* encodes a primary transcriptional regulator of the central regulatory pathway for development.

The pivotal role of *brlA* in controlling development is most dramatically demonstrated by results of experiments involving misscheduled activation of *brlA* in inappropriate cell types. An *A. nidulans* strain was constructed that allowed controlled transcription of *brlA* in vegetative cells by fusing the *brlA* coding region to the promoter for the *A. nidulans* catabolic alcohol dehydrogenase gene, *alcA(p)*. Because *alcA* transcription is induced by threonine or ethanol and repressed in the presence of glucose, *brlA* could be induced or suppressed in vegetative cells by simply changing the medium. When the *alcA(p)::brlA* fusion strain was transferred from medium with glucose as a carbon source to medium with threonine, hyphal tips immediately stopped growing and

196

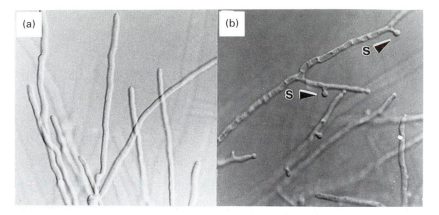

Figure 6.10a,b Overexpression of *brlA*. The *brlA* gene was placed under the control of the alcohol-inducible promoter *alcA*. The *alcA(p)::brlA* strain and a wild-type strain were grown for 12 h in liquid minimal media containing glucose to repress *brlA* expression from the *alcA* promoter then shifted to *alcA*-inducing media and observed. 3 h after the shift the *alcA(p)::brlA* strain produced spores from the tips of hyphae (b), no such structures were produced in the wild-type strain even 24 h after the shift (a).

differentiated into reduced conidiophores that produced viable conidia (Figure 6.10). In addition, this forced activation of *brlA* resulted in activation of the *abaA* and *wetA* regulatory genes, as well as additional developmentally specific transcripts with known and unknown functions.

Activation of *abaA* initiates a positive feedback loop

abaA encodes a developmental regulator that is activated by *brlA* during the middle stages of conidiophore development following sterigmata differentiation. The phenotype of *abaA* null mutants is described as 'abacus' because these mutants produce aconidial conidiophores that differentiate sterigmata but do not form sporogenous phialides. Instead, these mutants form branching sterigmata leading to long chains of cells that appear like beads on a string as in an abacus. Although the *brlA* transcript accumulates normally in *abaA* mutants, many other developmentally regulated mRNAs (including the *wetA* message) do not.

The predicted AbaA polypeptide contains a domain that is very closely related to the DNA binding domain of the SV40 enhancer factor TEF-1. This domain is also closely related to the yeast Ty1 enhancer-binding protein TEC1 and has been called the TEA (TEF-1 and TEC1, AbaA) or ATTS (AbaA, TEC1p, TEF-1 sequence) domain. AbaA also contains a potential leucine zipper for dimerization and, like *brlA*, *abaA* is required for transcriptional activation of numerous sporulation specific genes. Results from *in vitro* biochemical analyses showed that AbaA binds to the consensus sequence 5′-CATTCY-3′ (ARE). Multiple AREs are present in the regulatory regions for developmentally regulated genes including *brlA*, *wetA*, *yA*, *rodA*, and *abaA* itself. As with BREs, these *cis*-acting sequences are able to confer transcriptional activity in a heterologous *S. cerevisiae* gene expression system that is dependent on expression of the *Aspergillus* activator, *abaA*.

Forced activation of *abaA* from the *alcA* promoter in vegetative hyphae resulted in growth cessation and accentuated cellular vacuolization, but not in conidial differentiation. *abaA* induction also led to activation of several developmentally specific genes with known and unknown functions including *wetA*, and perhaps surprisingly, *brlA*. Thus, *brlA* and *abaA* are reciprocal inducers but *brlA* expression must occur before *abaA* expression for productive conidiophore development.

wetA is required late in development

The *wetA* gene is required late in development for synthesis of crucial cell wall components. The phenotype of a *wetA* null mutant is described as 'wet-white'. These mutants differentiate normal conidiophores but the conidia produced are defective in that they never become pigmented but instead autolyse.

wetA appears to encode a regulator of spore-specific gene expression. This hypothesis is based on the finding that *wetA⁻* mutants fail to accumulate many sporulation-specific mRNAs. In addition, forced activation of *wetA* in vegetative cells caused growth inhibition and excessive branching and resulted in accumulation of transcripts from several genes that are normally expressed only during spore formation and whose mRNAs are found in mature spores. *wetA* activation in hyphae did not result in *brlA* or *abaA* activation and never led to premature conidiation.

198

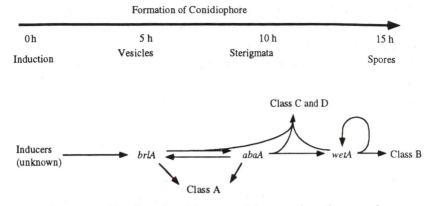

Figure 6.11 Timeline of conidiation and the central regulatory pathway. The genes in the central regulatory pathway for conidiation are predicted to activate other genes responsible for the production of conidiophores. The activation of the class A, B, C, and D genes give rise to the formation of a conidiophore with the timing indicated in the upper part of the figure.

Coordinated activation of gene expression by the central regulatory pathway

By examining patterns of RNA accumulation in a series of *brlA*⁻, *abaA*⁻, and *wetA*⁻ mutant strains in which *brlA*, *abaA*, or *wetA* were induced by *alcA* activation, the non-regulatory developmentally activated genes were divided into four categories. Class A genes are activated by either *brlA* or *abaA* or both, independent of *wetA* (Figure 6.11). These Class A genes are likely to be involved in early development events. *abaA* alone activates *wetA* which in turn activates Class B genes, independent of *brlA* and *abaA*. These Class B genes are likely to encode spore-specific functions. Class C and D genes require the combined activities of *brlA*, *abaA*, and *wetA* for their expression and have been proposed to encode phialide-specific functions. Class C and D genes are distinguished from one another by their expression patterns during normally induced development in wild-type and mutant strains. Accumulation of *wetA* mRNA requires *wetA*⁺ activity both during normal conidiophore development and in forced expression experiments bringing up the interesting possibility that *wetA* is autogenously regulated.

These results have been incorporated into the model describing the

199

genetic processes underlying the temporal and spatial control of gene expression during differentiation of conidiophores shown in Figure 6.11. In this model, activation of *brlA* expression initiates a cascade of events that are coordinated by the interactions of *brlA*, *abaA*, and *wetA*. The timing and extent of expression of the regulatory and non-regulatory genes is determined by intrinsic changes in the relative concentrations of the regulatory gene products in the various conidiophore cell types. There remains a great deal to learn regarding the molecular genetic processes that lead to *brlA* expression as well as the precise mechanisms through which *brlA*, *abaA*, and *wetA* interact in controlling their own expression and the expression of other developmentally regulated genes.

Activation of the central regulatory pathway occurs through multiple mechanisms

Regulation of *brlA* expression

As described above, it has been suggested that *brlAα* and *brlAβ* have evolved to provide separate means of responding to the multiple regulatory inputs controlling *brlA* expression. *LacZ* fusions have shown that *brlAα* and *brlAβ* are regulated in different ways (Figure 6.12). The *brlAβ* message is transcribed in vegetative cells before developmental induction but does not accumulate to appreciable levels presumably because translation of the *brlAβ* μORF represses BrlA translation blocking development. The importance of this translational block was demonstrated by eliminating the μORF AUG which led to *brlA* expression in hyphae and inappropriate sporulation. This result implies that one way *brlA* expression could be initiated following induction is through a controlled change in the choice of translational initiation sites from the μORF AUG to the BrlAβ AUG. However, it is also possible that increased transcription of *brlAβ* could provide sufficient template to allow inefficient BrlA translation from the internal AUG. In support of the latter hypothesis, it has been demonstrated that forced activation of *brlAβ* transcription in hyphae is sufficient to cause sporulation despite the presence of the μORF. In addition, there is a significant increase in *brlAβ* mRNA levels following developmental induction. Although this has not been shown to result from increased transcription, it has been demonstrated that

200

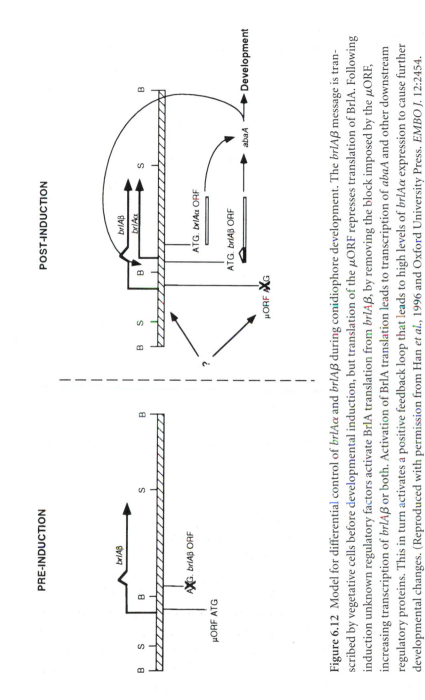

Figure 6.12 Model for differential control of *brlAα* and *brlAβ* during conidiophore development. The *brlAβ* message is transcribed by vegetative cells before developmental induction, but translation of the μORF represses translation of BrlA. Following induction unknown regulatory factors activate BrlA translation from *brlAβ*, by removing the block imposed by the μORF, increasing transcription of *brlAβ* or both. Activation of BrlA translation leads to transcription of *abaA* and other downstream regulatory proteins. This in turn activates a positive feedback loop that leads to high levels of *brlAα* expression to cause further developmental changes. (Reproduced with permission from Han *et al.*, 1996 and Oxford University Press. *EMBO J.* 12:2454.

specific fragments from the *brlAβ* regulatory region confer development-specific expression to an otherwise inactive promoter-*lacZ* fusion. In any case, the ultimate result of *brlA* activation is activation of other developmentally specific genes. Following BrlAβ translation, *brlAα* transcription is activated primarily through the *brlA*-dependent positive feedback loop by a mechanism that probably involves *abaA* and may also involve *brlA* directly. This prediction is supported by the observation that there are 5 AbaA and 12 BrlA binding sites found upstream of *brlAα*. Forced overexpression of *brlAβ* resulted in accumulation of *brlAα* mRNA whereas forced activation of *brlAα* did not result in *brlAβ* accumulation.

The role of *abaA* in feedback regulation of *brlA* expression is somewhat complicated by the observation that *brlA* expression actually increases in *abaA* null mutants. Thus, while overexpression of *abaA* in hyphae causes *brlA* activation, *brlA* is apparently overexpressed in *abaA* null mutants indicating that AbaA has both repressive and stimulatory activities towards *brlA*. Aguirre (1993) attempted to explain this paradox by proposing that AbaA functions as a transcriptional repressor of *brlA* when present at low concentrations but acts as an activator when present at high concentrations as would occur following *alcA* promoter-induced expression.

Analyses of fluffy mutant loci

In estimating the number of genes required for various stages in development, Martinelli and Clutterbuck proposed that 83 per cent of all sporulation mutants were altered in their ability to initiate development before formation of the conidiophore vesicle and presumably before activation of *brlA* expression. This result indicates that regulation of the decision to initiate conidiophore development is highly complex, requiring the activities of numerous genes. The most common phenotype observed among these proposed early conidiation mutants has been described as 'fluffy'. The term fluffy includes a diverse group of phenotypes, but in general all fluffy mutants grow as undifferentiated masses of vegetative hyphae forming large cotton-like colonies. This has led to the characterization of six fluffy loci, designated *fluG*, *flbA*, *flbB*, *flbC*, *flbD*, and *flbE*, which when mutant result in fluffy colonies with greatly reduced *brlA* expression (Figure 6.13a–d). Four of these genes, *fluG*,

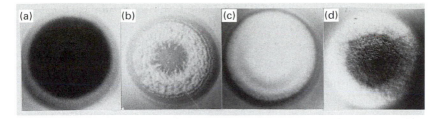

Figure 6.13 Phenotypic classes of fluffy mutants. Colonies were grown for 3 days on solid minimal media. Panel a shows a developmentally wild-type strain of *A. nidulans*. Panel c shows a *fluG* mutant. Panel b shows an *flbA* mutant and panel d shows a delayed conidiation mutant typical of *flbB,C,D* and *E* mutants.

flbA, *flbC*, and *flbD*, have been physically isolated and characterized in some detail.

fluG is required for production of an extracellular factor and is related to prokaryotic glutamine synthetase I

Although *fluG* mutants produce fluffy colonies (Figure 6.13c) and fail to activate *brlA* under normal growth conditions, they can be induced to conidiate in two ways. First, when *fluG* mutant strains (even a *fluG* deletion mutant) are grown on a solid surface under suboptimal nutritional conditions they produce some conidiophores. This result has been interpreted to indicate that *fluG* is required for a genetically programmed aspect of conidiophore development. In the absence of *fluG* it is possible to detect sporulation that occurs in response to environmental stress. This environmentally regulated response presumably also occurs in wild type but is overwhelmed by the more prolific *fluG*-dependent programmed development response. As predicted in this model, *fluG* mutants can respond to starvation stress in submerged culture by producing conidiophores. However, the response differs from that observed for wild type in two respects. First, although elimination of the carbon source causes *fluG* mutants to sporulate in submerged culture, the number of conidiophores observed is less than in wild type. Second, although elimination of the nitrogen source does cause morphological changes in the *fluG* mutant in submerged culture, it does not cause sporulation. Taken together, these results might indicate that there are two aspects to stress-induced sporulation in submerged culture: one that is *fluG* dependent and a second that is *fluG* independent. Because over-expression of *fluG* in submerged culture can cause development (see below), one aspect of stress-induced development could be increased activity of *fluG*.

The second way that *fluG* mutants can be induced to conidiate is by growing next to wild-type strains, or strains with mutations in different fluffy genes. In this case, a strong band of conidiation can be observed at the interface between two colonies suggesting that wild-type provides a missing factor to the mutant strain that can stimulate conidiation. This phenotypic rescue of the *fluG* mutant strain occurs even if the two strains are separated by dialysis membrane having a pore size of 6–8 kDa. These results have led to the proposal that *fluG* is required for synthesis of a low molecular weight factor that signals activation of the major, programmed pathway for conidiophore development, independent of the environmental conditions. Without *fluG*, development can be activated by a mechanism that involves sensing growth rate or nutrient status directly, bypassing the need for *fluG*.

The wild-type *fluG* gene was isolated and the FluG polypeptide (96 kDa) was found in the cytoplasm of cells grown vegetatively and following developmental induction. The carboxy-terminal 436 amino acids of FluG share about 30 per cent identity with prokaryotic glutamine synthetase I and there is also limited similarity to eukaryotic glutamine synthetases, particularly in those regions thought to comprise the active site of the enzyme. By contrast, the amino-terminal half of the protein shares no significant similarity with any proteins found in the current databases. Analysis of sequence changes in three distinct mutant *fluG* alleles showed that all three result from alterations in the glutamine synthetase-like domain. These results have led to the speculation that *fluG* could be involved in constitutive synthesis of an extracellular sporulation-inducing factor that is related to glutamine or glutamate, but we have not yet been able to isolate this compound. Overexpression of *fluG* in vegetative hyphae is sufficient to cause activation of *brlA* and sporulation in liquid medium, where conidial development is normally suppressed (Figure 6.14b). This result implies a direct role for *fluG* in controlling

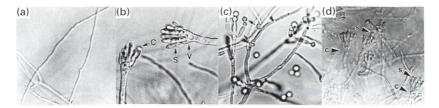

Figure 6.14 Overexpression of *fluG*, *flbA* and *flbD* causes conidiophore production in submerged culture. The alcohol-inducible promoter *alcA* was used to drive the expression of *fluG*, *flbA*, and *flbD*. The *aclA(p)::fluG*, *aclA(p)::flbA*, *aclA(p)::flbD* strains and a wild-type strain were grown 14 h in minimal media containing glucose to repress *alcA* and then shifted to *alcA*-inducing media and observed. Overexpression of *fluG* (b), *flbA*(c), and *flbD* (d) led to production of conidiophores in ~18 h, 9 h and 9 h, respectively after the shift while the wild-type strain (a) never produced spores.

development and supports the idea that the concentration of the FluG-supplied conidiation signal normally limits wild-type development in liquid medium.

flbA may function in intracellular signalling

A second group of fluffy mutants analysed is distinguished by the fact that by 3 days after inoculation, the centre of the colony begins to disintegrate and by 5 days post-inoculation, the entire colony has autolysed so that only a few hyphal strands remain (Figure 6.13b). All of the recessive mutant strains that have been identified as having this fluffy autolytic phenotype have been shown to result from mutations in a single locus designated *flbA*. The wild-type *flbA* gene was isolated through complementation of the mutant phenotype. The 3 kb mRNA was present at low levels in vegetative cells and remained relatively constant throughout the life cycle. The DNA sequence of the *flbA* is most significantly similar to that of the *S. cerevisiae* Sst2 polypeptide. This indicates that the gene is important in negatively controlling signalling of G-protein-mediated signal transduction pathways. Thus, one likely function for *flbA* is in regulating an intracellular signalling pathway for conidiation. For instance, *flbA* may be required for controlling the intracellular response to the proposed *fluG* signal described above. Whatever the role of *flbA*, it may function directly in activating development; overexpression of *flbA* in liquid culture is sufficient to induce *brlA* expression and sporulation (Figure 6.14c).

flbC and *flbD* encode nucleic acid-binding proteins
Another group of fluffy mutants can be distinguished by the observation that beginning 2–3 days after inoculation, conidiophores are produced in the centre of the colony while the margins remained fluffy (Figure 6.13d). All of the delayed-conidiation fluffy mutants characterized to date define four distinct complementation groups designated *flbB*, *flbC*, *flbD*, and *flbE*. The wild-type *flbC* and *flbD* genes were each isolated by complementation of the respective mutants. As described for *flbA* and *fluG*, the cloned *flbC* and *flbD* genes recognize RNAs that are present in vegetative cells and following developmental induction. *flbD* is predicted to encode a polypeptide similar to the DNA binding domain in the human protooncogene *c-myb* (55, 87). *flbC* is predicted to encode a 354 amino acid that, like *brlA*, has two polypeptide zinc fingers. Thus, both *flbC* and *flbD* are likely to function as DNA-binding proteins and could control the transcriptional activation of other developmental regulators. Again, overexpression of *flbD* in vegetative hyphae results in activation of *brlA* expression and sporulation (Figure 6.14d).

Future prospects

The ease with which *A. nidulans* can be manipulated both through classical and molecular genetics has made it possible to learn much about the genetic controls for asexual sporulation in this fungus; however, much remains to be learned. Work must be continued to determine the biochemical activities of the developmental regulatory genes isolated to date. New screens and selection methods will have to be implemented to identify genes that subtly modify the activities of the regulatory genes and formation of conidiophores. In addition, little is known about the internal and external signals that lead to the acquisition of developmental competence or for that matter the mechanisms that control spore dormancy and germination. Though genes have been identified that may function to transduce a signal for conidiophore development the nature of the signals themselves has not been determined. Finally, most work has concentrated on understanding the mechanisms involved in formation of a single conidiophore and has neglected the colony as a developmental unit. Future work should therefore address the question of the coordinate regulation of colony development.

It will also be interesting to learn if any of the signals and genes found to be important for asexual sporulation in *A. nidulans* are common for the same process in other filamentous fungi. While variation in conidiophore size and structure has been noted among the various aspergilli, homologues of *brlA*, *fluG*, and *flbA* have been found in other species of *Aspergillus*. However, to date no putative homologues of *A. nidulans* developmental regulatory genes have been reported in more distantly related Ascomycetes, though very little effort has been made in this direction. It is possible that the mechanisms and genes that control conidiation in other filamentous fungi will prove to be as diverse as the types of conidiophores produced by these organisms.

Further reading

Adams, J. N.-B.a. T. H. (1994). *Development of Conidia and Fruiting Bodies in Ascomycetes*. Springer Verlag, Berlin Heidelberg.

Adams, T. H. (1994). Pp. 367–382 in N. A. R. Gow and G. M. Gadd, eds. *The Growing Fungus*. Chapman & Hall, London.

Beever, R. E. and Dempsey, G. P. (1978). Function of rodlets on the surface of fungal spores. *Nature*. 272:608–610.

Champe, S. P., Kurtz, M. B., Yager, L. N., Butnick, N. J. and Axelrod, D. E. (1981). Pp. 63–91 in G. Turian and H. R. Hohl, eds. *The Fungal Spore: Morphogenetic Controls*. Academic Press, New York.

Champe, S. P. and Simon, L. D. (1992) Pp. 63–91 in E. F. Rossomando and S. Alexander, eds. *Morphogenesis: An Analysis of the Development of Biological Form*. Marcel Dekker, New York.

Clutterbuck, A. J. (1990). The genetics of conidiophore pigmentation in *Aspergillus nidulans. J Gen. Microbiol.* 136:1371–1378.

Cole, G. T. (1986). Models of cell differentiation in conidial fungi. *Microbiol. Rev.* 50:95–132.

Lee, B. N. and Adams, T. H. (1994a). The *Aspergillus nidulans fluG* gene is required for production of an extracellular developmental signal. *Genes Dev.* 8:641–651.

(1994b). Overexpression of *flbA*, an early regulator of *Aspergillus* asexual sporulation leads to activation of *brlA* and premature initiation of development. *Mol. Microbiol.* 14:323–334.

(1996). *fluG* and *flbA* function interdependently to initiate conidiophore development in *Aspergillus nidulans* through *brlAβ* activation. *EMBO J.* 15:299–309.

Mims, C. W., Richardson, E. A. and Timberlake, W. E. (1988). Ultrastructural analysis of conidiophore development in the fungus *Aspergillus nidulans* using freeze-substitution. *Protoplasma.* 44:132–141.

Oliver, P. T. P. (1972). Conidiophore and spore development in *Aspergillus nidulans. Gen. Microbiol.* 73:45–54.

Pontecorvo, G., Roper, J. A., Hemmons, L. M., MacDonald, K. D. and Bufton, A. W. J. (1953). The Genetics of *Aspergillus nidulans. Adv. Genet.* 5:141–238.

Skromne, I., Sanchez, O. and Aguirre, J. (1995). Starvation stress modulates the expression of the *Aspergillus nidulans brlA* regulatory gene. *Microbiology* 141:21–28.

Timberlake, W. E. (1990). Molecular genetics of *Aspergillus* development. *Annu. Rev. Genet.* 24: 5–36.

Wieser, J. and Adams, T. H. (1995). *flbD* encodes a myb-like DNA binding protein that controls initiation of *Aspergillus nidulans* conidiophore development. *Genes Dev.* 9:491–502.

Wieser, J., Lee, B. N., Fondon, J. W. and Adams, T. H. (1994). Genetic requirements for initiating asexual development in *Aspergillus nidulans. Curr. Genet.* 27:62–69.

7 Fungal cell division

C. PITT AND J. DOONAN

Introduction

An essential feature of living cells is their potential to replicate and produce exact copies of themselves. In single-celled organisms replication is achieved by cell division. In multicellular organisms programmed cell division is essential for growth and normal development. In this case, cell proliferation can be seen as a strategy for increasing overall size as well as permitting cells or groups of cells to specialize. Multicellular animals, plants and fungi, therefore, can develop different tissues each containing one or more cell types adapted to a particular function.

The coordination of the various processes required to bring about successful cell division is not trivial. Simply cutting a cell in two is unlikely to result in the production of two viable daughter cells (the products of a cell division). Mistakes at any stage during the entire division cycle can lead to mutant or non-viable daughter cells. To produce two viable cells from one, partitioning of all the necessary cellular components into the two daughters is essential. Each daughter must receive a complete copy of the genome of its mother cell (the cell undergoing division) if genetic integrity is to be maintained through subsequent generations. The *genome* contains the complete set of genetic information that encodes the proteins required for viability. This information is packaged into a set of chromosomes. The fidelity of replication of these chromosomes must be high in order to maintain genetic stability. Incomplete replication may lead to the loss of essential genes and, ultimately, death of daughter cells. Re-replication of parts or all of the genome will lead to an increase in the number of copies of some or all of the genes. The concomitant alteration in gene dosage may be detrimental to the cell. Thus the integrity of the duplicated chromosomes must be monitored before their physical separation takes place.

The dividing cell must have an apparatus for separating the two genomes so that each daughter cell receives one full set of chromosomes. Separation by this apparatus must take place only after the chromosomes have been faithfully duplicated. After separation of the chromosomes the cell can divide in two (a process known as *cytokinesis*), isolating the two genomes into two separate cells.

Even if cell division occurs correctly, inappropriate cell proliferation can have disastrous consequences. Single-celled organisms have mechanisms which prevent cell division under unfavourable conditions such as nutrient starvation or stress, when young cells might die. Multicellular organisms suffer from cancer when controls which normally limit cell proliferation fail to operate. With the reduction in other human diseases, coupled to the increase in life expectancy, cancer is now a major cause of death in the Western world. Over the past 30 years, the cellular and molecular basis of cell division has come under intense scrutiny because it may offer insight into the defects which arise during cancer.

The time interval from the point at which a daughter cell is formed until the point at which it forms two new daughter cells is called the *cell cycle*. There is a mitotic cell cycle and a meiotic cell cycle. The *mitotic cycle* is the kind of cell division that occurs in the somatic cells of multicellular organisms and in many unicellular organisms. This sort of division is *non-reductive*, in that the number of copies of each chromosome in the mother cell is maintained in the two daughter cells. Thus a *diploid* cell (i.e. has two copies of each chromosome) produces two daughter diploid cells, and a *haploid* cell (i.e. has one copy of each chromosome) produces two haploid daughter cells. The *meiotic cycle* defines the type of cell division that occurs in the germline cells of multicellular organisms. In this case the mother cell undergoes *reductive division* to produce two daughter cells, with half the number of chromosomes. These then divide again by a mitotic-like division to produce the haploid gametes. It was stated earlier that each daughter cell must receive a complete copy of its genome. In reductive division this is clearly not the case. However, the full genetic complement is restored in the zygote after fusion of the haploid genomes of the two gametes.

The three fungi that are the subjects of this chapter are all able to propagate as either haploids or diploids and are therefore useful systems for studying both the mitotic and meiotic cycles. Due to the complexity of both types of cell division, this chapter will be limited to following the events in the mitotic cycle. What follows is a generalized view that aims to give the reader a flavour of how three fungi have solved the various problems involved in ensuring that cell division results in two genetically equal and viable daughter cells. Cell division is not just about nuclear division, but also requires the segregation of other cellular components, such as mitochondria and Golgi bodies. Segregation of these components will not be addressed.

The identification of mammalian homologues for some of the fungal components involved in these processes highlights the immense value of working with these genetically amenable organisms.

Role of fungi in elucidating growth controls

In contrast to higher organisms, cytological observations of fungi reveal little about cell division. These microbes are too small for convenient study, and in most cases, do not display nuclear envelope breakdown at mitosis. However, the ease with which three particular fungi can be genetically analysed, together with their short generation time, makes them ideal organisms for dissecting the genetic basis of cell cycle progression. Two of these are unicellular yeasts, the budding yeast *Saccharomyces cerevisiae*, and the fission yeast *Schizosaccharomyces pombe*. The third is the multicellular filamentous fungus *Aspergillus nidulans*. In each organism, a series of strains have been generated and isolated that carry mutations in cell cycle-related genes.

Cell division in *Saccharomyces cerevisiae*

Saccharomyces cerevisiae, or budding yeast, is the best-characterized eukaryote. Decades of extensive genetic research have isolated numerous mutations that affect many processes, including cellular metabolism, growth and morphogenesis. The sequence of the entire genome is now available. Cell division occurs by budding (Figure 7.1a). The slightly elliptical cell first forms a bud at one pole. This bud expands as nuclear division occurs and one of the daughter nuclei enters it. This nucleated

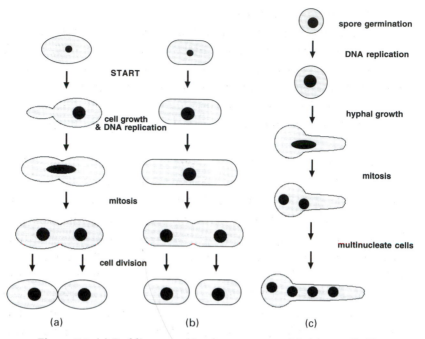

Figure 7.1 (a) Budding yeast (*Saccharomyces cerevisiae*) is so called because cell division takes place by budding. The mitotic spindle is oriented such that migration of sister chromatids to opposite poles ensures that only one set of chromosomes ends up in the bud. The bud eventually pinches off from the mother cell by formation of a septum to produce two daughter cells. (b) Fission yeast (*Schizosaccharomyces pombe*) are so called because cell division takes place by fission of the cell along the mid zone of the mother cell. This occurs after chromosome separation on the mitotic spindle. (c) *Aspergillus nidulans* forms multinucleate cells: septum formation occurs only after three nuclear divisions have taken place. As the spore germinates a germ tube is formed, down which the products of mitosis migrate.

bud will form one of the daughter cells as it separates from the original cell by the formation of a septum. The characteristic change in cell shape during the cell cycle has facilitated the isolation of a large collection of *cell division cycle* (*cdc*) mutants which have proven instrumental in providing insight into the molecular basis of cell division.

212

Cell division in *Schizosaccharomyces pombe*

Schizosaccharomyces pombe, or fission yeast, divide by the formation of a septum across the mid zone of the mother cell (Figure 7.1b). Cell cycle mutants were originally identified on the basis of changes in cell size because the species characteristically divides at a constant cell length. Mutants which failed to divide at the correct size were called cell division cycle mutants (*cdc*) or, in the case of one small-celled variety, *wee*.

Cell division in *Aspergillus nidulans*

Aspergillus nidulans is a multicellular filamentous fungus with several cell types. The mode of cell division varies with cell type, and can resemble that of either yeast depending on the stage of the life cycle. The filamentous hyphal stages (Figure 7.1c) are *multinucleate* (i.e. multiple nuclei are contained within a common cytoplasm). Cell division occurs by the production of a septum after about three nuclear divisions and resembles that of fission yeast. Septation and nuclear division are coupled thereafter. Asexual reproduction involves the formation of a multicellular structure called a conidiophore. Chains of haploid spores are produced at the apex of this structure in a manner that resembles the budding seen in *S. cerevisiae*. Nuclear division mutants were isolated from *Aspergillus* on the basis that they reduced the number of nuclei present in a hypha. Such mutants were classified into '*never in mitosis*' (*nim*) or '*blocked in mitosis*' (*bim*) based on the appearance of their *chromatin* (DNA together with its associated proteins).

Events occurring in one organism may not necessarily happen in precisely the same way, or with the same timing, as in another. For instance, the mechanism of cytokinesis is rather different between the two yeasts, but many of the control molecules are similar. Both yeasts are unicellular and can proliferate as either haploids or diploids. Under appropriate conditions, the diploid cells may undergo meiosis to produce haploid progeny. The haploid cells may mate or fuse to reestablish the diploid phase. Mating is an alternative to normal division and usually occurs under nutrient deprivation. This aspect of yeast biology has been useful in understanding how the cell's environment can influence cell division.

Mutant isolation

Typically, mutants are generated by UV irradiation or by incubation with mutagenic substances. As these organisms are cultured in the laboratory as haploids, i.e. have one set of chromosomes, then mutations must be *conditionally lethal.* In other words, there must be a set of conditions under which the mutation has no effect on cell viability, but under a different set of conditions the mutation exerts a lethal effect (wild-type cells are completely viable under the lethal conditions for the mutants). This must be the case, for if the mutation was lethal under all conditions then a strain could never be isolated. In a diploid organism the faulty gene may have no detrimental effect as a *wild-type* copy is present on the homologous chromosome. Many such conditionally lethal mutants are *temperature sensitive* (*ts⁻*), being able to grow within a certain temperature range (the *permissive temperature*), but die when the temperature is raised above this (the *restrictive temperature*). *Cold-sensitive* (*cs⁻*) mutants die when the temperature is lowered below the optimal range.

Another useful type of mutant is the promoter-replacement type. Using genetic engineering, a particular gene has its normal promoter removed and replaced with one that is sensitive to the presence or absence of particular compounds in its environment. For instance, the *alcA* promoter of *A. nidulans* induces the expression of the alcohol dehydrogenase gene when grown on ethanol-containing medium that lacks a more energy-efficient carbon source (such as glucose). The enzyme encoded by this gene is involved in the metabolism of ethanol. By placing a gene of interest under *alcA* control the gene can be switched on and off at will by controlling the carbon source. Growth medium that contains the carbon source that switches on expression (in this case ethanol) is called *inducing medium,* and that which switches off expression (in this case glucose) is called *repressing medium.* This may lead to under- and overexpression and/or improper timing of expression. The *phenotype* (i.e. the observable characteristics that define the mutant) of such a cell may be informative as to the role of this gene in cell cycle events. Similar mutants can be made in budding yeast with the *gal* promoter (gene expression is induced on galactose medium but repressed on glucose) and in fission yeast with the thiamine repressible promoter.

These mutants have a number of uses: first, many are '*loss of func-*

tion' mutants (i.e. the mutation prevents the gene from carrying out its normal function but does not lead to the acquisition of novel functions. If the latter were the case then the mutation would be described as a '*gain of function*') and so the arrest phenotype provides a strong indication as to the point in the cycle at which the gene product is normally required. Second, the availability of conditional mutants has allowed the isolation of the genes involved in cell cycle progression. These simple organisms can be transformed efficiently with DNA libraries, and if the *wild-type* gene is in the library, it will repair the defect. This complementing DNA fragment (i.e. the repairing fragment) can be subsequently isolated and analysed. DNA-mediated complementation of cell cycle mutants has led to the isolation of many important functions.

In order to isolate *ts*⁻ mutants, first the permissive temperature range has to be decided upon. Typically this will be from around 20–32 °C. After mutagenizing cells they are allowed to grow at the permissive temperature. Colonies that form are then replica-plated and placed at the proposed restrictive temperature, say 37 °C. The master plates, from which the replica plates are made, are kept at the permissive temperature. Cells that fail to grow at the restrictive temperature are selected from the original master plate and subjected to further analysis. In the case of budding yeast, mutations causing a block in the cell division cycle were identified by microscopic examination of the terminal cellular phenotype (the appearance of the dead cells). Cells from individual mutant colonies were grown in culture at the permissive temperature, resulting in a large population of cells carrying the same mutation. The cultures were then raised to the restrictive temperature for several hours. Should the mutation not affect cell division, then each population of mutant cells should arrest with a random cell morphology. That is to say, some cells will not be dividing when they die and will be unbudded whilst other cells will be in various stages of cell division and will display a range of bud sizes. Such mutants were disregarded. However, those cultures that produced populations of cells in which greater than 95 per cent displayed the same terminal cellular phenotype were retained and assigned a number prefixed with the letters cdc (for cell division cycle). The synchronous arrest given by the cdc mutants was due to progression through the cycle up to the point at which the mutagenized cell cycle gene would normally exert its

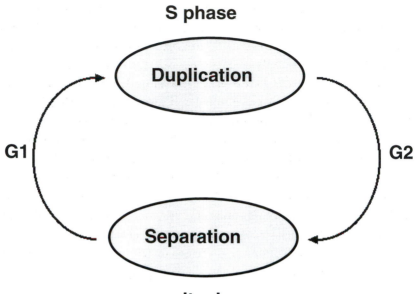

S phase

Duplication

G1

G2

Separation

mitosis

Figure 7.2 The cell cycle can be thought of as an alternation between genome duplication (S phase) and genome separation (mitosis). Processes occurring in G1 and G2 ensure the alternation between S phase and mitosis.

effect. As the mutation prevented progression beyond this point then all of the cells in the culture eventually accumulated at this block point.

Cell cycle progression

Traditionally, the cell cycle has been divided into four phases: G1, S, G2 and M, but it is convenient to think of it as an alternation between *genome duplication* and *genome separation* (Figure 7.2). The purpose of *S phase* is to ensure that the genome is completely and faithfully duplicated. The purpose of *mitosis* is to ensure that the resulting two genomes are segregated into two daughter nuclei, which will nucleate two daughter cells. The gap between S phase and mitosis (*G2*) provides an opportunity to ensure that duplication has been successful. Without this monitoring of S phase the cell may enter a lethal mitosis or generate non-viable daughter cells.

Other key events take place in G2 that are prerequisites for entry into mitosis. After a successful cell division the daughter cells undergo a period of growth. This growth is carefully monitored. A cell that is committed to divide must only enter S phase when a certain minimum size has been attained. A cell that is programmed to divide before obtaining this size will give rise to consecutive generations of cells that become increasingly smaller. Conversely, a cell that grows too large before dividing will lead to increasingly larger cells with each new generation. Observations of single-celled organisms in culture shows us that this is clearly not the case. This period of growth between the previous mitosis and the next S phase is called *G1*. Some cells are not committed to divide and leave the cell cycle at G1. Such cells are said to be in *G0*. Cells may remain in this quiescent state for a variable length of time, but are still able to re-enter the cycle at G1. Monitoring of cell size is still important in quiescent cells in order to maintain tissue integrity.

Genome duplication occurs in S phase

The initiation of cell division requires exit from G1 and entry into S phase. This is an irreversible step which is referred to as *START* in yeasts and the *restriction point* in other organisms. Progression beyond START requires the activation of *cyclin-dependent kinase* (cdk) through its association with G1 cyclins. The cdk is encoded by the $cdc2^+$ gene in fission yeast and was isolated by the complementation of a cdc^- mutant. Subsequent experiments led to the isolation of a human homologue which was able to complement the yeast mutant. This was a significant landmark in cell cycle research as not only did it show for the first time that a human gene was able to complement a yeast mutation, but also demonstrated the high degree of conservation of a key cell cycle regulator in two highly divergent species. Homologues of $cdc2^+$ have been found in all species in which this gene has been searched for. The protein is encoded by the *CDC28* gene in budding yeast and *nimX* in *Aspergillus*. The *CDC28* gene had been previously isolated as a cell cycle mutant in budding yeast. Temperature-sensitive mutations have been generated in *Aspergillus* cdk using reverse genetics. This involves identifying the mutations in yeast and recreating them in *Aspergillus* using molecular biological techniques. This has

217

allowed a greater understanding of cdk function in the latter system. *Cyclins* are so named because the level of these proteins fluctuates throughout the cell cycle, accumulating steadily during some phases whilst falling rapidly during others. *G1 cyclin* levels are low early in G1 but as the cell approaches a stage where it is competent to divide these activating subunits accumulate. Three genes for G1 cyclins have been identified in budding yeast, *CLN1*, *CLN2* and *CLN3*. They are *functionally redundant*. This means that they have overlapping functions so that should one, or even two, of the cyclins become inactivated through mutation, then the third cyclin is capable of carrying out the required function(s) alone. The proteins encoded by these genes were shown to accumulate and disappear in a cell cycle-dependent manner. *CLN1* and *CLN3* were deleted in a budding yeast strain (this entails using molecular biological techniques to remove the gene from the genome) and *CLN2* was placed under control of a galactose promoter. This strain was able to grow normally on medium containing galactose, but when grown on glucose medium the cells were able to complete one cell cycle but arrested prior to START in the next cell cycle. This demonstrated that the cyclin had been used up and had to reaccumulate in order to initiate a new round of cell division.

The activation of cdk by G1 cyclins is a good example of a positive feedback loop. The presence of small amounts of G1 cyclins leads to low-level activation of cdk. This activated cdk phosphorylates a complex of components that contains two proteins, Swi4 and Swi6. Phosphorylation of this complex activates it as a transcription factor. This leads to transcription of genes that contain a *cis-acting element* (this refers to a motif in the DNA that is situated on the same DNA strand as the gene to which it is associated) in their promoters called an *SCB element*. These genes are only transcribed when this particular transcription factor is activated. Proteins whose genes contain an SCB element include two of the G1 cyclins themselves, Cln1 and Cln2. The production of more G1 cyclins leads to the activation of more cdk and hence to even greater accumulation of G1 cyclins, and so on. The result is that there is a rapid build-up of activated cdk, which is sufficient to allow progression through START.

Another promoter element that confers periodic transcription of its gene at START is the *MluI cell cycle box* (MCB). This element is present in genes involved in DNA synthesis (including POL1 DNA polymerase) as well as a number of other cell cycle related genes. The MCB is recognized by a binding activity known as DSC1 (for DNA synthesis control). The regulation of DSC1 binding and transcriptional activation is unclear, but may be due to alteration of its activity in a cell cycle-dependent manner.

DNA replication takes place at specific regions of the chromosomes called *replication origins*. At these sites the DNA is made single stranded by the replication machinery. This machinery then moves along a single strand and uses it as a template to synthesize a complementary strand. Unreplicated origins can be thought of as activated for replication. Once the replication machinery has passed over a particular origin it is rendered inactive for replication and cannot be reactivated until completion of mitosis and entry into the next cell cycle. This ensures that all regions of the genome are replicated once and only once.

Genome separation occurs in mitosis

Entry into mitosis involves a number of key events: activation of cdk, DNA condensation, spindle pole body (SPB) duplication and separation. Mitotic exit requires the inactivation of the cdk and the removal of the phosphate groups added to its substrates.

Activation of cyclin-dependent kinase (cdk) as a promoter of mitosis is brought about by dephosphorylation at its tyrosine 15 residue (Tyr15) and by its association with a *cyclinB* subunit (*mitotic cyclin*). The resulting activated complex is referred to as *mitosis promoting factor* (MPF). MPF is believed to phosphorylate nuclear proteins, thereby changing their properties and leading to the construction of the mitotic spindle (see later), DNA condensation and ultimately chromosome separation. Entry into mitosis in *Aspergillus* also requires the activity of a second kinase, *nimA*. Strains carrying mutations in *nimA* have been shown to arrest just prior to mitosis with cdk active. Therefore, activation of cdk alone is insufficient to promote entry into mitosis in *Aspergillus*. Homologues of *nimA* have not been found in the two yeasts but have been identified in humans. This may reflect the extra levels of complexity required to control the cell cycle in multicellular eukaryotes.

During most of a cell's life the chromosomes are diffuse structures packed within the nucleus, and are not visible by light microscopy. Trying to separate chromosomes in this state would almost certainly lead to errors, as the thread-like strands of DNA would become entangled. *DNA condensation* heralds the first stage (prometaphase) of mitosis. This compaction of chromosomes into discrete structures enables the two genomes to be separated with relative ease. Condensed chromosomes can be visualized by light microscopy in certain cell types. Each pair of identical chromosomes is attached at a region called the *centromere*. At this stage the homologous chromosomes are referred to as *sister chromatids*. The centromere consists of non-coding repeat elements of DNA and a complex of proteins. Part of the complex of proteins forms a structure called the *kinetochore* and each pair of sister chromatids has two kinetochores situated on opposite sides of the centromere. Certain proteins within the centromere are responsible for the cohesive forces that hold the sister chromatids together.

Separation of the two genomes is carried out on a complex structure called the *mitotic spindle*. Prior to mitosis the *spindle pole body* (SPB), a structure associated with the outer nuclear envelope, undergoes duplication (Figure 7.3a). In prometaphase the two SPBs migrate to opposite sides of the nucleus, creating two poles (Figure 7.3b). The *cytoskeleton*, a cytoplasmic framework which maintains cell shape and consists of *microtubules* (MTs) polymerized from subunits of α and β tubulin, is dismantled and the SPBs use the tubulin subunits to construct the mitotic spindle. The MTs are very dynamic structures, being constantly assembled and disassembled. If the rate of subunit addition to an MT exceeds the rate of subunit removal, then the MT will grow. Conversely, if the rate of subunit removal exceeds that of addition, the MT will shrink. The growing end of a MT is generally referred to as the plus end, and the other, the minus end. The *tubA* and *benA* genes of *Aspergillus* encode α tubulin and β tubulin, respectively. Mutations in both genes lead to blocks in mitosis. The *benA33* mutant actually causes the hyperstabilization of MTs, thus showing that MT depolymerization is essential for mitotic progression. Another type of tubulin, γ tubulin, was first isolated in *Aspergillus*. This class of tubulin is also essential for MT function, but unlike α and β tubulin it is not a component of the MTs but of the SPBs (see later: checkpoint controls).

220

Fungal cell division

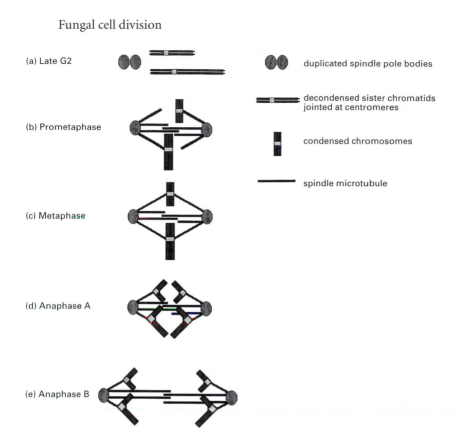

(a) Late G2 — duplicated spindle pole bodies

— decondensed sister chromatids jointed at centromeres

— condensed chromosomes

— spindle microtubule

(b) Prometaphase

(c) Metaphase

(d) Anaphase A

(e) Anaphase B

Figure 7.3 (a) By late G2 the spindle pole body (SPB) has duplicated but migration to form the poles of the nucleus has not occurred. Chromosomes are decondensed. (b) Early in mitosis (prometaphase) the SPBs migrate to form the poles and begin nucleating microtubules (MTs) that constitute the mitotic spindle. Chromosomes have condensed. Some kinetochores have captured an MT. (c) By metaphase all pairs of sister chromatids are attached to both poles by their kinetochore MTs. (d) Sister chromatids have lost their cohesion and kinetochore MTs depolymerize to move the sisters to opposite poles. (e) Interdigitating non-kinetochore MTs are polymerized and the two poles are moved apart. (Just two pairs of sister chromatids have been shown for clarity.)

Motor proteins interact with MTs and are able to utilize the energy released from ATP hydrolysis to move along the MTs. Two classes of motor proteins have been recognized, *kinesins* and *dyneins*. Kinesins are able to move in either direction along MTs whereas dyneins move towards the minus ends. The BIMC protein in *Aspergillus* is a kinesin-related protein and was isolated as a *ts*⁻ cell cycle mutant. This mutant arrests in mitosis due to the inability to set up a mitotic spindle. In such strains the SPB has duplicated and produced two half spindles but they fail to migrate to opposite poles. Thus motor protein activity is required to set up the mitotic spindle.

The half-spindles produced by each SPB interdigitate with the other (Figure 7.3b). Kinetochores are capable of capturing MTs and when an MT is captured it is referred to as a *kinetochore microtubule*, to distinguish it from those that are not attached. Should a kinetochore capture an MT from one pole, then it will tend to be pulled toward that pole due to shrinkage of its kinetochore MT. However, repulsion from the pole is also experienced by the chromatids, and this may be due to the growing non-kinetochore MTs. The net result is that pairs of sister chromatids tend to move back and forwards between the poles. Eventually, the unattached kinetochore will capture MTs emanating from the other pole (due to its opposite orientation to the kinetochore that is attached). The sister chromatids will now experience equal forces from both poles and will tend to spend most of their time in the central plane of the nucleus, equidistant from the poles. This movement of chromosomes towards the central plane is called *congression*. When all the pairs of sister chromatids are attached to both poles the sisters are ready to divide. This stage is referred to as the *metaphase plate* (Figure 7.3c). The pairs of sister chromatids are under equal tension from both poles and this tension is believed to be involved in generating a signal that leads to the dissolution of the cohesive forces that hold the sisters together. Dissolution of these forces also requires the physical degradation of certain, as yet unidentified, proteins that can be thought of as *anaphase inhibitors*. Once cohesion is lost the sister chromatids move to opposite poles due to the depolymerization of the kinetochore MTs. This movement is referred to as *anaphase A* (Figure 7.3d). The second part of anaphase, *anaphase B*, involves the polymerization of non-kinetochore MTs (Figure 7.3e). MTs from one pole move against the MTs from the other pole in their interdigitating regions. The net result is that the poles are moved apart. Progression through anaphase requires the activity of a type 1 protein phosphatase. This enzyme is encoded by the *bimG* gene in *Aspergillus* (and *dis2*$^+$ in *S. pombe*) and a *ts*$^-$ mutant is unable to resolve the genomes into two distinct nuclei. Soon after anaphase completion the nuclear envelope pinches off between the two genomes to yield two distinct nuclei (*karyokinesis*). The DNA decondenses and the nuclei enter the interphase state. At this time cytokinesis separates the nuclei into two daughter cells. The mitosis that takes place in the two yeasts and *Aspergillus* is called *closed mitosis*. This refers to the nuclear envelope which remains intact throughout the whole process. Higher eukaryotes undergo *open mitosis* in which the nuclear envelope breaks down into a number of vesicles. These vesicles eventually coalesce to form two separate nuclear envelopes around the two genomes once anaphase has been completed. Eventual mitotic exit is brought about by phosphatase activity (which presumably dephosphorylates the substrates of the mitotic kinases that initiated mitosis) and inactivation of MPF through destruction of cyclinB (see next section). There is also evidence that the NIMA kinase is destroyed as mitosis is traversed.

Alternation of genome duplication and separation

To maintain the *ploidy* (i.e. the number of copies of each chromosome) of an organism it is essential that genome duplication is always followed by genome separation, and that genome separation is always followed by genome duplication. In other words, S phase must be followed by mitosis, and vice versa. This is achieved by the biochemical cell cycle which regulates the activity of certain key enzymes, thus preventing the short-circuiting of the cycle. Only when cell division has been completed can the initiation of a new cycle commence. Commitment to genome duplication requires progression through START, which is dependent upon activation of cdk with G1 cyclins. Commitment to genome separation requires entry into mitosis which is dependent upon activation of cdk with cyclinB (mitotic cyclin) to form MPF. Thus alternation of genome duplication and separation depends upon the alternate interaction of cdk with different cyclin subunits. The mitotic cyclin,

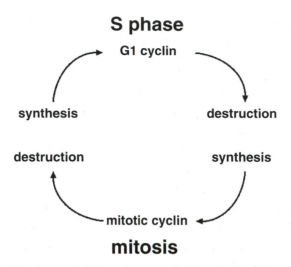

S phase
G1 cyclin
synthesis
destruction
destruction
synthesis
mitotic cyclin
mitosis

Figure 7.4 As S phase is approached synthesis of G1 cyclins leads to their accumulation. Association with these cyclins allows cdk to drive the cell past START. Downregulation of G1 cyclin transcription (by mitotic cyclin) and degradation of G1 cyclin protein allows the mitotic cyclin to replace its G1 counterparts as the cdk-interacting subunit. This activates cdk as mitosis-promoting factor (MPF) and allows mitosis to take place. At the end of mitosis the anaphase-promoting complex targets mitotic cyclin for destruction by the proteasome and the cell cycle returns to the early G1 state.

which represses transcription of the G1 cyclins, is expressed in G2 only. As the cell completes S phase and progresses towards mitosis the appearance of mitotic cyclin begins the downregulation of G1 cyclin transcription (Figure 7.4). Due to lowered G1 cyclin transcription and to protein turnover, the level of G1 cyclin protein begins to fall. Mitotic cyclin is now able to replace its G1 counterparts in the cdk–cyclin complex. Eventually, sufficient MPF (cdk–cyclinB) is accumulated to allow entry into mitosis.

Mitotic exit requires the degradation of cyclinB, which inactivates MPF. CyclinB is targetted for destruction by a complex known as the *anaphase promoting complex* (APC). This complex recognizes a short amino acid motif (called a *destruction box* or D box) in the N-terminal domain of its target proteins. Chains of a small peptide, called *ubiquitin*, are then added to the target protein (in this case cyclinB) by the APC. Genes encoding subunits for this complex have been isolated in all three of the fungal systems under discussion. The Cdc16p, Cdc23p and Cdc27p proteins in budding yeast are all components of the APC. Mutations in the genes encoding these proteins lead to arrest in mitosis. When the genes encoding BIMA and BIME in *Aspergillus* are mutated, cells arrest in metaphase. The interpretation of these phenotypes is that APC activity is also required to target the anaphase inhibitor for degradation (see previous section) and the inability to do this prevents anaphase entry. Ubiquitination of the target protein identifies it as a substrate for a proteolytic complex called the *proteasome*, which degrades it. This degradation resets the cell cycle to its G1 state, allowing G1 cyclins to begin accumulating again. It also removes any possibility of cdk being activated by a mitotic cyclin before START has taken place. Mitotic cyclin will not reappear until entry into G2. The requirement for BIMG/Dis2 type 1 phosphatase activity to progress through anaphase suggests that this enzyme could also help to reset the cell cycle by removing the phosphate groups added by MPF (and NIMA kinase in *Aspergillus*) to its substrates. However, the precise targets of BIMG/Dis2 are not known.

Checkpoint controls

The alternate interaction of cdk with G1 cyclins and then mitotic cyclin ensures the correct ordering to events in cell cycle progression. It is also of critical importance to ensure the successful completion of each phase before moving on to the next. This is achieved through *checkpoint controls*. Such controls are envisaged to monitor discrete cell cycle events, and prevent subsequent events taking place until the one being monitored has been successfully completed.

The *rad*⁻ mutants in budding yeast demonstrated the existence of one such checkpoint. Some of the *rad* class of genes were shown to be responsible for detecting DNA damage (as a result of UV irradiation) and prevented progression into mitosis until the damage had been repaired. Mutations in these genes did not prevent cell cycle progression and cells underwent mitosis with damaged DNA with lethal consequences. Subsequent experiments showed that these cells were still able to repair

DNA so that the mutations must lie in checkpoint components that monitor DNA integrity, rather than in the repair machinery itself.

The radiation damage checkpoint proteins prevent mitotic entry by regulating the activity of cdk (Figure 7.5). Even when cdk associates with cyclinB to form MPF it remains inactive until the inhibitory phosphate group on its tyrosine 15 residue is removed. The phosphate group is added by the Wee1 kinase in *S. pombe*, so called because a strain carrying a mutation in this gene was unable to inhibit the mitosis-promoting activity of MPF and entered mitosis before sufficient cell growth had taken place, resulting in small daughter cells (wee is a Scottish colloquialism for little and reflects the geographical location where this mutant was isolated). The reduced size of the *wee* mutant showed that growth control also regulates the cell cycle through the phosphorylation state of cdk. Removal of the phosphate group is carried out by the Cdc25 phosphatase in *S. pombe* and it is the activity of this enzyme that is regulated by the DNA damage checkpoint. In response to DNA damage Cdc25 is inactivated, preventing the removal of Tyr15 from MPF and its activation. This type of regulation of a protein is referred to as *post-translational modification.*

γ Tubulin (encoded by the *mipA* gene in *Aspergillus*) appears to be part of a checkpoint control that ensures that the mitotic spindle cycle is coupled to the DNA duplication/condensation cycle. Mutations in α and β tubulin lead to a block in mitosis. However, disruption of the *mipA* gene (the gene is removed from the genome using molecular biological techniques) does not cause a block in mitosis. The cells of such disrupted strains are not able to form a mitotic spindle, but DNA continues to condense and decondense and undergo increases in ploidy at similar kinetics to a wild-type strain dividing normally. The ability of these cells to undergo sequential rounds of DNA duplication and condensation without genome separation suggests that a checkpoint (involving γ tubulin), which normally prevents progression through the cycle in the absence of mitotic spindle function, has been disabled.

Another checkpoint exists in mitosis where progression from metaphase to anaphase is held in check until correct alignment of sister chromatids on the mitotic spindle has taken place (see genome separation). This *metaphase checkpoint* is thought to be sensitive to the tension exerted

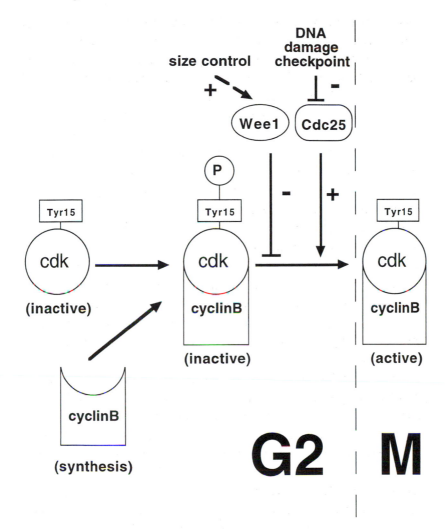

Figure 7.5 Activation of cdk requires association with cyclinB and dephosphorylation of its Tyr15 residue. The synthesis of cyclinB in G2 only ensures that S phase takes place before mitosis. Tyr15 phosphorylation is carried out by the Wee1 kinase. Growth controls are able to prevent mitosis taking place before a suitable size has been attained by regulating the activity of Wee1. Dephosphorylation at Tyr15 is mediated by Cdc25 phosphatase and this activates cdk as a kinase. The DNA damage checkpoint prevents mitosis taking place before the damage has been repaired by regulating the activity of Cdc25.

on the pairs of sister chromatids by the kinetochore MTs attached to the poles. Only when all the pairs of sisters are under equal tension from both poles does the checkpoint release its inhibition of anaphase. A decrease in phosphorylation at the kinetochores is concomitant with this release and this may play a part in signalling that a kinetochore is attached to the spindle.

Summary and outlook

Cell division requires accurate and complete genome duplication followed by genome separation. It is essential that the processes required to achieve this occur in the correct order and with high fidelity. This is achieved by post-translational modification and protein turnover of key cell cycle components. The consequences of improper progression through the cycle are potentially disastrous, leading to uncontrolled proliferation (in cancer) or cell death.

Three genetically amenable fungi have provided a simple means of dissecting the control pathways involved in cell division, thus contributing greatly to current knowledge. Many of the controls isolated are now understood to the extent that they can be used to manipulate the structure of the cell. Several notable gaps remain in our understanding of the division cycle and here we highlight those that are active areas of research.

We have only a vague idea of how sister chromatids are held together until metaphase and how they separate during early anaphase. The powerful fungal genetic systems are already beginning to provide clues as to how these processes are achieved. Another area of active research includes protein localization during the cell cycle. A given protein may be restricted to one cell compartment at one stage of the cycle but move into new regions at other stages, allowing it to interact with previously inaccessible proteins. Such movements could be envisaged to take place across the nuclear envelope, and there is already evidence that the NIMA mitotic kinase is regulated in such a spatial manner. External *mitogenic signals* (extracellular signals that regulate commitment to cell division) need to be relayed across the plasma membrane, and ultimately between the nucleus and cytoplasm, in order for the cell to respond to environmental cues. These cues can affect the rate of cell division in single-celled organisms (such as the yeasts) and influence development in multicellular organisms (such as *Aspergillus* and higher eukaryotes). In the latter, different cell types may have distinct cell division cycles. Thus a complete understanding of the control of cell division will require the integration of research in the areas of cell cycle, cell signalling and developmental biology.

The complete DNA sequence of the budding yeast genome is now known, allowing systematic function searches by means of gene disruption. Many of the genes involved in the regulation of fungal cell division have already been isolated. Subsequent searches for these genes in mammals have revealed the highly conserved nature of many of these components, and point to a common ancestor from which the eukaryotes have diverged. The conserved nature of these cell cycle regulators will allow us to make hypotheses as to the function of related genes and proteins isolated in higher eukaryotes. Ultimately this may identify the aberrations that lead to cancer and chromosome mis-segregation (such as in Down syndrome), thus allowing the rationalization of drug and therapeutic design.

Bibliography
General

Murray, A. and Hunt, T. (1993). *The Cell Cycle: an Introduction.* Oxford University Press, Oxford.

Science 274:1577–1804 (Several critical reviews of ongoing cell cycle research).

Reviews relevant to this chapter

Doonan, J. H. (1992). Cell division in *Aspergillus. J. Cell Sci.* 103:599–611.

Fisher, D. and Nurse, P. (1995). Cyclins of the fission yeast *Schizosaccharomyces pombe. Semin. Cell Biol.* 6:73–78.

Li, R. and Murray, A. (1991). Feedback control of mitosis in budding yeast. *Cell* 66:519–531.

Murray, M. (1994). Cell cycle checkpoints. *Curr. Opin. Cell Biol.* 6:872–876.

   (1995). Cyclin ubiquitination: the destructive end of mitosis. *Cell* 81:149–152.

Nurse, P. (1994). Ordering S phase and M phase in the cell cycle. *Cell* 79:547–550.

O'Connell, M. J. and Nurse, P. (1994). How cells know they are in G1 or G2. *Curr. Opin. Cell Biol.* 6:867–871.

Osmani, S. A. and Ye, X. S. (1996). Cell cycle regulation in *Aspergillus* by two protein kinases. *Biochem J.* 317:633–641.

Rudner, A. D. and Murray, A. W. (1996). The spindle assembly checkpoint. *Curr. Opin. Cell Biol.* 8:773–780.

Woollard, A. and Nurse, P. (1995). G1 regulation and checkpoints operating around START in fission yeast. *BioEssays* 17:481–490.

Wuarin, J. and Nurse, P. (1996). Regulating S phase: CDKs, licensing and proteolysis. *Cell* 85:785–787.

Further reading

Chang, F. and Nurse, P. (1996). How fission yeast fission in the middle. *Cell* 84:191–194.

Gorbsky, G. J. (1995). Kinetochores, microtubules and the metaphase checkpoint. *Trends Cell Biol.* 5:143–148.

Kelly, T. J., Nurse, P. and Forsburg, S. L. (1993). Coupling DNA replication to the cell cycle. *Cold Spring Harbor Symp. Quant. Biol.* LVIII:637–644.

Martín-Castellanos, C. and Moreno, S. (1997). Recent advances on cyclins, CDKs and CDK inhibitors. *Trends Cell Biol.* 12:95–98.

Miyazaki, W. Y. and Orr-Weaver, T. L. (1994). Sister-chromatid cohesion in mitosis and meiosis. *Ann. Rev. Genet.* 28:167–187.

Nicklas, R. B. (1997). How cells get the right chromosomes. *Science* 275:632–637.

Norbury, C. and Nurse, P. (1992). Animal cell cycles and their control. *Annu. Rev. Biochem.* 61:441–470.

Page, B. D. and Snyder, M. (1993). Chromosome segregation in yeast. *Annu. Rev. Microbiol.* 47:231–261.

Stern, B. and Nurse, P. (1996). A quantitative model for the cdc2 control of S phase and mitosis in fission yeast. *Trends Genet.* 12:345–350.

Yanagida, M. (1995). Frontier questions about sister chromatid separation in anaphase. *BioEssays* 17:519–525.

8 Sexual development of higher fungi

S.-W. CHIU AND D. MOORE

Introduction

Fungal mycelia will continue to grow and invade new substrates for as long as satisfactory conditions prevail (Chapter 9), typically producing numerous asexual spores (Chapter 7) and other mitotically derived invasive, reproductive and/or resistant structures (strands, rhizomorphs, sclerotia, stromata, etc.). Except for the Mycelia Sterilia (Deuteromycotina, Chapter 2), under particular conditions a fungus enters a sexual pathway, resulting in genetic segregation and production of recombinant progeny.

Since most fungi seem to be haploid for most of their life cycles, the first step in this process is to bring together two haploids so that nuclei can coexist in the same cytoplasm, undergo karyogamy followed by the meiotic division, and then generate and distribute progeny spores. These processes are considered in this chapter (see also Carlile and Watkinson, 1994; Elliott, 1994; Moore, 1998).

Sex: what and why?

Most fungi produce abundant asexual spores which are extremely effective in dispersing the organism. We have to ask why so many fungi invest more resources in a more complex sexual reproduction. There are, indeed, many fungi which only reproduce asexually but the majority still have a sexual cycle. Sex must have selective advantage if sexual stages are not to be replaced by asexual ones entirely (Maynard Smith, 1978).

The crucial point which provides the contrast with asexual reproduction is fusion of nuclei derived from different individuals. If the individuals differ in genotype, the fusion nucleus will be heterozygous and the products of the meiotic division can have recombinant genotypes. Thus, in one sexual cycle, new combinations of characters can be created.

231

This is the most usual 'explanation' for sex, namely that it promotes genetic variability through out-crossing and that variability is needed for the species to evolve to deal with competitors and environmental changes. Copious evidence exists to show that out-crossing certainly does promote variability, and that asexual lineages change little in time, in apparent support of the view that variability in the population enables the organism to survive ecological and environmental challenges.

However, this is a 'group selectionist' interpretation (variation generated in an *individual* meiosis is seen as benefiting the *group* or population to which the individual belongs) and current theory emphasizes, instead, that selection acts on individuals, so any feature which is argued to be advantageous in selection must be so because it benefits either the individual itself or its immediate progeny (Dawkins, 1976; Carlile, 1987). Bernstein *et al.* (1985) suggested that repair of damaged DNA is the crucial advantage of the meiotic sexual cycle, damaged DNA (caused by mutation or faulty replication) in one chromosome being repaired by recombination with the normal chromosome provided by the other parent.

Out-crossing might also give rise to heterozygous advantage, where the heterozygous phenotype is better than either of its homozygous parents, which has frequently been demonstrated in plants and animals and has also been demonstrated in the yeast *Saccharomyces cerevisiae.*

These alleged advantages of the sexual cycle (generation of variation and heterozygous advantage) are not mutually exclusive, nor of equal value; rather, they are themselves phenotypic characters which may or may not have selective value for the organism concerned. Different species have different life cycles and experience different evolutionary challenges and may therefore make use of, enhance or dispense with various aspects of sexual reproductive processes for any one or more than one of the interpretations outlined above. Generally, experimental evidence for the advantage of sex in the wild is lacking and fungi, which have both asexual and sexual cycles, are ideal organisms with which to perform such experiments.

Getting it together

Hyphal anastomosis (fusion between hyphae; or conjugation [cell fusion] in the unicellular fungi, yeasts) is the essential first step in the sexual cycle of most fungi. The process involves breakdown of two hyphal (cell) walls and union between two separate plasma membranes to bring the cytoplasms of the fusing hyphae into continuity with each other. To maximize the advantage of sexual reproduction the parental nuclei should be genetically as different as possible, whereas safe operation of the cell requires that cytoplasms which are to mingle must be as similar as possible.

The problem at the cytoplasmic level is that hyphal anastomosis carries the risk of exposure to contamination with alien genetic information from defective or harmful cell organelles, viruses, transposons or plasmids. Protection against alien DNA is provided by expression of vegetative compatibility which is controlled by one to several nuclear genes which limit completion of hyphal anastomosis between colonies to those which belong to the same vegetative compatibility group (v-c group). Members of a v-c group possess the same vegetative compatibility genes (or alleles).

Cytoplasmic compatibility and the individualistic mycelium

Fungal isolates from nature usually show interactions which demonstrate self/non-self recognition. If the colonies involved are not compatible the cells immediately involved in anastomosis are killed. Vegetative compatibility (also called somatic or heterokaryon incompatibility) will prevent formation of heterokaryons except between strains which are sufficiently closely related to belong to the same v-c group.

Fusion incompatibility (where the compatibility system determines the ability to fuse) occurs in slime moulds, but the type of vegetative compatibility which is most usual in fungi is called post-fusion incompatibility. Hyphal anastomosis is consequently promiscuous, but compatibility of the cytoplasms determines whether the cytoplasmic exchange process will progress beyond the few hyphal compartments involved in the initial interaction. This strategy prevents transfer of nuclei and other organelles between incompatible strains, but if the incompatibility reaction is slow, a virus or cytoplasmic plasmid may be communicated to adjacent undamaged cells before the incompatibility reaction kills the hyphal compartments where anastomosis occurred.

It is these compatibility reactions (including vegetative and sexual) which define in real life what constitutes the fungal individual. In yeasts each cell is clearly an individual but a mycelial individual is not so obvious. Spores are individuals and colonies developed from single spores must also be individuals. Whether a heterokaryon is an individual, rather than a chimera or mosaic of two or more individuals, is more debatable. When individuals do exchange nuclei it is the mating systems which then regulate sexual exchange between the vegetatively compatible mycelia.

Sexual mating systems

Mating systems (or breeding systems) depend on nuclear genes that prevent mating between genetically identical mycelia. A mycelium which possesses a system to prevent mating between identical gametes will be self-sterile and, since *different* mycelia must come together for a successful mating to occur, the system is called *heterothallism*. The alternative, where there is nothing to prevent a single mycelium completing a successful mating, is called *homothallism*. A number of mating type systems have been recognized in fungi, they all regulate the sexual process so that meiosis occurs only if the two mycelia concerned carry different mating type factors.

In many fungi there are only two mating types. The common 'bread-mould' *Neurospora crassa*, the brewer's yeast *Saccharomyces cerevisiae*, the grass rust *Puccinia graminis* are examples. In these cases the 'mating type' of a culture depends on which allele it has at a *single* mating type locus (hence the name *unifactorial* incompatibility) and successful mating can only take place between (yeast) cells or mycelia that have different alleles at the mating type locus. The diploid nucleus which is eventually formed is, of course heterozygous at the mating type factor, and meiosis produces equal numbers of progeny of each of the two mating types (hence the alternative name bipolar heterothallism). A modification of this system is seen in *Coprinus comatus* ('Lawyer's wig' or 'Shaggy-cap' mushroom) which has a large number of mating type alleles at its single mating type locus. When there are only two alleles the likelihood that two unrelated individuals will be able to mate (which is the outbreeding potential) is 50 per cent. But if there are *n* mating types in a

234

population the outbreeding potential would be $[1/n \times (n-1)] \times 100$ per cent. The greater the number of mating type alleles, the greater the outbreeding potential.

Many of the Basidiomycotina have *two* unlinked mating type factors (designated A and B). This, of course, is called a *bifactorial* incompatibility system. In this case also, compatibility requires that two mycelia have different alleles, but this time *both* mating type factors must differ. The diploid nucleus will therefore be heterozygous at the two mating type loci and meiosis will generate progeny spores of four different mating types (so tetrapolar heterothallism is the alternative name of this system). The classic examples of this mating type system are the Hymenomycetes *Coprinus cinereus* and *Schizophyllum commune*. In both organisms the wild population contains many different A and B alleles, so the outbreeding potential is about 100 per cent. The advantage over the unifactorial system is that the inbreeding potential of bifactorial incompatibility (the likelihood of being able to mate with a sibling) is 25 per cent (because there are four different mating types among the progeny of a single meiosis) whereas it is 50 per cent in unifactorial incompatibility where there are only two mating types among the progeny. The percentage values quoted here are simplifications. Change in the mating type (by recombination or mating type switching) can increase inbreeding potential and is discussed below.

In self-fertile (homothallic) fungi the sexual process can occur between genetically identical cells or hyphae, but this does not mean that mating type factors are not involved. Primary homothallism does indeed occur in species completely lacking heterothallism, but secondary homothallism occurs in species which have an underlying heterothallism which is bypassed in some way. The best examples are found in *Neurospora tetrasperma* and *Agaricus bisporus*. In both cases, fewer spores are formed than there are post-meiotic nuclei with the result that spores are binucleate and heterozygous for mating type factors. When the spores germinate they form heterokaryotic vegetative mycelia which are able to complete the sexual cycle alone; i.e. they act like homothallics. A different example is provided by the yeast *Saccharomyces cerevisiae*, in which most strains are heterothallic with two mating types, a and α (see below, section: mating type factors in *S. cerevisiae*). However, in some strains mating occurs between progeny of a single haploid ancestor; that is, the culture appears to be 'homothallic'. The apparent 'homothallism' results from a switch from one mating type to the other (see below) so that the clone comes to contain cells of both mating types.

Mating type factors in *Saccharomyces cerevisiae*

The yeast *Saccharomyces cerevisiae* has a life cycle which features an alternation of a haploid phase with a true diploid phase, and in this respect differs from other Ascomycotina in which the growth phase after anastomosis is a heterokaryon. Haploid yeast cells have one of two mating types which are symbolized a and α. Nuclear fusion (karyogamy) follows fusion of cells of opposite mating type and the first bud after these events contains a diploid nucleus.

Most natural cultures of yeast are diploid because the haploid meiotic products mate soon after meiosis while they are in close proximity. The diploid state is maintained by mitosis and budding until specific environmental conditions (deficiency in nitrogen and carbohydrate, but well aerated and with acetate or other carbon sources which favour the glyoxylate shunt) induce sporulation when the entire cell is converted into an ascus in which meiosis occurs and haploid ascospores are produced. Ascospore germination re-establishes the haploid phase, which is itself maintained by mitosis and budding (Figure 8.1).

The mating type factors in yeast are responsible for production of peptide hormones (pheromones called α- and a-factors; Figure 8.2) and pheromone-specific receptors. The pheromones organize the mating process by binding to pheromone receptors on the surface of cells of opposite mating type (Figure 8.1) acting through GTP binding proteins to alter metabolism and (i) causing recipient cells to produce an agglutinin, so that cells of opposite mating type adhere; (ii) stopping growth, so that cells are blocked in the G1 stage of the cell cycle; (iii) changing wall structure and consequently cell shape.

Both pheromones cause their target cells to elongate into projections but have no effect on cells of the same mating type or on diploids. Fusion occurs between the projections (Figure 8.1). The elongated cells are called 'shmoos'.

Mating phenotypes in *S. cerevisiae* are controlled by a complex

236

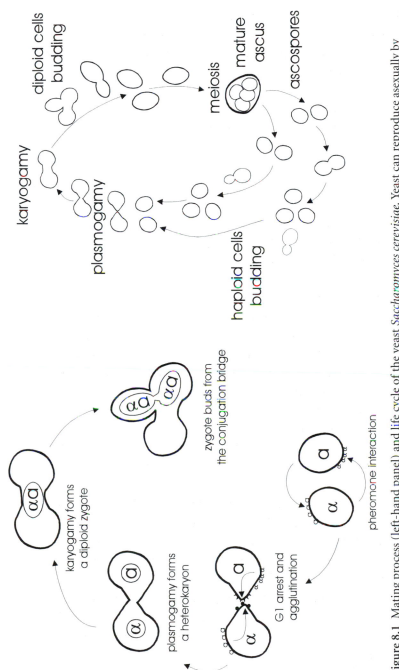

Figure 8.1 Mating process (left-hand panel) and life cycle of the yeast *Saccharomyces cerevisiae*. Yeast can reproduce asexually by budding. Haploid cells of different mating types fuse to form dumbell-shaped zygotes, which can themselves bud to establish a diploid clone. Well-nourished diploid cells which are exposed to starvation conditions enter meiosis, forming a four-spore ascus. Ascospores germinate by budding. In the laboratory, ascospores can be separated to form haploid clones but in nature ascospores usually mate immediately, so the haploid phase is greatly reduced.

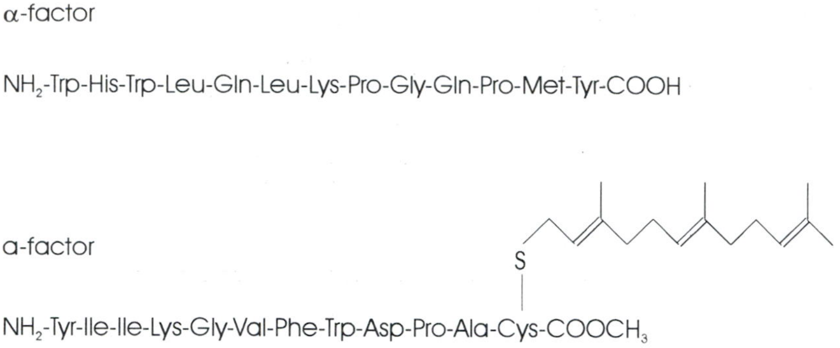

α-factor

NH$_2$-Trp-His-Trp-Leu-Gln-Leu-Lys-Pro-Gly-Gln-Pro-Met-Tyr-COOH

a-factor

S

NH$_2$-Tyr-Ile-Ile-Lys-Gly-Val-Phe-Trp-Asp-Pro-Ala-Cys-COOCH$_3$

Figure 8.2 Simplified chemical structures of yeast pheromones. The a-factor contains a farnesyl group.

genetic locus called *MAT* where two linked genes are harboured (a1, a2 for mating type a and α1, α2 for mating type α). The *MAT*a locus encodes the divergently transcribed a1 and a2 polypeptides (Figure 8.3), and *MAT*α encodes polypeptides α1 and α2 (no function has been identified for the a2 gene product). Heterozygosity at *MAT* signals diploidy and eligibility to sporulate (i.e. even partial diploids carrying *MAT*a/*MAT*α will attempt to sporulate). The α2 polypeptide is a repressor of transcription of a-factor (in α-cells), whilst a1 represses αspecific genes in a-cells. The α1 protein activates transcription of genes coding for αpheromone and a-factor surface receptor. In a/α diploids, a1 and α2 polypeptides form a heterodimer which represses genes specific for the haploid phases, including a gene called *RME*1 (Repressor of Meiosis) which suppresses meiosis and sporulation (Figure 8.3).

    S. cerevisiae is heterothallic but a clone of haploid cells of the same mating type frequently sporulates, producing equal numbers of a and αprogeny. Originally thought to result from mutation, this was eventually found to result from mating type switching controlled by the gene *HO* (HOmothallic) which exists in two allelic forms; rare occurrence of mating type switching (about once in 10^5 divisions) is the phenotype of allele *ho*, whereas in strains carrying *HO* the switch occurs at every cell division. *HO* is a gene encoding an endonuclease creating a double-strand break at locus *MAT* and its transcription is repressed by the a1/α2 heterodimer. The mating type of the yeast is determined by the expressed allele in the *MAT* locus, but on either side of this (and on the same

Saccharomyces cerevisiae

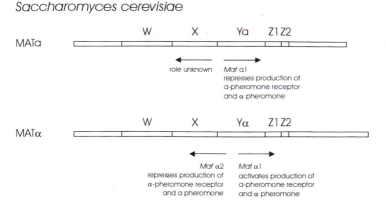

In a/α diploids, the MATα1/MATα2 heterodimer protein activates meiotic and sporulation functions, and represses haploid functions (turning off α-specific functions by repressing MATα1, a-specific functions being repressed by MATα2 alone).

Neurospora crassa

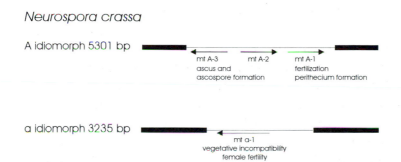

Figure 8.3 Functional regions of mating type factors in *Saccharomyces cerevisiae* (top) and *Neurospora crassa* (bottom). In both panels the arrows indicate direction of transcription and the legends beneath the arrows indicate functions of the gene products. In *S. cerevisiae* of mating type a, a general transcription activator is responsible for production of a-pheromone and the membrane-bound α-pheromone receptor. In the *N. crassa* illustration the black bars represent the conserved DNA sequences which flank the idiomorphs, the latter shown as lines. These diagrams are oriented so that the centromere is on the left, consequently the centromere-distal sequence is on the right.

239

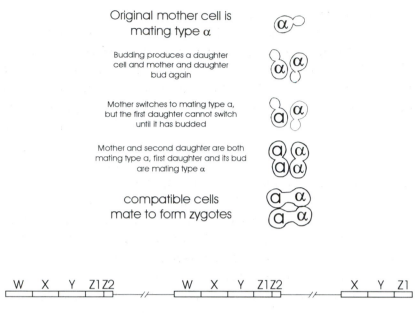

Original mother cell is
mating type α

Budding produces a daughter
cell and mother and daughter
bud again

Mother switches to mating type a,
but the first daughter cannot switch
until it has budded

Mother and second daughter are both
mating type a, first daughter and its bud
are mating type α

compatible cells
mate to form zygotes

| W | X | Y | Z1 Z2 | | W | X | Y | Z1 Z2 | | X | Y | Z1 |

HML   *MAT*   *HMR*

Figure 8.4 Top: pattern of mating type switching in *Saccharomyces cerevisiae* showing the consequences of a mating type switch in one mother cell. Bottom: the three loci involved in mating type switching, *HML*, *MAT* and *HMR*, are located on the same chromosome.

chromosome) there are storage but silent loci, one for each mating type, *HML* and *HMR*. Switching involves replacement of information at the *MAT* locus by that at either *HML* or *HMR* (Figure 8.4; see Schmidt and Gutz, 1994) by an intrachromosomal recombination event. *HO* is expressed in G1 phase of the cell cycle under the positive regulation of SWI5 protein which is synthesized in late S and G2 phases. As a result, the asymmetrical distribution of SWI5 protein in the mother cell, which has undergone one complete cell cycle, grants the cell the developmental potential of mating type switching (Figure 8.4).

Mating type switching also occurs in the distantly related fission yeast *Schizosaccharomyces pombe* but mating types in filamentous fungi tend to be far more stable although unidirectional switching of mating type has been reported in *Chromocrea spinulosa*, *Sclerotinia trifoliorum*, *Glomerella cingulata* and *Ophiostoma ulmi*.

Mating type factors in *Neurospora* species

The genus *Neurospora* has at least four markedly contrasting mating strategies: (i) bipolar heterothallism with mating types A and a (e.g. *N. crassa, N. sitophila, N. intermedia* and *N. discreta*), but unlike the yeast *Saccharomyces cerevisiae*, the mating type genes are present in a single copy per genome, and the two alleles do not share homology (hence the name idiomorph); (ii) secondary homothallism (e.g. *N. tetrasperma* which produces asci containing four ascospores each containing compatible nuclei); (iii) primary homothallism in which each haploid genome carries genetic information of both mating types (e.g. *N. terricola, N. pannonica*); and (iv) primary homothallism in which genetic information for only one mating type is detected (e.g. *N. africana* which possesses only the *A* idiomorph which shares 88 per cent similarity with that of *N. crassa*). In the primary homothallic species, meiosis produces a linear eight-spored ascus in which all progeny are self-fertile.

In bipolar heterothallic species, the mycelium of each mating type is hermaphroditic. Under nitrogen starvation, strains of either mating type develop female structures (protoperithecia and trichogynes). Mycelia and asexual spores (macroconidia or microconidia) of the strain of the opposite mating type can serve as the male in a sexual cross and secrete mating type-specific pheromones which orient trichogynes to cells of the opposite mating type. Mating is followed by the formation of perithecia which contain asci (see Figure 8.13). In *N. crassa,* the mating type locus is one of the 10 heterokaryon incompatibility (*het*) loci active during vegetative growth, preventing the formation of viable $a + A$ heterokaryons (see section: Cytoplasmic compatibility, above). Yet this mating type heterokaryon incompatibility is suppressible by an unlinked wild-type suppressor, *tol*$^+$ (tolerant).

The mating type genes *A* and *a* in *Neurospora crassa* have been cloned and sequenced (see Figure 8.3). In all the known *Neurospora A* and *a* idiomorphs and flanking regions, the centromere-proximal flank contains species-specific and/or mating type-specific DNA sequences. Immediately adjacent to the centromere-proximal are the variable regions, which are very different between species. Next to the variable regions are the idiomorphs themselves which are highly conserved between species but are completely dissimilar between the two mating types within the species. These are then followed by a 'mating type common region' of 57–69 bp which separates an idiomorph from its nearby variable region and is highly similar between all species and between both mating types. These flanking regions may provide positional information for the proper expression of the idiomorph enclosed.

The *A* idiomorph of 5301 bp in length gives rise to at least three transcripts (A-1, A-2 and A-3), with the first two transcribed in the same direction (Figure 8.3). The first 85 amino acids at the N-terminal region of the mating type *A* product are minimally sufficient for female fertility. A region from position 1 to 111 confers vegetative incompatibility while amino acids from position 1 to 227 are required for male-mating activity. This mating type-specific mRNA is expressed constitutively in vegetative cultures, and during the sexual cycle both before and after fertilization. Transcript A-1 which shows a high degree similarity to *MATα*1 of *Saccharomyces cerevisiae* is essential for fertilization and fruiting body formation and the other two transcripts are involved in post-fertilization functions including ascus and ascospore formation. The *A*-3 transcript is a HMG (high-mobility group) polypeptide with DNA-binding ability, indicating its potential function as a transcriptional factor and shares 50 per cent similarity to the mating type polypeptide (*Mc*) of yeast *Schizosaccharomyces pombe*.

In *N. crassa*, bases 2923–4596 of the *a* idiomorph give rise to a single mt *a*-1 transcript which encodes a polypeptide (288 amino acids) belonging to high-mobility group proteins (the HMG box) with DNA-binding activity. The HMG box is the domain for mating function whereas amino acids 216–220 of mt *a*-1 act in vegetative incompatibility. The separation of vegetative incompatibility from both mating and DNA binding indicates that vegetative incompatibility functions by a biochemically distinct mechanism. All *a* idiomorphic DNA sequences between 1409 and 2923 are non-essential. Unlike the case with *Saccharomyces cerevisiae* (see section: mating type factor in *S. cerevisiae*), the genes downstream to and regulated by the *A* and *a* transcripts, however, have not been characterized.

Mating type factors in *Ustilago maydis*

Ustilago maydis causes the smut disease of maize. It has a tetrapolar mating system. Haploid, yeast-like sporidia can be grown on synthetic

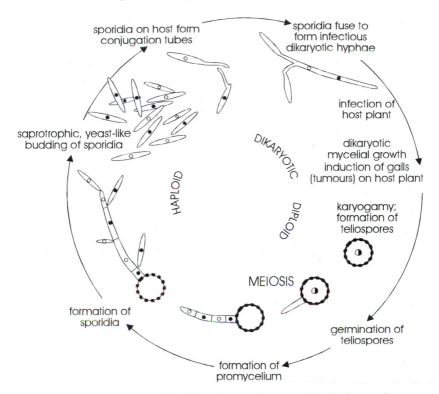

sporidia on host form
conjugation tubes

sporidia fuse to
form infectious
dikaryotic hyphae

saprotrophic, yeast-like
budding of sporidia

infection of
host plant

DIKARYOTIC

HAPLOID

DIPLOID

dikaryotic
mycelial growth
induction of galls
(tumours) on host plant

karyogamy;
formation of
teliospores

MEIOSIS

formation of
sporidia

germination of
teliospores

formation of
promycelium

Figure 8.5 Diagram of the life cycle of *Ustilago maydis*. (Redrawn after Kämper *et al.*, 1994.)

media and are non pathogenic for the host plant (Figure 8.5). When sporidia of opposite mating types are mixed, conjugation tubes are formed and fusion of these produces the dikaryon which grows as a filamentous fungus. The dikaryon is the pathogenic stage, however, and its growth is strictly dependent on the host plant. The hyphal cells grow within the plant, causing tumour (or gall) formation. Within the galls, virtually all the hyphal segments differentiate into diploid teliospores which have thickened black walls. Eventually the host epidermis ruptures and galls break open, releasing teliospores. Teliospores germinate, undergo meiosis, and form a promycelium that usually consists of four haploid cells, each of which produces numerous sporidia by successive budding.

Fusion of sporidia is controlled by the biallelic '*a*' mating-type locus, a contrast with Homobasidiomycetes, where hyphal fusion is not controlled by mating type factors, but by vegetative compatibility genes. The heterozygous $a1/a2$ genotype is required for the generation of structures similar to conjugation tubes for the proper transition between the yeast and filamentous growth form. The multiallelic 'b' locus determines true filament (hyphal) growth form, pathogenicity and stops cells fusing after the formation of the diploid.

Formation of conjugation tubes is induced by the action of diffusible, mating type-specific pheromones released by haploid cells. Sporidia of each mating type secrete a pheromone specific to their mating type and sense the presence of pheromone from cells of opposite mating type with mating type-specific pheromone receptors. These pheromones are short hydrophobic lipopeptides. The peptide consists of 11 to 15 amino acids; they all have a C-terminal cysteine residue that is post-translationally modified. The cysteine's carboxyl group is methylated and a farnesyl group (a 15-C isoprenyl moiety) is attached to the sulfhydryl group. The latter makes the pheromone so hydrophobic that it is nearly insoluble in water and it may act as a membrane anchor.

The $a1$ allele comprises 4.5 kb of DNA, and the $a2$ allele 8 kb. Two mating type-specific genes have been identified in these regions: $mfa1$ (in $a1$) and $mfa2$ (in $a2$) code for precursors of the farnesylated pheromones and $pra1/pra2$ which encode pheromone receptors (Figure 8.6).

The b mating type factor in $U.$ $maydis$ contains a pair of genes (separated by a 260-bp spacer region) which are transcribed in opposite directions. The genes are called bE and bW (East and West) and their coding sequences are equivalent to polypeptides of 473 and 629 amino acids, respectively. There are a number of conserved sites with similarities to the homeodomain or DNA-binding regions known in regulatory proteins of many other eukaryotes. In different alleles the coding sequence shows a high degree of variation at the amino terminal end, whereas the carboxy-terminal ends are highly conserved.

The variable regions at the amino terminal end determine the allele specificity at this multiallelic locus; whilst the highly conserved region, and particularly the homeodomain, provide for its function in regulating sexual development. The current notion is that bW and bE proteins form a heterodimer which might act as an activator of genes required for the sexual cycle and/or repressor of haploid-specific genes. It must be

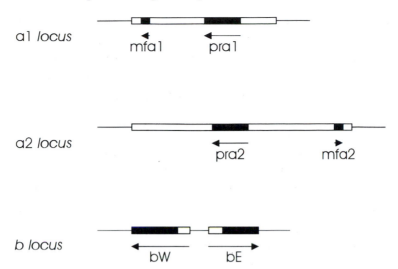

Figure 8.6 Schematic representations of the structures of the a and b mating type loci of *Ustilago maydis*. Alleles of the a locus consist of mating type-specific (i.e. variable) DNA sequences (4500 base pairs in a1, 8000 base pairs in a2), here shown as open boxes, within which are the genes for mating (*mfa* and *pra*). The b locus has two reading frames, bW and bE, which produce polypeptides containing domains of more than 90 per cent sequence identity (shown as black boxes) and variable domains (open boxes) which show 60 to 90 per cent identity. Arrows indicate the direction of transcription.

assumed, of course, that the complex comprised of bE & bW from the same allele is always inactive (perhaps the homeodomains are not properly exposed); only when the proteins come from different alleles will the heterodimer function properly.

Mating type factors in *Coprinus cinereus* and *Schizophyllum commune*

These two homobasidiomycetes have been the main subjects in studies of mating in saprotrophic basidiomycetes over many years. The attraction has been that in both organisms mating results in a major change in mycelial morphology (Figures 8.7 and 8.8) and growth pattern as it converts the sterile parental homokaryons (monokaryon) into a fertile heterokaryon (dikaryon). In *Coprinus cinereus* the monokaryon

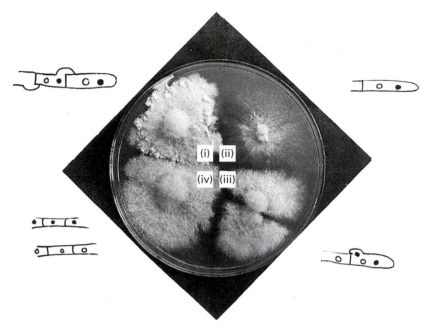

Figure 8.7 Mating reactions in *Schizophyllum commune*. (i) + reaction, compatible 'A different, B different' heterokaryon; (ii) flat reaction, common A heterokaryon; (iii) barrage reaction, common B heterokaryon; (iv) – reaction, common AB heterokaryon.

has uninucleate cells and produces abundant uninucleate arthrospores (oidia). The dikaryon has binucleate cells with characteristic clamp connections (Figure 8.8) at each septum. It does not produce oidia but under appropriate environmental conditions (nutrition, temperature and illumination) it does produce the mushroom fruit bodies. Hyphal fusions occur between any monokaryons (i.e. it is promiscuous, see section: Cytoplasmic compatibility, above), the mating type factors recognize compatibility intracellularly after hyphal fusion. Because of the clear phenotypic differences between homokaryons, heterokaryons and dikaryons the mating type factors have been subjects for classical genetic studies for many years and have recently been subjected to detailed molecular analysis.

C. *cinereus* and S. *commune* exhibit tetrapolar heterothallism which is determined by two mating type factors, called *A* and *B*. The natural

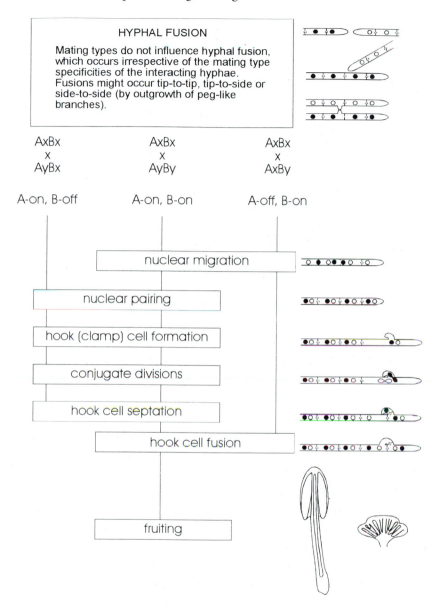

Figure 8.8 Flow chart diagram of A and B mating type factor activity in the homobasidiomycetes *Coprinus cinereus* and *Schizophyllum commune*. See text for details. (Revised and modified after Wendland *et al.*, 1995.)

S.-W. CHIU AND D. MOORE

Coprinus cinereus

Schizophyllum commune

Figure 8.9 Linkage maps of the chromosomes which carry the A mating type factor in *Coprinus cinereus* and *Schizophyllum commune*.

population contains many different mating types which behave in crosses as though they are multiple alleles at the two mating type loci. In fact, each mating type factor is a complex genetic region (which is why we refer to mating type *factors* rather than genes). The mating type factors are located on different chromosomes, and even conventional genetic analysis was able to demonstrate Aα, Aβ, Bα and Bβ subloci, the subloci being relatively far apart in *S. commune* but much closer together in *C. cinereus* (Figure 8.9). The subloci also exhibit multiple allelism and recombination between subloci can generate new mating type specificities.

In both fungi, a compatible mating with the clamp connections (hook) and conjugate nuclei in the mated hyphae requires heterozygosity at both *A* and *B* (A-on, B-on). In terms of cytological observations, mating type factor *A* controls nuclear pairing, clamp cell formation and synchronized (conjugate) mitosis whereas mating type locus *B* controls nuclear migration and clamp cell fusion (Figure 8.8). Nuclei migrate through the existing mycelium of the recipient and this requires breakdown of the septa between adjacent cells of that mycelium. Heterokaryons can also be formed in matings in which one of the mating type factors is common (i.e. homozygous). When the *A* factors are the same (A-off, B-on), nuclear migration occurs but no clamp connections form. A mating between strains carrying the same *B* factor (A-on, B-off) forms a heterokaryon only where the mated monokaryons meet because

248

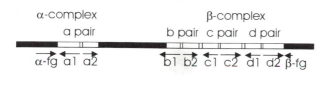

Figure 8.10 Schematic representations of the structures of parts of the A mating type factors in *Coprinus cinereus* and *Schizophyllum commune*. Arrows show the direction of transcription. The (predicted) archetypal A factor from *Coprinus cinereus* has four pairs of functionally redundant genes (a, b, c and d) which feature the homeodomain 1 (HD1 in a1, b1, c1 and d1) and homeodomain 2 (HD2 in a2, b2, c2 and d2) sequences. Interaction between HD1 and HD2 proteins is the basis of the compatible reaction (see Figure 8.11). A factors examined in different strains of *C. cinereus* isolated from nature contain different combinations, and different numbers, of these genes. In *Schizophyllum commune* the mating type genes are called X and Y and carry HD1 and HD2 respectively. Again, different alleles are found in different natural mating types; indeed, the Z gene is absent in the Aα1 mating type. The sequences shown as mep and α-fg are homologous, encoding a metalloendopeptidase; X is a flanking gene of unknown function.

nuclear migration is blocked. Terminal cells of heterokaryotic hyphae initiate clamp connections and nuclei divide but the hook cell fails to fuse with the subterminal cell and its nucleus remains trapped.

Molecular analysis of the *A* mating type factors showed that they are composed of many more mating type genes/gene pairs than the classical genetic analysis revealed (Figure 8.10). These gene pairs encode two

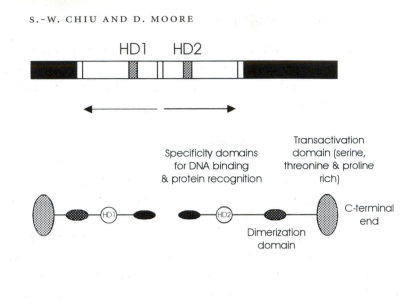

HD1/HD2 interaction

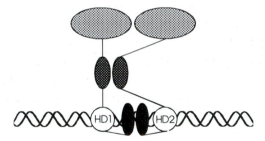

Figure 8.11 Schematic diagram showing a model of homeodomain protein production, structure and interactions involved in A mating type factor activity in *Coprinus*. (Redrawn and adapted after Casselton & Kües, 1994.)

families of proteins (HD1 and HD2) with homeodomain regions which may encode transcriptional factors. The compatibility reaction required for sexual development is triggered by heterodimerization between HD1 and HD2 proteins from the different *A* mating type factors of compatible individuals (Figure 8.11). The N-terminal regions of these proteins are essential for choosing a compatible partner but not for regulating gene transcription.

Although most work has been aimed at determining the structure

250

of the *A* mating type factors, recently the B sequences have been cloned and sequence analysis has shown that the multiallelic B mating type factor encodes several pheromone and receptor genes which might be involved in controlling the growth of the clamp connection.

Overview of mating type factors

Not all fungi possess mating type genes. In those that do, mating type genes may encode or positively regulate the transcription of pheromone and pheromone receptors involved in some aspect of the mating process (ranging from recognition between sexually competent cells in yeast to governing growth of clamp connections in homobasidiomycetes). Also, heterodimerization of homeodomains from different alleles is employed to trigger further aspects of sexual development and may at the same time repress haploid-specific events. In terms of sequence and functionality, the mating type alleles are usually conserved within the same family (e.g. mating type gene of *Neurospora crassa* in other members of the Sordariaceae). The sequences of the mating type alleles are highly dissimilar and have been interpreted as equivalent to the highly variable region in major histocompatibility loci in mammals as part of a self/non-self recognition system. The N-terminal region of the proteins is essential for such self/non-self recognition while the other region codes for DNA-binding. As some of the mating type loci show a series of linked genes, some of the genes are transcribed in opposite directions to prevent intragenic recombination.

Superior to the simple biallelic system, the functional redundancy of the multiallelic system is clearly to promote outbreeding. For instance, an estimated 28 000 mating types for *Schizophyllum commune* result from the combination of 9, 32, 9 and 9 different specificities found in the world-wide population for Aα, Aβ, Bα and Bβ, respectively. In *Saccharomyces cerevisiae*, the mating type genes are known to act as master genes (producing transcription factors) controlling the expression of multiple genes (downstream regulation) which then impact on the sexual development pathway such as fertilization and sporulation. However, in *Podospora*, the meiocyte pathway (meiosis and sporulation) does not require heterozygous mating type alleles. In addition the formation of apparently normal fruiting bodies by haploid cultures is not

251

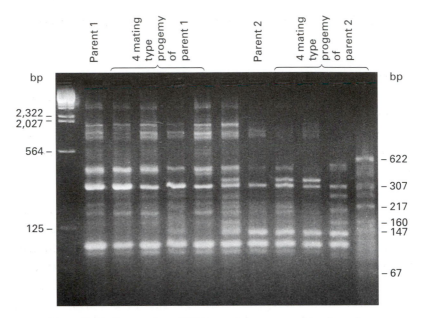

Figure 8.12 Segregation of DNA markers generated by the polymerase chain reaction in the sexual progenies of the basidiomycete, *Lentinula edodes* which is a tetrapolar heterothallic species. Some of the DNA bands are found in both parent and progeny while some bands are polymorphic (present/absent in/from some of the progeny).

uncommon in fungi and fruit body formation can usually be separated from other parts of the sexual pathway by mutation (discussion in Moore, 1994, 1998). Therefore, the significance of mating type factors in regulating events beyond the initial mating reaction is uncertain. Even the target genes within that mating reaction are unknown and it is also uncertain how the products of the multiple genes which make up the complex mating type factors avoid producing active multimers in unmated mycelia. Much remains to be learned about these mating type factors.

The sexual cycle

The outcome of mating is the recombination and segregation of the genetic elements which were brought together (Figure 8.12). The progeny have new genotypes but to achieve the expression of these the nuclei must be 'packaged' into progeny spores which can be distributed into the environment.

Sporulation in higher fungi

Unlike most animals and plants, after mating a persistent and independent heterokaryotic or diploid phase exists in fungi. Only under particular environmental triggers is the sexual cycle initiated. Taking a filamentous ascomycete as an example: hyphal fusion or similar mating between male and female structures results in nuclei moving from the male into the female to form an ascogonium in which male and female nuclei may pair but do not fuse (dikaryon). Ascogenous hyphae grow from the ascogonium. Most cells in these hyphae are dikaryotic, containing one maternal and one paternal nucleus, the pairs of nuclei undergoing conjugate divisions as the hypha extends. In typical development, the ascogenous hypha bends over to form a crozier. The two nuclei in the hooked cell undergo conjugate mitosis and then two septa are formed, creating three cells (Figure 8.13). The cell at the bend of the crozier is binucleate but the other two cells are uninucleate. The binucleate cell becomes the ascus mother cell, in which karyogamy takes place. In the young ascus meiosis results in four haploid daughter nuclei, each of which divides by mitosis to form the eight ascospore nuclei (Figure 8.13). Formation of ascospores results from the infolding of membranes around the daughter nuclei so that eight ascospores are delimited. A spore wall then forms around each ascospore. Ascus cytoplasm left outside the spores, called the epiplasm, may provide nutrients to the maturing spores, contribute to the outer layers of the spore wall, or contribute to the osmotic potential of the ascus to aid subsequent spore discharge. Ascus and ascospore morphogenesis has been reviewed by Read and Beckett (1996).

Typically, eight uninucleate and haploid ascospores are formed, though there are variations on this theme. In some species a further mitotic division forms binucleate spores, in others the immature spores become multinucleate prior to being divided up by septa. Some ascomycetes form only four ascospores as a result of the spore membranes enclosing a pair of nuclei rather than just one nucleus, while others may produce fewer spores in each ascus as a result of nuclear disintegration or spore abortion.

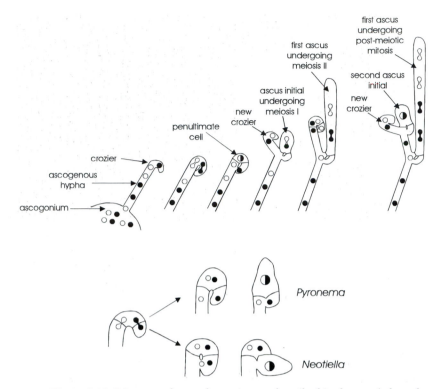

Figure 8.13 Diagram of ascus formation, as described in the text (adapted and redrawn from Webster, 1980). Karyogamy may occur in the penultimate cell of the crozier (top diagram and *Pyronema* in the lower diagram) or the terminal or stalk cell (*Neotiella* in the lower diagram, which is adapted and redrawn from Read and Beckett, 1996).

In basidiomycetes, karyogamy and meiosis take place in the basidium and basidiospores (usually four) are produced externally on outgrowths of the basidial wall which are called sterigmata (Figure 8.14). The walls of the sterigma and of the early spore initial are continuous and homologous. Starting with the spherical growth of the spore initial and nuclear migration into the maturing spore, a basidiospore grows further to attain the species-specific form and dimension, and the spore protection is increased by further wall layers formed with/without pigmentation, ornamentation and germ pore.

All of these spore formation processes include steps which require precise nuclear positioning and/or active nuclear movement. Various

254

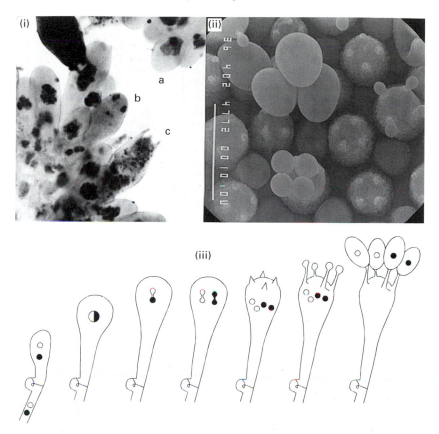

Figure 8.14 Meiosis and sporulation in the homobasidiomycete, *Volvariella bombycina*. (i) Light micrographs showing meiosis. a, prophase I; b, meiosis II; c, interphase II and basidium with sterigmata; (ii) scanning electron micrograph showing the initiation of sterigmata in the final spore maturation process; (iii) diagram of basidium formation: in a 'classic' homobasidiomycete the basidium arises as the terminal cell of a dikaryotic hyphal branch (see also Figure 8.16) which inflates and undergoes karyogamy and meiosis. At the conclusion of the meiotic division four outgrowths (sterigmata) emerge from the basidial apex and inflation of each sterigma tip produces the basidiospore (which is an exospore, produced outside the meiocyte in contrast to the endospores of ascomycetes). Nuclei then migrate from the basidium into the newly formed basidiospores. Mitosis may take place within the basidiospores before they are discharged. Comparison of this diagram with Figure 8.13 will indicate readily how tempting it is to suggest some evolutionary relationship between crozier formation and the early stages of basidium and clamp connection formation.

cytoskeletal structures have been associated with nuclear division, nuclear migration and cytokinesis. Benomyl-resistant (*ben*) mutants of *Coprinus cinereus* have defects in structural genes for α- or β-tubulin. The *ben* mutants were blocked in migration of nuclei during formation of dikaryotic hyphae with clamp cells, but migration of nuclei into developing spores was unaffected. This may indicate that there are at least two nuclear movement systems, only one being dependent on tubulin. A key functional difference may be that microtubules are involved in the pairing of the two conjugate nuclei in the dikaryon.

The spindle pole body (the centrosome in other organisms; SPB in abbreviation) shows a duplication cycle in meiosis and mitosis. Using mutant analysis, it has been deduced that structural modification of SPB in meiosis II is required to nucleate the endospore-wall formation in *Saccharomyces cerevisiae* and *Schizosaccharomyces pombe* (Esposito and Klapholz, 1981). In homobasidiomycetes, the SPB always leads the daughter nucleus migrating into the maturing basidiospore.

Meiosis

Although there are some inevitable modifications, karyogamy and meiosis in heterothallic fungi go through stages fairly typical for eukaryotic haploids. In particular the major round of DNA replication is premeiotic, occurring before karyogamy. Indeed, it was research with the ascomycete *Neottiella* which first demonstrated this aspect of meiosis (Westergaard and von Wettstein, 1970). Meiosis takes place in the meiocytes (basidia or asci) and chromosome behaviour in meiosis follows the Mendelian laws of segregation, independent assortment, linkage and crossing over.

Until recently meiosis has been described primarily as comprising the sequence synapsis, recombination and segregation. It is now known that chromosome pairing and synapsis are distinct processes in terms of both mechanism and timing. Premeiotic DNA replication and formation of the synaptonemal complex are independent events. Synapsis does not require DNA homology. At prophase I, homologous chromosomes are aligned prior to appearance of tripartite synaptonemal complex, and at this time double-strand breaks (providing sites for meiotic recombination) usually occur. The fission yeast, *Schizosaccharomyces pombe*, has

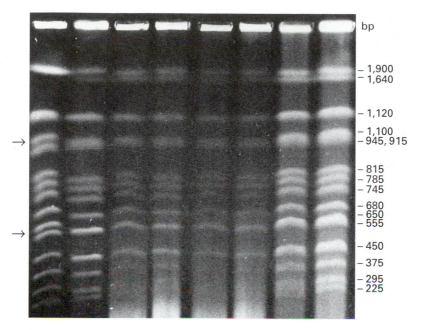

Figure 8.15 Electrophoretic karyotypes of two strains of *Saccharomyces cerevisiae* revealed by pulsed field gel electrophoresis showing chromosomal length polymorphisms. Arrows point to the polymorphic chromosomal DNA bands.

normal levels of meiotic recombination but does not make any synaptonemal complex, though structures resembling the axial core of the synaptonemal complex do occur. Mutants of *Saccharomyces cerevisiae* defective in synaptonemal complex structure showed defects in chromosome condensation but still underwent some meiotically induced homology pairing, revealing the independence of homology pairing and meiotic recombination. In meiotic recombination, heteroduplex DNA (hDNA), which refers to the hybrid DNA formed following strand exchange and with one or more mismatched base pairs, is an essential intermediate. The frequency of homologous recombination is 100- to 1000-fold higher during meiosis than during mitosis.

The synaptonemal complex, if formed, converts several sites of alignment and exchange into a functionally intact bivalent, consisting of kinetochores, which are sites for attachment to fibres of the division spindle, and the four strands connected by chiasmata. Chiasmata are the cytologically visible connections between homologous chromosomes which have long been regarded as being equivalent to genetic cross-overs formed in the recombination process. Yet the number of chiasmata observed in meiosis of some organisms is much less than that of cross-over events deduced from progeny analysis, so recombination events and chiasmata may not be numerically equivalent. Rather, chiasmata may be only those cross-overs which are required to balance mechanical forces exerted by the division spindle on the kinetochores to ensure that homologues move away from each other in the first meiotic division (Hawley and Arbel, 1993; Moens, 1994).

Mutants defective in DNA repair have been demonstrated to have defects in meiotic chromosome condensation, synapsis and recombination in meiosis, perhaps culminating in defects such as formation of non-viable spores or abortion in sporulation. Using mutant analysis, it has been found that recombination events are not sufficient in themselves to ensure disjunction (Hawley and Arbel, 1993; Pukkila, 1994). On the other hand, experiments with artificially constructed chromosomes in *Saccharomyces cerevisiae* have shown that neither chromosome size nor homologous DNA sequence had much effect on chromosome disjunction. The crucial factor was that when a cross-over occurred the chromosomes nearly always disjoined at meiosis I. Clearly, much more work needs to be done to establish the precise relationships between, on the one hand, synapsis, sequence homology and synaptonemal complex formation, and on the other hand between molecular recombination events, genetic cross-overs and chiasmata.

Duplicated DNA sequences in *Neurospora crassa* are extensively point mutated during the sexual cycle by a process known as repeat-induced point mutation (RIP)(Davis, 1995). The effect occurs at high frequency during pre-meiosis; 50 to 100 per cent of linked, direct duplications can be RIPed and many deleted as well. RIP and recombination are correlated but not interdependent, indicating that the two processes are distinct but perhaps one provokes the other. Although the likely function of RIP is to protect against the potential instability in chromosomal interactions which direct duplications threaten, the effect can extend for one to four kilobases from the duplication boundary. Single-copy

258

sequences adjacent to a duplicated sequence are consequently more prone to mutation. Although other fungi detect and modify (by methylation) duplicated sequences, RIP is only known to occur in *Neurospora*.

Unlike most plants and animals, fungi carry out meiosis with the nuclear membrane remaining intact in prophase I. Meiosis I is a reductional division and meiosis II, an equational division. The meiotic II division is mitosis-like, sharing the same machinery with mitosis. At least in *Saccharomyces cerevisiae* following proper environmental triggers, the a1/α2 heterodimer activates IME1 (Inducer of Meiosis) to synthesize a transcriptional factor which, in turn, switches on various meiotic genes (Figure 8.3). Yeast is the only fungus for which the physiological, biochemical and molecular controls of meiosis are sufficiently well known for some understanding to emerge; it may or may not be representative.

In fungi, the leptotene of prophase I may be very brief or even absent. Also, in contrast to plant and animal systems, fungal karyotype analysis is usually done at the pachytene stage when chromosomes are paired and appear as long and thick threads rather than at metaphase I because fungal chromosomes are usually so small at the latter stage. Lu (1993) developed a method of spreading and staining chromosomes with silver nitrate which greatly improves conventional examination of synaptonemal complex and chromosomal rearrangements by both light and electron microscopy. Fluorescence *in situ* hybridization (FISH technique or 'chromosome painting') has been applied to chromosome spreads of *Saccharomyces cerevisiae*. With a probe derived from the homologous genomic library, the technique allows the study of specific individual chromosomes during meiosis. The karyotypes of most fungi can be resolved electrophoretically by pulsed field gel electrophoresis (PFGE) and Southern hybridization. Southern hybridization with homologous probes can establish the ploidy levels of different isolates and reveal gene amplification during differentiation. This technique can also reveal the loss of supernumerary chromosomes, which are usually less than one million base pairs in size and are dispensable, and generation of novel-sized chromosomes (chromosome-length polymorphisms; Zolan, 1995). Chromosome length polymorphisms are widespread in both sexual and asexual species, revealing a general genome plasticity (Figure 8.15). Tandem repeats, e.g. repeats of rRNA genes, frequently vary in length and dispensable chromosomes and dispensable chromosome regions also occur. Many karyotype changes are genetically neutral, others may be advantageous in allowing adaptation to new environments (Zolan, 1995).

Spreading it around

In yeasts producing naked asci the mother cell becomes the ascus. There is no specialized dispersal apparatus and the ascospores formed endogenously are simply released by rupture of the ascus (mother cell) wall. Yeasts are not alone in producing 'naked' asci, but the majority of ascomycetes produce their asci in multicellular fruiting bodies called ascomata (see next section). Basidiomycetous yeasts also lack a fruiting body and after meiosis produce basidiospores by budding (sporidia; see section: Mating type factors in *U. maydis* as an example) or show pleomorphism, forming pseudohyphae or true hyphae and bearing naked basidia, e.g. *Cryptococcus neoformans*.

Fungal multicellular structures – ascomata and basidiomata

Macroscopic fungal structures are formed by hyphal aggregation. When seen in microscope sections fungal tissues appear to be comprised of tightly packed cells and often resemble plant tissue (Figure 8.16). It is crucial to remember, though, that fungal tissue is made up of a community of hyphae. Primary septa in fungal hyphae usually have a pore which may be elaborated with the parenthesome apparatus in most basidiomycetes or be associated with Woronin bodies in ascomycetes; in either case the movement or migration of cytoplasmic components between neighbouring compartments is under effective control. So the hypha is separated into compartments which can exhibit contrasting patterns of differentiation. So fungi can quite reasonably be considered to be cellular organisms able to produce differentiated tissues comprised of cells which derive from an initial cell community which is induced to start

Figure 8.16 Scanning electron micrographs of hyphal tissues in the fruit body of *Volvariella bombycina*. (i) Highly branched hyphae in the gill at a young stage; (ii) the closely packed hymenium with hyphal tips differentiated into different cell types (a, basidium; b, cystidium).

Sexual development of higher fungi

(i)

(ii)

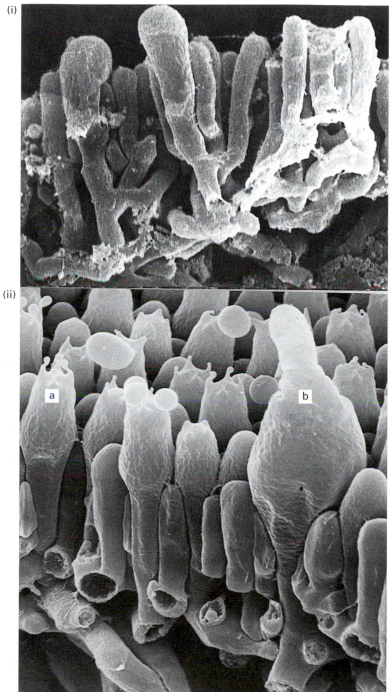

multiplication and differentiation. Nevertheless, there are fundamental differences between the fungal cell concept and one that might be applied to plants because the hypha grows only at its tip and cross walls form only at right angles to the long axis of the hypha. Consequently, fungal morphogenesis depends on hyphal branching. Proliferation in communities of fungal hyphae requires branching, and formation of a particular structure demands that the position of branch emergence and direction of growth are controlled (Moore, 1994, 1996, 1998; Watling and Moore, 1994; Chiu and Moore, 1996). Unfortunately, the cellular machinery involved in creating a new hyphal tip in a new lateral branch and determining its position, orientation and direction of outgrowth from the parent hypha is completely unknown at present.

Making ascomata

Ascogenous hyphae may branch repeatedly to form clusters of asci. Often the crook cell elongates into a new crozier rather than becoming an ascus mother cell immediately, and the tip and basal cells of the crozier fuse, producing another dikaryotic crozier beside the first (Figure 8.13). As a result, asci may arise singly, scattered within the fruiting body or may be in 'bunches' (called a fascicle). When asci form a distinct layer it is called a hymenium. Sterile hyphae interspersed between the asci are called paraphyses (singular paraphysis) and aid ascospore discharge.

Most ascomycetes form asci in a fruiting body called an ascoma (plural ascomata). Ascomata might be completely closed (a cleistothecium); more or less closed, but with an opening (ostiole) through which the ascospores escape when mature (a perithecium); completely open (an apothecium); or a cavity (locule) within a larger mass of tissue called a stroma (the whole structure is called an ascostroma or pseudothecium) (Figure 8.17). In some ascomycetes, an 'embryonic' ascoma develops before mating and the sexual organs are formed from hyphae within the developing ascoma (e.g. the protoperithecium of *Neurospora crassa*). In others, mating stimulates ascoma development by prompting mycelial hyphae around the ascogonium to grow and branch to generate the tissues of the ascoma.

In both cases, the ascoma develops from hyphae (most often maternal hyphae) that have not been involved in the mating process. Thus

262

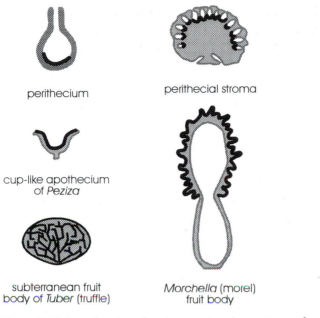

perithecium   perithecial stroma

cup-like apothecium
of *Peziza*

subterranean fruit   *Morchella* (morel)
body of *Tuber* (truffle)   fruit body

Figure 8.17 Line drawings showing construction patterns of some ascomata in the form of simplified diagrammatic sectional drawings (redrawn
after Burnett, 1968); in each case the hymenial tissue is represented by the
black line.

formation of the ascoma and coordination of its formation with sexual
reproduction must involve important signalling processes between the
different hyphal populations involved; nothing is known about the nature
of these signals (Novak Frazer, 1996).

Making basidiomata

The fruiting bodies of Basidiomycotina, the mushrooms, toadstools,
bracket fungi, puff-balls, stinkhorns, bird's nest fungi, etc., are all examples
of basidiomata (singular basidioma) which bear the sexually produced
basidiospores on basidia. Simplified diagrammatic drawings of some of the
different types of basidiomata are shown in Figure 8.18. The majority of
these fruit bodies, which vary in size from a few milligrams to tens of kilo-
grams fresh weight, are modifications of the basic theme of an umbrella-
shape, made up of a cap on top of a stem, so that the spore-release

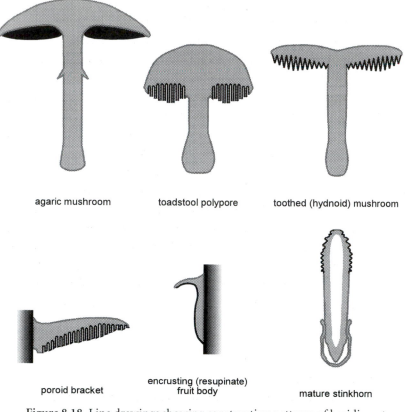

agaric mushroom   toadstool polypore   toothed (hydnoid) mushroom

poroid bracket   encrusting (resupinate) fruit body   mature stinkhorn

Figure 8.18 Line drawings showing construction patterns of basidiomata in the form of simplified diagrammatic sectional drawings with the hymenial tissue represented by a black line (redrawn after Burnett, 1968).

mechanism is protected from rain (Watling and Moore, 1994; Watling, 1996).

The tissue patterns in these structures are established very early in development. In *Coprinus cinereus*, fruit body initials only 800 μm tall, are clearly differentiated into cap and stem though this is only 1 per cent of the size of a mature fruit body (Moore, 1994, 1996). Examples like this bring to mind how early the basic body plan is established during development of an animal embryo and the concept of mushroom 'embryology' can be used to show that processes known to occur during animal embryo development have their analogues during basidioma development. These

264

include formation of inhomogeneous cell (hyphal) populations from homogeneous ones; regional specification of tissues (pattern formation) directed by organizers producing morphogens; specification and commitment of particular cells to particular fates; cell differentiation; and regulation of gene activity in ways specifically geared to morphogenesis. All of these processes are so well researched in animals (and, increasingly, in plants too); that other great Kingdom, Fungi, is a fertile area for research. So while the occurrence of these processes during fungal morphogenesis may be implied by, or can be inferred from, indirect observations, there are very few specific examples of research aimed directly at understanding how fungal multicellular structures develop (Chiu and Moore, 1996). Taxonomically, families are characterized by specific morphological features of fruit bodies, implying strict temporal and spatial control of morphogenesis. However, basidiome plasticity is also well known (IGURE 8.19) showing that the developmental programme has great flexibility to adapt. The key function of the multicellular structure, regardless of its form, is to ensure spore dispersal even under stress.

Conclusions

In the sections above we have given some indication of the diversity of sexual behaviour in fungi. At one extreme of the spectrum of behaviour are fungi which are completely asexual organisms (e.g. the deuteromycetes) but which nevertheless generate variation by modifying genetic expression or adapting mitotic processes to produce recombinants or segregants as asexual propagules. At the other extreme are the diverse forms of sexuality, ranging from bipolar (unifactorial) incompatibility systems to (bifactorial) incompatibility systems, with mating type switching possible in some organisms. The uniqueness of fungi is the possession of complex multiallelic system in the mating type gene and the redundancy of linked functional mating type genes in an incompatibility locus. The highest expression of this is in the basidiomycetes with mating type systems which can generate thousands of compatibility genotypes, producing multicellular fruiting structures capable of releasing hundreds of millions of basidiospores.

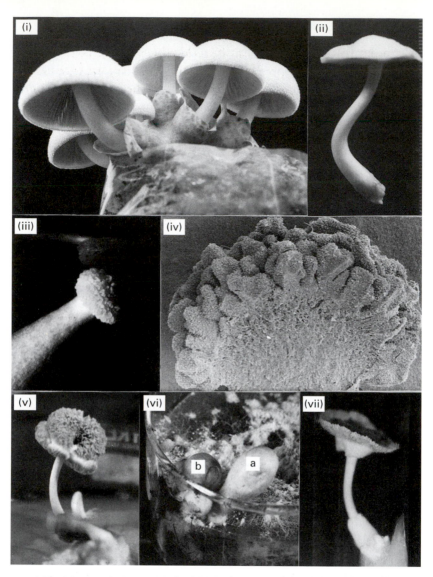

Figure 8.19 Spontaneous basidiome polymorphism in the basidomycete *Volvariella bombycina*. (i) Normal mushroom showing the genus-specific basal volva; (ii) spontaneous fruit body variant without volva; (iii) morchelloid fruit body variant; (iv) scanning electron micrograph of the longitudinal section of the morchelloid fruit body showing the convoluted hymenium without the formation of cap; (v) a fruit body variant with additional hymenium; (vi) a gasteroid fruit body variant (a) with gill maturation completed without rupture of the universal veil, next to a fruit body variant (b) which developed a hole in the universal veil early in development; (vii) a fruit body variant with inverted cap.

Sexual development of higher fungi

Most fungi have polyphasic life cycles, consisting of various phases such as haploid (N), dikaryotic (N + N) (or heterokaryotic with the two nuclear types in random ratio) and diploid (2N). In most species, reproduction is both sexual and asexual. Under a stable environment, haploidy and diploidy show equal selective advantage (Jenkins, 1993). The prolonged dikaryotic or heterokaryotic stage, however, may show hybrid vigour and can segregate into homokaryons if adverse selection pressure is imposed. Meanwhile, the cost (in metabolism and cellular resource) of producing a multicellular fruiting structure is in most cases greater than that of the simple conidiophore (for example) used for asexual reproduction. A sexual or asexual propagule, once separated from the parent, restarts the life developmental programme again. Thus, it seems that the best strategy to be adopted by fungi is to produce some offspring sexually for generating novel genotypes to adapt the unpredictable and fluctuating environment, and the rest asexually to rapidly colonize the favourable environment (once it is found) as well as to establish territorial control in a competitive world.

The mating type genes are involved in outcrossing and sexual development. The advantages of sexual reproduction include, obviously, the generation of genetic variation, but less obviously the process offers an escape from DNA parasites and a means to repair DNA damage. Sex may, therefore, be an important means to enhance the overall rate of adaptation (Hurst and Peck, 1996).

The beauty of a multicellular fruiting structure is not only the amplification in production of meiotic progeny widely differed in genetic make-up, but also the protection of the spore production against the fluctuating environment as well as the adaptation of the fungus towards its habitat. Regardless of its form, then, the multicellular fruiting structure, serves the same functions: to protect, to produce and to maximize the dispersal of the meiotic progeny for species propagation. In some cases, such as gasteromycetes bearing enclosed fruit bodies and hemiascomycetes producing naked asci, the spores are liberated by the 'mother structure' breaking open. In other cases the fruit body may be a complex mechanism which actively contributes to spore dispersal.

Rain splash is employed for spore dispersal in the bird's nest fungi in which the fruit body is adapted to redirect incoming raindrops so that they pick up the propagules and splash out, well away from the parent fruit body. Aquatic fungi, such as *Saprolegnia* produce flagellated zoospores which are able to swim through the water to find new territory. Insects are important vectors for some fungi. For instance, the stinkhorn autolyses after spore maturation to produce a sugary, sticky, stinking solution together with the spore mass to attract flies, which then take the sticky fungal spores with them when they leave.

Other fungal spores are ornamented making them rough or hooked which may be used for surface attachment or the outer layers of their walls may be wet and sticky to ensure adhesion to the surface. Active discharge of spores is widespread in fungi. Ascospores of many ascomycetes are forcibly discharged from the ascus by a pressurised extrusion process. In many basidiomycetes the basidiospores are actively discharged as ballistospores. The mechanism of this discharge has been a topic of hot debate for most of this century. It now appears that as a ballistospore matures, hygroscopic substances (including mannitol) are secreted at the base of the spore, causing condensation of water from the atmosphere to collect into a drop ('Buller's drop'). When this drop grows to a critical size it finally attempts to spread over the spore but the hydrophobicity of the wall amplifies surface tension and causes the spore to recoil away from the advancing liquid and lift off into the air, each successfully discharged spore representing a triumph of surface tension over gravity.

Acknowledgements

SWC thanks the British Council for an award under the HK/UK joint research scheme which enabled her to visit Manchester during the final stage of preparation of this chapter. A Research Grant Award from the Leverhulme Trust enabled DM to visit the Chinese University of Hong Kong for discussion of this chapter.

References and further reading

Bernstein, H., Byerly, H. C., Hopf, F. A. and Michod, R. E. (1985). Genetic damage, mutation and the evolution of sex. *Science* 229:1277–1281.

268

Sexual development of higher fungi

Birdsell, J. and Wills C. (1996). Significant competitive advantage conferred by meiosis and syngamy in the yeast *Saccharomyces cerevisiae*. *Proc. Natl. Acad. Sci. USA* 93:908–912.

Bourne, A. N., Chiu, S.-W. and Moore, D. (1996). Experimental approaches to the study of pattern formation in *Coprinus cinereus*. Pp. 126–155 in S.-W. Chiu and D. Moore, eds. *Patterns in Fungal Development*. Cambridge University Press, Cambridge, UK.

Burnett, J. H. (1968). *Fundamentals of Mycology*. Edward Arnold, London.

Carlile, M. J. (1987). Genetic exchange and gene flow: their promotion and prevention. Pp. 203–213 in A. D. M. Rayner *et al.*, eds. *Evolutionary Biology of the Fungi*. Cambridge University Press, Cambridge, UK.

Carlile, M. J. and Watkinson, S. C. (1994). *The Fungi*. Academic Press, London.

Casselton, L. A. and Kües, U. (1994). Mating-type genes in homobasidiomycetes. Pp. 307–321 in J. G. H. Wessels and F. Meinhardt, eds. *The Mycota, vol. I, Growth, Differentiation and Sexuality*. Springer-Verlag, Berlin.

Chiu, S.-W. (1996). Nuclear changes during fungal development. Pp. 105–125 in S.-W. Chiu & D. Moore, eds. *Patterns in Fungal Development*. Cambridge University Press: Cambridge, UK.

Chiu, S.-W. and Moore, D. (eds.) (1996). *Patterns in Fungal Development*. Cambridge University Press, Cambridge, UK.

Davis, R. H. (1995). Genetics of *Neurospora*. Pp. 3–18 in U. Kück, ed. *The Mycota, vol II, Genetics and Biotechnology*. Springer-Verlag, Berlin.

Dawkins, R. (1976). *The Selfish Gene*. Oxford University Press, Oxford.

Elliott, C. G. (1994). *Reproduction in Fungi. Genetical and Physiological Aspects*. Chapman & Hall, London.

Esposito, R. E. and Klapholz, S. (1981). Meiosis and ascospore development. Pp. 211–287 in J. N. Strathern *et al.*, eds. *The Molecular Biology of the Yeast Saccharomyces cerevisiae, vol. 1, Life Cycle and Inheritance*. Cold Spring Harbor Laboratory Press, New York.

Hawley, R. S. and Arbel, T. (1993). Yeast genetics and the fall of the classical view of meiosis. *Cell* 72:301–303.

Hurst, L. D. and Peck, J. R. (1996). Recent advances in understanding of the evolution and maintenance of sex. *Trends Ecol. Evol.* 11:79–82.

Jenkins, C. D. (1993). Selection and the evolution of genetic life cycles. *Genetics* 133:401–410.

Kämper, J., Bölker, M. and Kahmann, R. (1994). Mating-type genes in Heterobasidiomycetes. Pp. 323–332 in J. G. Wessels and F. Meindhardt, eds. *The Mycota, vol. I, Growth, Differentiation and Sexuality*. Springer-Verlag, Berlin.

269

Lu, B. C. (1993). Spreading the synaptonemal complex of *Neurospora crassa*. *Chromosoma* 102:464–472.

Maynard Smith, J. (1978). *The Evolution of Sex*. Cambridge University Press, Cambridge, UK.

Metzenberg, R. L. (1990). The role of similarity and difference in fungal mating. *Genetics* 125:457–462.

Metzenberg, R. L. and Glass, N. L. (1990). Mating type and mating strategies in *Neurospora*. *BioEssays* 12:53–60.

Metzenberg, R. L. and Randall, T. A. (1995). Mating type in *Neurospora* and closely related ascomycetes: some current problems. *Can. J. Bot.* 73(Suppl. 1):S251–S257.

Moens, P. B. (1994). Molecular perspectives of chromosome pairing at meiosis. *BioEssays* 16:101–106.

Moore, D. (1994). Tissue formation. Pp. 423–465 in N. A. R. Gow & G. M. Gadd eds. *The Growing Fungus*. Chapman & Hall, London.

Moore, D. (1996). Inside the developing mushroom: cells, tissues and tissue patterns. Pp. 1–36 in S.-W. Chiu and D. Moore, eds. *Patterns in Fungal Development*. Cambridge University Press, Cambridge, UK.

Moore, D. (1998). *Fungal Morphogenesis*. Cambridge University Press, Cambridge, UK.

Novak Frazer, L. (1996). Control of growth and patterning in the fungal fruiting structure. A case for the involvement of hormones. Pp. 156–181 in S.-W. Chiu and D. Moore, eds. *Patterns in Fungal Development*. Cambridge University Press, Cambridge, UK.

Pukkila, P. J. (1994). Meiosis in mycelial fungi. Pp. 267–281 in J. G. H. Wessels and F. Meinhardt, eds. *The Mycota, vol. I, Growth, Differentiation and Sexuality*. Springer-Verlag, Berlin.

Read, N. D. and Beckett, A. (1996). Centenary review. Ascus and ascospore morphogenesis. *Mycol. Res.* 100:1281–1314.

Schmidt, H. and Gutz, H. (1994). The mating type switch in yeasts. Pp. 283–294 in J. G. H. Wessels and F. Meinhardt, eds. *The Mycota, vol. I, Growth, Differentiation and Sexuality*. Springer-Verlag, Berlin.

Watling, R. (1996). Patterns in fungal development – fruiting patterns in nature. Pp. 182–222 in S.-W. Chiu and D. Moore, eds. *Patterns in Fungal Development*. Cambridge University Press, Cambridge, UK.

Watling, R. and Moore, D. (1994). Moulding moulds into mushrooms: shape and form in the higher fungi. Pp. 270–290 in D. S. Ingram and A. Hudson, eds. *Shape and Form in Plants and Fungi*. Academic Press, London.

Webster, J. (1980). *Introduction to Fungi*, Second edition. Cambridge University Press, Cambridge, UK.

Wendland, J., Vaillancourt, L., Hegener, J. *et al.* (1995). The mating-type locus Bα1 of *Schizopyllum commune* contains a pheromone receptor gene and putative pheromone genes. *EMBO J.* 14:5271–5278.

Westergaard, M. and von Wettstein, D. (1970). Studies on the mechanism of crossing over. IV. The molecular organization of the synaptinemal complex in *Neottiella* (Cooke) Saccardo (Ascomycetes). *Comtes Rendus des Travaux du Laboratoire Carlsberg* 37:239–268.

Zolan, M. E. (1995). Chromosome-length polymorphism in fungi. *Microbiol. Rev.* 59:686–698.

9 Lignocellulose breakdown and utilization by fungi

M. PENTTILÄ AND M. SALOHEIMO

Introduction

Fungi are the main decomposers of dead plant material and many fungal species are known as plant pathogens. Saprophytic fungi are able to get energy and nutrients from a large variety of organic compounds and therefore they are found in practically all terrestrial habitats in which there is life. Fungi can colonize new locations efficiently by spreading through their spores. A saprophytic fungus is always competing for nutrients against bacteria and other fungal species in its habitat. In many cases fungi benefit from their ability to degrade complex plant polymers more completely than bacteria and they also compete by antagonism against other saprophytic microbes. Yeasts are not generally able to decompose plant cell walls with polymeric components and thus saprophytic fungi are mainly filamentous. Saprophytic fungal species can be found in all fungal classes.

The saprophytic fungi are divided into three groups according to their mode of action on plant material. Soft-rot fungi cause softening of wood, degrading preferably plant polysaccharides but are often capable of slow lignin degradation as well. These species mostly belong to *Ascomycetes*. Brown-rot fungi degrade plant polysaccharides and leave behind a brown, modified lignin residue. White-rot fungi invade the lumens of wood cells and cause progressive thinning of the cell wall. They can degrade both plant polysaccharides and lignin efficiently. The members of the two latter groups mostly belong to *Basidiomycetes*.

Plant polymer structures

A major part of organic carbon on earth is bound to plant material. Cellulose is the most abundant carbon compound in plant material

272

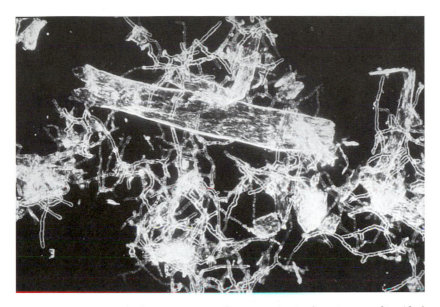

Figure 9.1 *Trichoderma reesei* mycelium growing in the presence of purified cellulose particles.

constituting 40–45 per cent, followed by hemicellulose (20–30 per cent) and lignin (15–30 per cent). Starch is an energy storage compound found only in certain parts of plants. Cellulose, the various hemicelluloses, starch and lignin are all complex polymeric structures. Cellulose, hemicellulose and starch are polysaccharides consisting of tens to thousands of sugar units linked together to form long and often branched chains. Lignin is a complex polymer formed of aromatic subunits linked by various linkage types. Since the plant materials consist primarily of sugars, they provide an excellent carbon source for microbial growth.

Cellulose can be considered as the chemically most simple of these plant polysaccharides because it consists in its pure form of only glucose units linked together by β-1,4-linkages (Figure 9.2). The long cellulose chains are then packed together to form sheets, which in turn form cellulose microfibrils. In its native form cellulose is largely crystalline and highly insoluble in water. β-Glucans on the other hand, have both β-1,4- and β-1,3-linked glucose units in the polymer, which results in kinks in the chain and give solubility to the polymer. Starch is yet another type of

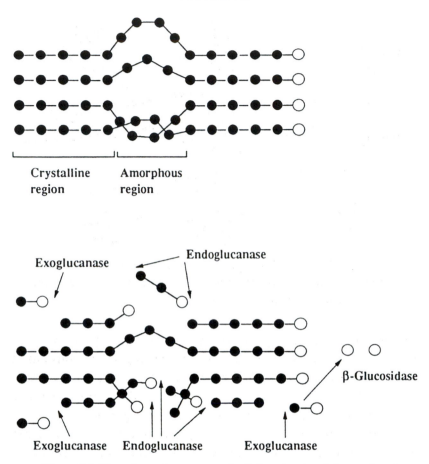

Figure 9.2 The schematic structure of cellulose (top) and the modes of action of different cellulase types (bottom). Cellulose is a linear polymer of glucose units (spheres, reducing chain ends are shown in white) linked by β-1,4-bonds.

glucose polymer but the linkages are now α-1,4-linkages with occasional branches formed by α-1,6-linkages (Figure 9.3). Hemicelluloses are highly variable in their structures depending on the plant species and tissue. The backbone chain can consist of, for instance, mannose, xylose, arabinose or galactose sugars, or their mixtures, or their mixtures with glucose (Figure 9.4). Furthermore, the main chain is substituted with sugar side chains of variable length which causes the overall structure of

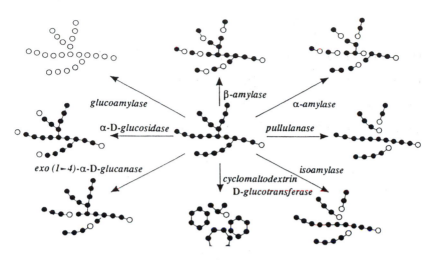

Figure 9.3 Starch consists of glucose units linked with α-1,4-bonds with some α-1,6-linkages forming branches to the main chain. The modes of action of starch-degrading enzymes are shown.

hemicelluloses to be amorphous and capable of forming viscous water solutions. Lignin is a polymer formed of phenylpropane subunits linked by ether and carbon–carbon bonds, of which the ether bonds are more common. The inter-subunit linkages can be formed in various ways and this results in a very complex molecular network (Figure 9.5).

Starch is the major energy storage compound in plants, found mainly in seeds, roots and tubers. Cellulose, hemicellulose and lignin together form the structures of plant cell walls. Cellulose microfibrils form the backbone of the structure and are cemented together by hemi-celluloses and lignin that are chemically bonded together. The different polymers are found in plants in variable proportions and form compact and rather resistant structures as shown by their use as building material (wood, hemicellulosic gums), textiles (cotton, cellulose) or raw materials for packaging (cellulose, starch).

Fungal extracellular glycosyl hydrolases

Living cells, including microbes, are able to take up from their environment only monomeric sugars or disaccharides (also mannotriose)

275

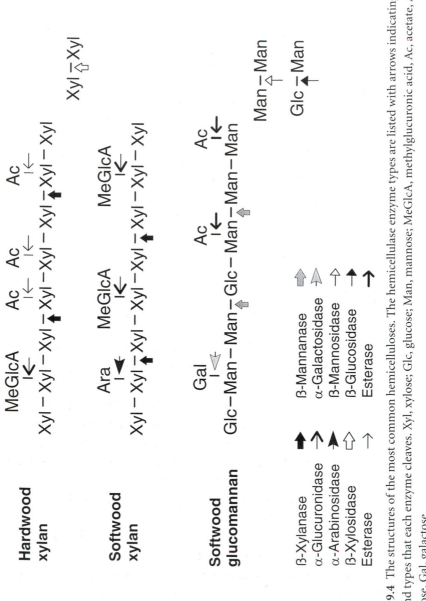

Figure 9.4 The structures of the most common hemicelluloses. The hemicellulase enzyme types are listed with arrows indicating the bond types that each enzyme cleaves. Xyl, xylose; Glc, glucose; Man, mannose; MeGlcA, methylglucuronic acid, Ac, acetate, Ara, arabinose, Gal, galactose.

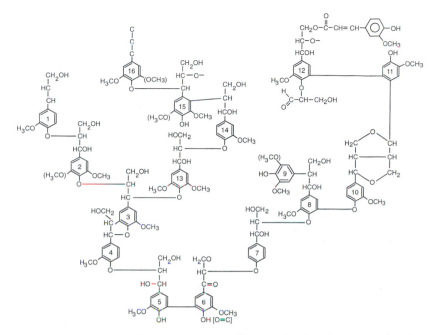

Figure 9.5 The schematic structure of lignin. The phenylpropene subunits are linked together with ether and carbon–carbon bonds.

through specific membrane-bound transporter proteins. Some fungi, such as the baker's yeast *Saccharomyces cerevisiae*, are unable to use most plant polysaccharides as a source for carbon and prefer habitats where free sugars are available such as in berries and fruits. Most fungi, however, produce enzymes which are secreted outside the cell and which hydrolyse plant polysaccharides to oligosaccharides and finally to di- and mono-meric sugars. These can be utilized by the fungi which can consequently live on even one of the polymers as a sole carbon source. Due to the great abundance of plant polysaccharides, fungi thus contribute significantly to the carbon turnover in nature.

Fungi grow on solid media and attach themselves to plant material. Secretion of extracellular enzymes is believed to occur at the apical or subapical regions of the growing hyphae and the enzymes are brought after synthesis and secretion directly into contact with the substrate. The many complex and variable structures of the polysaccharides, their compact structures and partial high crystallinity sets a challenge for the saprophytes. The polymeric structures are not usual substrates for enzymes. For efficient enzyme action the substrate needs to fit tightly to the active site of the enzyme. Due to the many different sugars, glycosyl linkages and tertiary structures found at different parts of the polysaccharide molecules and their higher ordered structures, it is expected that many different enzymes are needed for efficient and complete hydrolysis of plant polysaccharides.

Even for the hydrolysis of the chemically more simple polysaccharides, cellulose and starch, several different enzymes are produced by one organism. There are so-called endo-acting enzymes which can cut the linkages between sugar residues in the middle of the chain, and exoenzymes which release monomers, dimers or trimers at the end of the chains. A third class of enzymes cleaves the oligosaccharides released by the endo- and exoenzymes finally to monomeric sugars. In case of polymers such as hemicelluloses which are substituted, these side chain substituents are cleaved off from the main chain with different sets of enzymes. The different enzymes act synergistically. Cellulases, glucanases and hemicellulases are trivial names given for enzymes hydrolysing the respective substrates, and starch-degrading enzymes are called amylolytic enzymes. Some of the enzymes and their action towards the various substrates are shown in Figures 9.2, 9.3 and 9.4.

Structure and function of glycosyl hydrolases

The polysaccharide hydrolysing enzymes are usually 20–80 kDa in size and are glycosylated both with N- and O-linked sugars in the secretory pathway, which in fungi is expected to resemble in general features the secretory pathway of mammalian cells. Biochemical studies first indicated that the cellulases would consist of two functional domains, one responsible for the actual catalysis and the other contributing to binding of the enzyme to the substrate. This structure was further confirmed by cloning of the genes which enabled deduction of the amino acid sequences.

A large number of genes encoding cellulases, hemicellulases, glucanases and amylolytic enzymes as well as those hydrolysing various oligosaccharides have now been cloned from fungi. The best-studied

Figure 9.6 Schematic presentation of the published *Trichoderma reesei* cellulase sequences. With two exceptions, the enzymes have a cellulose-binding domain (shown in black) and a linker region (shaded) in their N- or C-termini. CBH, cellobiohydrolase; EG, endoglucanase; BGL, β-glucosidase.

organisms are *Trichoderma reesei, Aspergillus aculeatus, Phanerochaete chrysosporium* and *Agaricus bisporus* in respect of cellulases and hemicellulases, several *Aspergillus* species (*A. niger, A. awamori*) in respect of amylolytic enzymes, pectinases and hemicellulases such as arabinases, and *Humicola* species that produce enzymes with somewhat higher temperature and pH optima than the other species.

As an example, the schematic structures of the cellulase genes of *Trichoderma reesei* are presented in Figure 9.6. For the moment two exo-acting cellobiohydrolases and five endo-acting endoglucanases as well as a β-glucosidase hydrolysing oligosaccharides have been characterized from this fungus at the gene level. It can be seen that all except two of the enzymes share a common terminal domain of approximately 36 amino acids, the cellulose-binding domain (CBD). This region is found at either end of the cellulase enzyme. When the CBD is removed by proteases or through genetic engineering of the cellulase gene, the enzyme can still cleave glycosyl linkages from small oligosaccharides but binds to cellulose

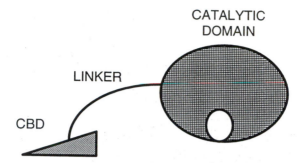

Figure 9.7 Presentation of the domain structure of a cellulase enzyme. The linker region takes an extended conformation to give spatial separation to the cellulose-binding domain (CBD) from the catalytic domain. A similar modular structure is found in some hemicellulases and starch-degrading enzymes.

much more weakly and consequently its action towards intact crystalline cellulose is impaired. The CBDs are linked to the catalytic core through a linker region which is rich in prolines, and serines and threonines which are O-glycosylated, keeping the linker region in an extended conformation. This spatial separation between the catalytic core and the CBD seems to be important for the most efficient hydrolysis of natural cellulose (Figure 9.7).

The three-dimensional structure of a CBD has been determined by NMR (nuclear magnetic resonance) spectroscopy which shows that it takes a compact structure through cysteine bridges and has one flat surface formed of aromatic amino acids (Figure 9.8). Site-directed mutagenesis of these amino acids has shown that they contribute to the binding of the domain to the substrate, and it is believed that these residues make a contact with the glucose units in cellulose. It is interesting that throughout the fungal kingdom (including the Basidiomycete mushroom *Agaricus bisporus*) the structure of the CBD is similar (except in rumen fungi). This is an indication for domain shuffling in evolution resulting in the same functional domain being attached to several different enzymes either to the carboxyl or the amino terminus of the enzymes. Cellulose-binding domains can also be found in bacterial enzymes but their structures are more variable and different from the fungal ones.

Interestingly, it has now become evident that some of the

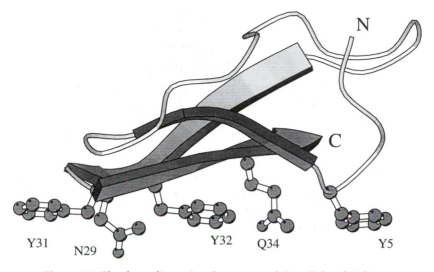

N

C

Y31   N29   Y32   Q34   Y5

Figure 9.8 The three-dimensional structure of the cellulose binding domain of a *Trichoderma reesei* cellulase (cellobiohydrolase I). The amino acid side chains forming a flat surface thought to bind on the cellulose surface are shown. The tyrosines found on this surface (Y5, Y31 and Y32) probably make contact with glucose units in the cellulose chain. (Figure: Markus Linder)

hemicellulases also contain a CBD which is attached to the catalytic core by a linker region although the enzymes have no activity towards cellulose. For instance a CBD can be found in the *Trichoderma* mannanase (a glucomannan-degrading endo-type enzyme) and acetyl xylan esterase (an enzyme removing acetyl substitutions from xylan), and in *Humicola* xylanases (a xylan-degrading endoenzyme). On the other hand, the characterized *Trichoderma* xylanases do not have a CBD and neither does the *Humicola* mannanase. The importance of the CBD for enzyme action is yet to be resolved and especially their role for the fungus in natural habitats. Since hemicelluloses are often heavily substituted with side chains it is not conceivable that the CBDs would have a significant role in binding of the enzymes to these substrates. However, cellulose and hemicelluloses are found in nature in close connection to each other and it could be beneficial if the hemicellulases are brought near to the actual substrate through the capability to bind cellulose.

The fungal amylolytic enzymes, glucoamylases and some other enzyme types have a similar domain structure in having a substrate-binding domain attached to the catalytic core with a linker region. However, this starch-binding domain (SBD) is about 100 amino acids long and is completely different in amino acid sequence from the CBDs. The three-dimensional structure of the *Aspergillus* glucoamylase SBD has been determined by NMR spectroscopy. It seems that as in the case of CBDs of cellulases, aromatic amino acids are important for the starch binding of the SBD.

The carbohydrate-degrading enzymes catalyse hydrolytic reactions, i.e. a glycosidic bond is cleaved and a water molecule is added to the new chain ends. The bond cleavage occurs through acid catalysis which involves the acidic amino acids glutamate or aspartate that are found in the active sites of the enzymes. Despite a similar catalytic mechanism there is no need for overall amino acid similarity between the catalytic cores of the different enzymes, not even in the case of, for instance, cellulases that all share a chemically similar substrate. For instance the three-dimensional (3D) structure of the catalytic core of the *Trichoderma* cellobiohydrolase I reveals a β-sandwich structure whereas cellobiohydrolase II has a α/β-barrel structure. On the other hand some enzymes with similar function seem to have evolved from a common ancestor by gene duplication as, for example, some fungal xylanases. Even though the enzymes produced by one organism might not show any significant sequence similarity to each other, similar enzymes in terms of activity and amino acid similarity are often found in other species. Based on primary sequence analysis, and recently also on 3D crystal structures, the glycosyl hydrolases have been grouped into total of 60 families, fungal enzymes belonging currently to 28 of these.

In addition to the interesting multidomain structure of many of the fungal hydrolases some of them also have an unusual active site architecture. The crystal structure of the cellobiohydrolase II of *Trichoderma reesei* (Figure 9.9a) shows a tunnel-shaped active site that is so tight that it can incorporate only one cellulose chain. This probably explains why the enzyme cannot cleave linkages in the middle of the chain and a cellulose chain rather enters into the active site from its end. The result of the catalysis is mostly the release of cellobiose (two glucose units) from the

(a)

(b)

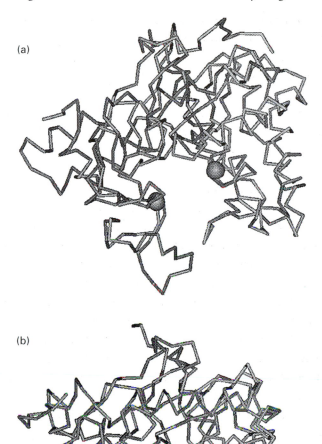

Figure 9.9 The three-dimensional structures of exoglucanase and endoglucanase catalytic domains. (a) *Trichoderma reesei* cellobiohydrolase II
(CBHII). (b) A bacterial endoglucanase with homology to CBHII. The
active site residues are shown with spheres. The exoglucanase active site is
formed inside a tunnel. The endoglucanase active site is more open,
enabling the enzyme to cleave a free cellulose chain at any position. (Figure:
Olle Teleman)

chain end. The active sites of endoglucanases, on the other hand, are more open, and it can be easily envisaged how the enzyme can grasp the cellulose chains from the middle (Figure 9.9b). The active sites of polymer-degrading enzymes have several subsites for binding of the sugar units in the chain, thus positioning the substrate tightly and correctly in respect to the amino acid residues participating in the catalysis (Figure 9.10). As in the case of the CBD binding to the glucose units in cellulose, aromatic amino acids in the subsites mediate binding of the sugars of the cellulose chain to the active site of the enzyme. Similar overall active site architecture with subsites can be found in the polymer-degrading amylases with the exception that tunnel-shaped active sites have not been found in any member of this group.

Lignin breakdown

The most efficient and best-known lignin degraders are the so-called white-rot fungi. Most wood-rotting fungi show the white-rot type of decay. These saphrophytes are able to degrade both lignin and the plant polysaccharides efficiently. In laboratory cultivations, however, none of the species studied is able to grow with polymeric lignin as the only carbon source. Thus it seems likely that in nature the white-rot fungi degrade lignin only when they need to gain access to the plant carbohydrates. The best-studied lignin-degrading fungi include *Phanerochaete chrysosporium*, *Trametes versicolor*, *Phlebia radiata* and *Pleurotus ostreatus*.

The complexity of the lignin molecule places stringent demands on the degradative enzyme systems. Lignin has no chains of repeating units and repeating linkages as the polysaccharides do but rather is a complicated network of subunits joined with various linkages. For its degradation, the white-rot fungi utilize a limited number of enzymes of low specificity, principally non-specific oxidative enzymes. The ligninolytic enzymes oxidize the phenylpropane subunits of lignin, which causes breaking of intersubunit linkages and turns the subunits into unstable cation radicals. These radicals can either form new linkages with other lignin subunits, which results in polymerization, or turn into quinones or ring cleavage products. These compounds can probably be catabolized

284

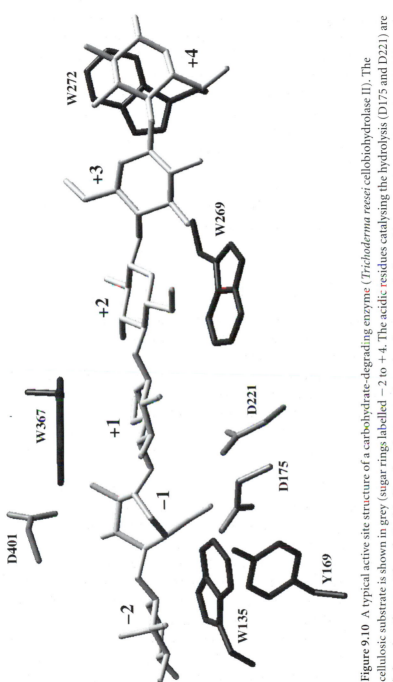

Figure 9.10 A typical active site structure of a carbohydrate-degrading enzyme (*Trichoderma reesei* cellobiohydrolase II). The cellulosic substrate is shown in grey (sugar rings labelled −2 to +4. The acidic residues catalysing the hydrolysis (D175 and D221) are below the substrate. The subsites binding the substrate to the active site are mostly formed by tryptophans and tyrosines (W135, Y169, W367, W269, W272). The glycosidic bond cleaved by the enzyme is between subsites −1 and +1. (Figure: Gerd Wohlfahrt)

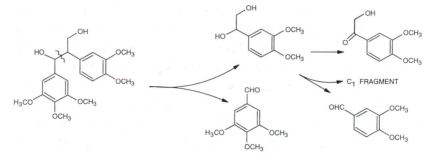

Figure 9.11 An example of a lignin model compound and a cleavage reaction catalysed by lignin peroxidase. This compound has the most abundant linkage type found in lignin (β-O-4 ether bond).

further or taken up by the fungus, which shifts the reaction balance towards depolymerization. It is difficult to examine reactions catalysed by ligninolytic enzymes using polymeric lignin as a substrate, and therefore monomeric or oligomeric lignin model compounds mimicking certain structures in the lignin molecule are used (an example is shown in Figure 9.11).

According to the electron acceptor used, the oxidative ligninolytic enzymes can be divided into peroxidases (H_2O_2 as the acceptor) and laccases (polyphenoloxidases, O_2 as the acceptor). Lignin peroxidases (LiP) react directly with aromatic substrates and manganese-dependent peroxidases (MnP) use Mn ions as mediators in their reaction with the actual substrate. These peroxidases are 40–50 kDal glycoproteins that have a haem (protoporphyrin IX) as a prosthetic group. During a reaction with the substrate, the enzyme is oxidized by two electrons, one of which is extracted from the haem iron atom and the other from the haem ring.

Both lignin peroxidases and Mn-dependent peroxidases are produced by the white-rot fungi as multiple isoenzymes. At least in *Phanerochaete chrysosporium* and *Trametes versicolor* the multiplicity of the enzymes is accounted for by multiple genes. Many of the lignin peroxidase genes are clustered in the same chromosomal region in *Phanerochaete*. The ligninolytic peroxidases need H_2O_2 for their activity. This is produced from molecular oxygen by oxidases secreted by the fungus, e.g. glyoxal oxidase.

The lignin peroxidases and Mn-dependent peroxidases show primary amino acid sequence conservation with each other and also with other peroxidases and other haem proteins. The three-dimensional structures of lignin peroxidase and Mn-dependent peroxidase from *Phanerochaete chrysosporium* have been resolved by X-ray crystallography. They both have a fold similar to the yeast cytochrome c peroxidase, where the haem group is embedded between two major protein domains formed mainly by α helices. The environment of the haem iron atom is very similar in the different peroxidases, with one histidine serving as a ligand for the iron and a second histidine at the opposite side of the haem plane that participates in the catalysis. The locations of the substrate-binding sites in LiP and MnP have been proposed. Expression of the Mn-dependent peroxidase as a recombinant product has been carried out in several systems, including *Phanerochaete* itself. For lignin peroxidases this has proved more difficult, and has complicated structure–function studies on these enzymes.

Laccases oxidize phenolic compounds with a broad substrate specificity. One molecule of O_2 is reduced to water when four substrate molecules are oxidized by one electron by the laccase. The role of laccases in lignin degradation has been disputed, since some white-rot fungal species do not secrete laccase during ligninolysis and are still capable of efficient lignin degradation. However, most species do secrete laccases that are able to catalyse degradative reactions with small lignin model compounds similarly to the ligninolytic peroxidases. Lignin degradation is not the only function of fungal laccases; some fungi produce laccases for the purpose of pigment formation, morphogenesis and plant pathogenesis.

The most-studied white-rot fungi in respect of laccase enzymes and their genes are the *Trametes* species, *Pleurotus ostreatus* and *Phlebia radiata*. The ligninolytic laccases are typically about 65 kDal glycoproteins that have four copper atoms in their active sites. Isolation of laccase genes from white-rot fungi has revealed sequence conservation of 60–90 per cent within the group and clearly lower similarity with laccases of other fungal groups. Laccases typically produce the distinct absorbtion spectrum of blue copper oxidases. From this enzyme family, the three-dimensional structure is known for the ascorbate oxidase of zucchini.

This oxidase is composed of three sequential domains with the β-barrel fold. The active site with four copper atoms is located in the space between two of these domains. The laccases have low overall sequence conservation with the ascorbate oxidase. The active site regions and most importantly the amino acids serving as ligands for the four copper atoms of the ascorbate oxidase are conserved (with one exception) in the laccases. Thus the laccases most probably share the basic architecture and a well-conserved active site structure with the ascorbate oxidase.

Regulation of enzyme production

Efficient hydrolysis of plant polysaccharides demands expression of a large number of gene products. Fungi often produce the different classes of enzymes simultaneously although their relative levels might vary depending on the growth conditions. Some industrial mutant strains secrete several tens of grams of extracellular hydrolases, which demonstrates the intrinsic high capacity of fungi to synthesize and secrete these enzymes. A single enzyme such as glucoamylase of *Aspergillus* or cellobiohydrolase I of *Trichoderma* can constitute a major part of all protein secreted. Production and secretion of the enzymes obviously demands a lot of energy and it is thus natural that expression of the genes is regulated.

The presence of the easily metabolizable carbon source glucose in the culture represses generally expression of many of the genes. The main regulatory protein responsible for glucose repression, CRE, has been characterized from *Aspergillus* and *Trichoderma* (CREA and CRE1, respectively). DNA binding of this protein to the hydrolase gene promoters is mediated by zinc fingers which are similar to those found in the glucose repressor protein Mig1 of *Saccharomyces cerevisiae*. Other amino acid similarities with the yeast protein are not obvious. Mutagenesis of the consensus-binding sites for this type of zinc finger proteins in the promoters of the hydrolase genes or analysis of *cre* mutants has revealed that the CRE protein is at least partly responsible for glucose repression of many of the genes. The characterized ones include those encoding xylanases and arabinosidases of *Aspergillus,* and cellulases, xylanases and many hemicellulose side chain-cleaving enzymes of *Trichoderma*. Consequently, regulation of the expression of the extracellular hydrolase genes representing one of the major mechanisms for fungi to obtain carbon for growth, seems to be under the same or similar carbon catabolite repression mechanisms that regulate utilization of mono- and disaccharides and other carbon sources in fungi.

Although glucose represses expression of many of the hydrolase genes, and some of them very tightly, it is, however, possible that some genes are not under this common regulation. They maybe also be repressed to varying extents, which might depend on the external concentration of glucose. For instance, *Aspergillus* glucoamylase is produced in the presence of glucose albeit less than in starch-containing cultures. The possible extent of constitutive expression and its role in natural conditions still remains to be investigated. Low basal levels of expression have been suggested to occur in order to explain how fungi can start growth in the presence of polymeric substrates and release soluble oligosaccharides from the polymers, which would further serve as inducers for higher levels of expression. On the other hand, significant expression levels of most of the cellulases and various hemicellulases occur in *Trichoderma* without any external addition of an inducing compound once glucose has been consumed from the culture. This type of release of repression after carbon source depletion could be sufficient to explain the survival of fungi in nature. Other possibilities are the presence of hydrolytic enzymes on the conidia, or simply that small inducing sugars might often be present in natural conditions.

Complex plant polysaccharides generally provoke the strongest expression of most of the enzyme genes. Cellulose, various hemicelluloses and starch result in a relative increase in expression of the respective enzymes involved in the hydrolysis of the particular polymer. It is generally believed that soluble small oligosaccharides or some monosaccharides (excluding glucose) released from the polymers serve as inducing compounds. Expression of cellulases can be observed in the presence of cellobiose, xylanases with xylobiose and xylose, α-galactosidases with galactose and amylolytic enzymes with maltose. Some of the extracellular hydrolases also have transglycosylation activity, and it has been proposed that transglycosylation products of the hydrolysis reactions could serve as inducing compounds. A good example is sophorose, a β-1,2-linked disaccharide of glucose, which can be formed from the β-1,4-linked cellulose oligosaccharides by some cellulases. This compound provokes very high levels of cellulase expression even in 1mM amounts in *Trichoderma*, and moderate expression also of many hemicellulase genes. If formed in nature, it is possible that this molecule could serve as an inducing compound due to its slow utilization by microbes and strong inducing power in small amounts. That true induction mechanisms are involved in hydrolase expression has been demonstrated for instance by growing the fungus on a 'neutral' carbon source such as glycerol, which neither represses nor induces expression as such, but if sophorose is added to glycerol cultures, strong expression of (hemi)cellulase genes can be obtained. Some evidence for promoter regions responsible for activation of the fungal α-amylase, glucoamylase, cellulase and xylanase genes has been presented, and it is expected that the first genes encoding some of the activator proteins will be presented in the near future.

Expression of plant material-hydrolysing enzymes, and the utilization of the sugars liberated, presents some interesting regulatory and metabolic consequences. These are, for instance, the fact that strong cellulase expression occurs in the presence of cellulose but the sugar released from this polymer, glucose, strongly represses expression of the same genes. Thus, it would be of interest to investigate the balances between glucose repression and de-repression. Another interesting point is that the pentose sugars released from hemicellulose, xylose and arabinose, are metabolized by fungi through the pentose phosphate pathway (PPP) whereas glucose, galactose and mannose enter the upper part of glycolysis. Thus it would be of interest to study the possible successive uptake of the various sugars, their inductive and repressive effects and the balances between glycolysis and PPP. In this respect, it is of interest that *Aspergillus* mutants, which are defective in arabinose utilization and accumulate the intracellular intermediate arabitol, produce elevated levels of secreted hemicellulases such as arabinosidases. Also arabitol addition to culture medium provokes gene expression. This has been observed also in *Trichoderma* in which arabitol induces expression of many hydrolase genes including that of cellulases.

The ligninolytic enzymes are produced by white-rot fungi in the secondary metabolic state, sometimes known as the idiophase. Their synthesis can be induced by nitrogen, carbon or sulfur deficiency of the culture. Of these, nitrogen limitation has the most pronounced effects. Apparently no specific inducer such as lignin or small aromatic compounds is generally needed for the induction. Ligninolytic enzymes are also produced in a high-glucose medium under nitrogen limitation. After isolation of the first genes encoding ligninolytic enzymes, it was shown that lignin peroxidases, Mn-dependent peroxidases and laccases are all regulated at the mRNA level. The onset of the ligninolytic growth stage is preceded by a peak of intracellular cAMP concentration in *Phanerochaete chrysosporium* and it has been postulated that cAMP would be involved in their regulation. Consensus-binding sites of the cAMP responsive regulatory protein AP-2 have been found in the promoter regions of the Mn-peroxidase genes and some of the lignin peroxidase genes but the significance of these sequences has not been shown. The Mn-peroxidase gene promoters also have similarity with the consensus eukaryotic heat shock element and metal response element, and it has been shown that the genes are inducible by heat shock and regulated by Mn concentration.

Concluding remarks and future prospects

The fungal systems for lignocellulose breakdown are under study by numerous research groups involved both in basic and applied research. The molecular mechanisms of the enzymes degrading cellulose, hemicelluloses, starch and lignin have been revealed to some extent by structure–function studies but many interesting questions still remain to be answered. For instance, what is the mechanism of hydrolysis of crystalline substrates, what are the structure–function requirements for the specificities towards various polymeric and oligomeric substrates, and to what extent are the enzymes capable of polymerization or transglycosylation? Analysis of mechanisms of gene regulation involved in lignocellulose utilization is an important and demanding task for the future. For example, elucidation of the induction mechanisms of the carbohydrate-degrading enzymes and finding the links between gene regulation and carbon or nitrogen starvation are of importance. A broader understanding of fungal physiology is needed when trying to understand how the regulation of the degradative systems is connected to growth control. Secretion of extracellular enzymes is very efficient in saprophytic fungi and is attracting increasing attention of researchers. Secretion is thought mainly to take place at the tip regions of the fungal mycelium and therefore only a small part of the fungal cell mass is involved in this efficient process. Correlations between secretion and the mycelial mode of growth and how changes in mycelial morphology affect secretion are points of future research.

The plant polymer-degrading enzyme systems have significant applications in many industrial processes including, for example, pulp and paper, food, animal feed and textile industries. Therefore, a major research effort has been invested in the production of individual enzymes or their optimized mixtures and finding new applications for them. Xylanases can be used in aiding pulp bleaching, amylolytic enzymes in preparation of various syrups, pectinases in extraction of fruit juices, cellulases and glucanases in predigestion of animal feed, and certain cellulases in textile polishing and instead of stones in stone-washing of jeans. The removal of lignin from the plant cell wall structure is of importance in the pulp and paper industry and many of the anticipated industrial applications of ligninolytic enzymes are in this sector. They include pulp bleaching, purification of waste effluents, detoxification of hazardous compounds and catalysis of organic synthesis reactions.

Bibliography

Key references

Gold, M. H. and Alic, M. (1993). Molecular biology of the lignin-degrading basidiomycete *Phanerochaete chrysosporium*. *Microbiol. Rew.* 57:605–622.

Henrissat, B. and Bairoch, A. (1996). Updating the sequence-based classification of glycosyl hydrolases. *Biochem. J.* 316:695–696.

Nevalainen, H. and Penttilä, M. (1995). Molecular biology of cellulolytic fungi. In: H. Kück (ed.) *The Mycota II. Genetics and Biotechnology*. Springer-Verlag, Berlin.

Reddy, C. A. (1993). An overview of the recent advances on the physiology and molecular biology of lignin peroxidases of *Phanerochaete chrysosporium. J. Biotechnol.* 30:91–107.

Svensson, B. (1994). Protein engineering in the α-amylase family: catalytic mechanism, substrate specificity, and stability. *Plant Mol. Biol.* 25:141–157.

Thurston, C. F. (1994). The structure and function of fungal laccases. *Microbiology* 140:19–26.

Tomme, P., Warren, R. A. J. and Gilkes, N. R. (1995). Cellulose hydrolysis by bacteria and fungi. *Adv. Microb. Physiol.* 37:1–81.

Warren, R. A. J. (1996). Microbial hydrolysis of polysaccharides. *Annu. Rev. Microbiol.* 50:183–212.

Visser, J. and Voragen, A. G. J. (eds.) (1996). *Pectins and pectinases.* Progress in Biotechnology 14. Elsevier, Amsterdam.

Further reading

Divne, C., Ståhlberg, J., Reinikainen, T., *et al.* (1994). The three-dimensional crystal structure of the catalytic core domain of cellobiohydrolase I from *Trichoderma reesei. Science* 265:524–528.

Ilmén, M., Saloheimo, A., Onnela, M.-L. and Penttilä, M. E. (1997). Regulation of cellulase gene expression in the filamentous fungus *Trichoderma reesei. Appl. Environ. Microbiol.* 63:1298–1306.

Poulos, T. L., Edwards, S. L., Wariishi, H. and Gold, M. H. (1993). Crystallographic refinement of lignin peroxidase at 2 Å. *J. Biol. Chem.* 268:4429–4440.

Rouvinen, J., Bergfors, T., Teeri, T., Knowles, J. K. C. and Jones, T. A. (1990). Three-dimensional structure of cellobiohydrolase II from *Trichoderma reesei. Science* 249:380–386.

Törrönen, A. and Rouvinen, J. (1995). Structural comparison of two major endo-1,4-xylanases from *Trichoderma reesei. Biochemistry* 34:847–856.

10 Plant disease caused by fungi: phytopathogenicity

P. BOWYER

The importance of fungal plant pathogens

Fungal plant diseases have profound effects on food production and society in general. Modern industrial society has evolved a number of methods for the control of plant disease comprising hygenic agricultural practice, breeding plants for disease resistance and the use of anti-fungal chemicals. Such measures are needed to produce crops of consistent quality and yield but contribute greatly to the cost of food production. Despite these advances, some 5–10 per cent of crop plants in Europe and North America are still lost to disease. Fungal races which can overcome resistance and which are insensitive to fungicides are constantly arising. Most farmers in underdeveloped countries are unable to afford either fungicides or the latest resistance-bred seeds and losses of 50 per cent are not unusual in these areas. Increasing human populations and concern over the excessive use of many fungicides has emphasized the need to understand fully the biochemical and genetic interactions between plant and pathogen in order to find new strategies to combat disease.

Fungal plant disease occurs because some fungi have evolved the ability to penetrate and colonize plant tissue. Fungal pathogens have evolved along with their host plants for many millions of years and have had a major impact on crop production since the first attempts at organized agriculture. In addition to pathogens, there are many fungal species that interact intimately with plants symbiotically. These include the very important mycorrhizal symbionts, which are believed to assist mineral nutrient assimilation in exchange for carbohydrates from the plants (Smith and Smith, 1997; Newsham, Fitter and Watkinson, 1995). These will not be considered further.

As well as possessing the normal metabolic systems found in other,

294

free-living, fungi, plant pathogens have developed specialized sets of enzymes to enable them to penetrate plant cell walls, overcome plant defences and complete their life cycle on the plant. In this chapter we will concentrate on these specialized systems and discuss our current understanding of the offensive and defensive strategies employed by fungi in disease.

History

Although the phenomena of plant disease had been observed by the Greek philosophers and natural scientists, the ability to observe the true nature of fungal plant disease began with Leeuwenhoek's invention of the microscope in 1675. The first experiments demonstrating that fungi were transmittable agents of plant disease were performed by Tillet in 1755, who showed that the black dust from infected wheat could cause similar disease on freshly sown plants. Widespread understanding of the causes of crop loss did not really arise until the disastrous epidemic of vine downy mildew (*Plasmopara viticola*) in Europe in the 1870s which spurred further investigation of the causal agents. Prior to this many devastating fungal epidemics, such as the Irish potato blight epidemic of the 1840s which resulted in more than a million deaths, were not recognised as such until some years after the event. With hindsight many of the famines in recorded history were probably caused by fungal disease. More dramatic effects have included the madness and death caused by consumption of rye bread made from plants infected with *Claviceps purpurea*. Entire villages were struck by 'St Anthony's fire', which also accounted for the deaths of a large proportion of the Russian army in 1772. An interesting example of the effects of a plant pathogen on crop utilization is that of the coffee rust outbreak in Sri Lanka in the 1870s causing the predominant coffee crop to be replaced by tea, leading to the English conversion from coffee to tea drinking.

Gastronomy

As well as their ability to change the drinking habits of a nation or to decimate armies, fungal plant pathogens are deliberately eaten for their own sake. The otherwise economically damaging disease of grapes, *Botrytis cinerea*, is allowed to infect vines in the Sauternes region of Bordeaux. The diseased grapes are partially digested and desiccated,

Table 10.1. *Estimated 1982 world crop losses*

	Production (megatonnes)	Percentage of crop lost to:			Total crop loss
		Diseases	Insects	Weeds	
Cereals	1695	9.2	13.9	11.4	34.5%
Potatoes	255	21.8	6.5	4.0	32.5%
Other root crops	556	16.7	13.6	12.7	43.0%
Sugarbeets	319	10.4	8.3	5.8	24.5%
Sugarcane	811	19.2	20.1	15.7	55.0%
Legumes	45	11.3	13.3	8.7	33.3%
Vegetables	368	10.1	8.7	8.9	27.7%
Fruits	302	12.6	7.8	3.0	23.4%
Coffee/cocoa/tea	8	17.7	12.1	13.2	42.4%
Oilseeds	240	9.8	10.5	10.4	30.7%
Fibre crops	40	11.0	12.9	6.9	30.8%
Tobacco	6	12.3	10.4	8.1	30.8%
Rubber	4	15.0	5.0	5.0	25.0%

resulting in wines of intense flavour and sweetness much sought after by connoisseurs world-wide. *Ustilago maydis*, the corn smut pathogen, causes infected maize to produce grossly distorted cobs. These cobs are regarded as a delicacy (huitlacoche) by American Indians and Mexican people and have a delightful taste and odour.

Indiscriminate harvesting of plants means that many fungal pathogens are present in our diet. In general there is little adverse effect from consumption of these organisms but there are certain exceptions, notably *Fusarium graminearum* which produces tricothecine toxins during infection of wheat.

Crop losses

Virtually every plant has fungal pathogens. Every crop plant has a number of fungal pathogens which cause varying amounts of damage from year to year; as a general rule of thumb, crop losses caused by fungal disease are 5–15 per cent in developed countries and may rise as high as 40–50 per cent in undeveloped countries. Table 10.1 shows crop losses caused by pests in a typical year.

Some definitions

As with many specialized branches of biology, the study of plant–pathogen interactions has its own specialized jargon and terminology.

Fungi are termed *pathogenic* when they are capable of causing disease. *Pathogenicity* therefore is the capability to cause disease and *aggression* (also referred to as *virulence*) is the degree or severity of disease caused by the fungus. Hence isolates of a single fungal species might be *pathogenic* (disease causing) or *non*-pathogenic (unable to cause disease) and the *pathogenic* isolates might be divided into highly *virulent* (causing severe disease) and *non-aggressive* or *avirulent* isolates (causing limited disease). The process of fungal disease and plant defence is often referred to as an *interaction*. When a fungus successfully infects and causes disease the *interaction* is *compatible* and when the fungus cannot successfully cause disease the interaction is referred to as *incompatible*.

Biotrophic fungi grow only on living plant tissue in contrast to the *necrotrophic* fungi which kill plant tissue in order to facilitate feeding or to pre-empt plant defence. *Biotrophs* are nevertheless capable of causing severe crop losses. When fungi can grow only on living tissue they are referred to as *obligate biotrophs*; such fungi cause many of the most severe crop losses and are proving very difficult to study as they cannot be cultured in the laboratory. Individual plant pathogens can usually infect only a limited number of plant species. The set of plant species susceptible to a particular fungus comprises the *host range* of that fungus.

Some *cultivars* of a plant might be susceptible to only a few *races* of a pathogen. Races causing disease are said to be *virulent*. Such race–cultivar specificity is often controlled by dominant plant *resistance* genes and dominant fungal *avirulence* genes, according to the *gene-for-gene* hypothesis.

How plant pathogen interactions are studied
Diagnosis ('what is it?')

A major problem in the study of disease in the field is that many fungal diseases look rather similar. Different infection processes and symptoms may also be caused by closely related fungi. Knowledge of which particular pathogen is causing disease is vital to the farmer or researcher, so that farmers can take appropriate action and so that researchers can identify their target organism. Great progress is being made in this area with the introduction of ribosomal gene sequencing which relies on similarities and differences in the ubiquitous ribosomal RNA genes to establish similarity or difference to known pathogens. Simpler techniques involve the use of highly specific monoclonal antibodies for identification or related groups of pathogens.

Physiology ('what does it do?')

The first step in the study of a plant pathogen is a microscopic examination of the physical progress of the disease. Fungal hyphae can be particularly difficult to see inside plant cells using conventional microscopy so the favoured method is to cut sections through the plant at regular intervals during infection and to submit these sections to transmission electron microscopy. Scanning electron microscopy can be used to detect events that occur on the plant surface and light microscopy can be used in conjunction with hyphae-specific stains such as trypan blue. Where antibodies are available against particular proteins, immunocytochemistry can be used to localize these proteins in the fungus during infection. In this way it is possible to determine how the fungus penetrates the plant surface and what 'life style' it has within the plant.

Biochemistry ('how does it do it?')

When a fungus penetrates a plant surface, it must produce enzymes to degrade plant cell walls. Plant surfaces and cell walls consist of layers of wax and polysaccharide fibrils. Most plant pathogenic fungi produce multiple forms and large quantities of enzymes which can break down specific polysaccharides and cutin. Though current research indicates that individual enzymes are not necessary for pathogenesis it seems likely that the concerted action of these enzymes is required for infection to occur. Another consideration is that the pathogen must adjust its metabolism to utilize the metabolites available in the plant. Certain amino acids may be abundant within infected tissue but others may need to be synthesized *de novo*.

Genetics

The application of genetics to the study of plant–pathogen interactions has been an important advance in our understanding of the field. Mutants are made in either plant or pathogen and are tested for alteration in disease progress. Depending on the manner in which the interaction has changed, conclusions can be drawn about what is actually happening during the interaction. For instance, fungal mutants which provoke an enhanced plant reaction must have lost either the ability to suppress plant reaction or have gained the ability to provoke one. Genetics has provided insight into host–pathogen interactions that could not be provided by biochemical or physiological studies as, for example, the demonstration of previously unknown systems (see Flor, 1955). Above all genetics has shown us the real extent of the communication, recognition and response events that occur during any infection.

Molecular biology

Although genetics has given much information on the types of systems which exist in an interaction, molecular biology provides the means for finding out exactly what the genetic mutations are. The genomes of plant or pathogen can be manipulated to analyse how they interact and how they might be exploited by man. Molecular biology provides a means of converting genetic systems in to precisely understood biochemical networks and is the most powerful tool currently available for analysis of disease. Gene cloning from plant or pathogen is well established. A particularly powerful tool in fungi is targeted gene disruption where the high integration frequencies of homologous DNA during transformation can be used to inactivate specific genes (Figure 10.1, Table 10.2). It is worth remembering however, that gene sequence or gene disruption commonly yield no information as to the identity or function of the gene, at which point the researcher must delve more deeply into genetics, biochemistry and physiology (see Oliver and Osbourn, 1995; Knogge, 1996).

A typical research project might take the form of making mutants in a pathogenic fungus, using genetics to find which are interesting (e.g. have altered virulence), cloning the genes which harbour the mutation and reconfirming the observed phenotype by gene disruption, sequencing the

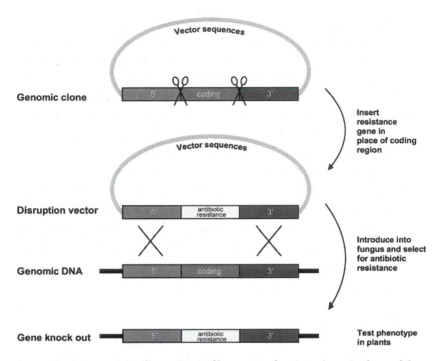

Figure 10.1 Gene disruption in filamentous fungi. A genomic clone of the gene of interest is prepared and the coding region of the gene replaced by a marker gene. The construct is transformed into the fungus where homologous recombination events insert the construct into the genome so replacing the native copy of the gene.

gene to find its function (comparison with existing sequence databases), using biochemistry to confirm that function and use of microscopy to investigate the precise nature of the phenotype.

The infection process

The infection process generally begins with the arrival of the fungus at the plant surface and continues with the formation of a series of specialized structures, each adapted for a particular purpose. It is very important to realize that environmental factors such as moisture or temperature play a crucial role in the establishment of infection though such factors are not discussed in detail here.

Getting to the plant

In general fungi are not able to propel themselves large distances and thus most rely on passive transport such as wind or water to carry them to a plant they can infect. Soil-living pathogens can generally grow a limited distance through soil to get to a plant. The most common inoculum in fungal disease is the spore. Spores are small, tough propagules containing nuclei, mitochondria and stored nutrients. Spores are produced in vast numbers at the end of the infection process and can then be carried from plant to plant by wind or rain dispersal. Spores can survive for several years in soil to form the initial inoculum for the season's disease. Fungi which do not readily sporulate can survive from year to year as hyphae on the remains of the previous years crop (e.g. stubble). Many fungi can also exist saprophytically in the absence of a plant on soil or rotting vegetation, simply waiting for a susceptible plant to come along or to be carried to plants in the mud, on tractor wheels, boots or animal feet.

Getting into the plant

Having arrived at the plant surface, a fungus then faces the formid-able task of getting into the nutrient-rich interior. The plant (leaf) surface is an inhospitable cuticular wax/lipid expanse with occasional openings (e.g. the stomata). Underlying the cuticle are the thick, physically tough walls of the epidermal cell layer. There are three ways in: penetration of the stomata, direct penetration of the cuticle and cell wall and penetra-tion of damaged plant tissue.

Fungi that penetrate through stomata are somehow able to recog-nize the presence of an opening in the plant surface and to grow through it; entry through the stomata bypasses the physical barrier of the cuticle allowing direct access to the interior of the plant. In some pathogens there seems to be some ability to grow actively towards the stomata.

Direct penetration of the cuticle is performed in a variety of ways. Fungi may penetrate either between epidermal cells or directly into them depending on the species. Generally a spherical structure know as the *appressorium* is produced at the site of penetration. The best understood example of appressorium formation is from the *Magnaporthe grisea* (rice blast)–rice interaction (Figure 10.2). The *M. grisea* spore germinates on the plant surface in a water droplet and the hypha grows until it recognizes a

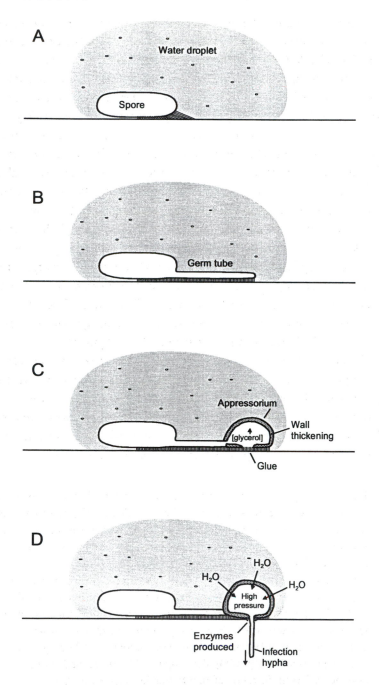

good penetration point. The specific means of recognition are not precisely known but the surface must be hydrophobic and above the cytoplasm of the epidermal cell. Certain long chain alcohols or lipids are also able to trigger appressorium formation. Recognition of hydrophobic surfaces is known to be controlled by a cyclic AMP-dependent protein kinase (Kronstad, 1997). Once the signal for appressorial development is given, the hyphal tip swells up and seals itself off from the hypha by deposition of a new fungal cell wall. The appressorium glues itself to the plant surface with secreted glycoproteins (10 times stronger bonding than superglue) and simultaneously strengthens its cell wall by the deposition of pigmented melanin compounds. Glycerol is synthesized in large quantities inside the appressorium until a 3–4 molar concentration is reached (de Jong *et al.*, 1997). As the appressorium forms inside a water droplet, osmotic pressure begins to build up inside; pressures of 8 Mpa (40 times typical car tyre pressure) are theoretically possible. At the base of the appressorium a hyphal peg forms which secretes cuticle and plant cell wall degrading enzymes. This infection hypha penetrates the plant cuticle and cell wall through a mixture of applied pressure and destruction of the physical barrier.

Certain pathogenic fungi secrete a range of enzymes capable of degrading plant cutin and cell walls in sufficient amounts so that they may penetrate the plant surface without appressorium formation. It is uncertain what the advantages if any of appressorium formation may be; perhaps it is a particularly energy-efficient method of penetration given the high metabolic cost of production of extracellular enzymes.

Figure 10.2 Appressorium formation in *Magnaporthe grisea* (rice blast)– rice interaction. A Spore lands on plant surface and adheres using proteinaceous glue. B The spore germinates to form a hypha, which can sense plant surface features, perhaps by means of secreted proteins or by the ability to sense the surface topology. C When the hypha senses that conditions are right for penetration, a quasi-spherical structure, the appressorium, is formed. The appressorium is strongly attached to the plant surface and has thickened walls, often containing melanin. D The appressorium seals itself off from the spore and germ tube. The high internal glycerol concentration creates an osmotic potential and water diffuses into the appressorium leading to high internal pressure. This, possibly also coupled with the production of lytic enzymes, drives the insertion of an infection hypha into the plant.

Many apparently less specialized pathogens are unable to penetrate intact plant barriers and rely on physical damage caused by other (pioneer) fungi, animal pests or harvest damage to effect entry.

Growth in the plant

Once inside the plant fungi grow in a variety of ways. Some penetrate cells directly and grow through them whereas others grow under the cuticle or only within the cell walls or intercellular spaces. The interior of the plant cell is nutrient rich and able to support vigorous pathogen growth but sugars may also be obtained from the breakdown of the plant cell wall or from intercellular fluid. Fungi feed on plant cells in one of two ways: by direct uptake of nutrients into invading hyphae or by formation of a specialized feeding structure known as the *haustorium* (Figure 10.3). Some fungi such as *Cladosporium fulvum*, the grey leaf mould of tomato, are even able to exist in the intercellular spaces of the plant without direct penetration into plant cells.

Necrotrophic pathogens typically ramify through plant tissue as hyphae which secrete a variety of enzymes to break down the cell contents into compounds usable by the fungus. This results in death of the plant cell and forces the fungus to penetrate further plant cell walls and invade new plant cells to continue feeding.

Biotrophic pathogens typically invade plant cells by penetration of the cell wall followed by extrusion into the cell of a haustorium. Although the haustorium penetrates the cell wall it does not breach the cell plasma membrane. It is presumed that the haustorium secretes factors into the plant cell to induce the plant to make nutrients available. This feeding mechanism may be the basis for the biotrophic life style in that it requires plant cells to be alive to provide the invading pathogen with nutrition.

Fungal hyphae spread through the plant forming an area of stressed, damaged or destroyed plant tissue known as a *lesion*. Lesions are the visible symptoms of plant disease and range from completely blackened, destroyed tissue to slight chlorosis. Lesions may extend beyond tissues that are directly colonized by the fungus due to secretion and diffusion of fungal toxins. Lesions are first seen as pinhead-sized areas of brown or lighter green tissue which spread according to the progress of the disease.

304

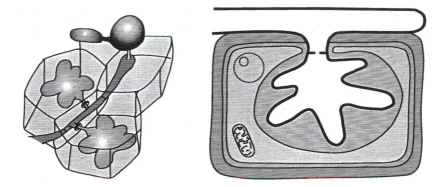

Figure 10.3 Generalized view of the morphology of haustoria. Many biotrophic pathogenic fungi feed on plant cells by the production of haustoria. These structures are formed inside plant cells, although the plant plasma membrane is never breached. The fungus is thought to feed on the plant cell through the medium of the extra-haustorial matrix, which is presumed to contain proteins that can direct the plant cell to export nutrients. This idea has proved difficult to test due to the difficulty of making viable mutants in obligate biotrophic pathogens.

After a period of growth within the plant, the fungus will either undergo sexual reproduction to form sexual spores (ascospores or basidia) or re-emerge from the plant for production of asexual spores (see Chapter 8), which are then dispersed to start new infections. Typically asexual spores are formed a few days post-infection whereas sexual spores are formed some weeks post-infection. The corn smut pathogen *Ustilago maydis* only causes disease after a haploid yeast-like phase mates to form a pathogenic diploid. In this organism one of the mating type genes, the b locus, controls pathogenicity. Some pathogens may go through many cycles of disease in a single growing season and since a single infectious spore may give rise to many millions of spores, these diseases can be devastating.

Barriers to infection

Given that the majority of fungi use closely related strategies to infect plants and that plants themselves are constructed of and contain similar structural and nutrient molecules it is perhaps surprising to discover that

305

Table 10.2. *Fungal genes that have been functionally tested for a role in pathogenicity*

Nature of process	Fungus	Molecular determinant	Reduction in pathogenicity?
Overcoming physical barriers	*Cochliobolus carbonum*	Endo-xylanase	No
		Endopolygalacturonase	No
		Exo -1,3 glucanase	No
		Polygalacturonase	No
	Penicillium olsonii	Cutinase	No
	Magnaporthe grisea	Cutinase	No
	Nectria hematococca	Cutinase	Yes
Detoxification of			
(a) pre-formed inhibitors	*Gaeumannomyces graminis*	Avenacinase	Yes
	Gleocercospora sorghi	Cyanide hydratase	No
(b) phytoalexins	*Nectria hematococca*	Pisatin demethylase	No?
Toxin production			
(a) non-specific	*Gibberella pulicaris*	Trichodiene synthase	Yes
(b) host-specific	*Cochliobolus carbonum*	HC toxin synthase	Yes
Avirulence	*Cladosporium fulvum*	*Avr4*	No
		Avr9	No?
	Rynchosporium secalis	*Nip1*	Yes
	M. grisea	*AVR2-YAMO*	No
		PWL1	No
		PWL2	No
Others	*Ustilago maydis*	Mating type	Yes
		Chitin synthase	No
	M. grisea	MPG1 hydrophobin	Yes
		CPK1 kinase	Yes
		MAP kinase	Yes

Figure 10.4 Composition of a 'typical' cell wall.

an individual pathogen can only infect a very limited set of plants. This limitation is so severe that in many cases an individual pathogen isolate can only infect certain members of a particular species of plant. The set of plant species that can be infected by a particular pathogen is called the *host range* of that pathogen. The factors that limit host range or that are responsible for the appearance of new pathogenic races are discussed below (Table 10.2).

Physical barriers

Plants have evolved a battery of pre-formed obstacles to fungal growth. The most obvious of these is the physical structure of the plant, consisting of a surface layer of wax, cutin and thick physically robust cell wall. Wax and cutin are formed from monomers and polymers of C16 or C18 fatty acids. The cell wall itself is more complex, being composed of layers of cellulose (polyglucose) hemicellulose (cellulose with side chains of arabinose or other sugars) and pectin (polygalacturan). Figure 10.4 shows the composition of a 'typical' plant cell wall. The composition of cellulose and pectin is far from homogenous with both being subject to many modifications such as inclusion of other sugars or methylation of the core sugars. In addition to the chemical modifications, the sugar chains themselves are branched and woven into a complex high-tensile fabric, making physical disruption of plant cells a difficult task. Thus to break through a plant cell wall fungi must utilize not one but a large array of cellulose- and pectin-degrading enzymes each suited to the particular modification of the sugar chain that is present. Lack of a key enzyme might prevent fungi gaining access to the cell or at least slow the fungus sufficiently so that plant defences can disable it. The importance of such mechanisms in determining host range remains unknown; targeted gene disruption of individual fungal cellulase or pectinase genes seems to have little effect on overall virulence (Table 10.2). Disruption of sets of genes (e.g. all the cellulases) is currently being attempted in several laboratories and such 'group-mutation' approaches may well yield clearer results.

Chemical barriers

Plants are known to produce a wide array of compounds that are toxic to fungi. These include saponins (glycosylated steroids), cyanogenic glycosides, and phenolic compounds. These compounds are often found in epidermal cells and at plant surface in sufficient quantities to inhibit fungal growth and it is thought that the differing toxicity of these compounds to various fungal species or the presence or lack of a toxin may determine host range in many cases. Toxins are classed as either *phyto-anticipins* (*pre-formed inhibitors*), which are present in the plant at all time regardless of the presence of a pathogen, or *phytoalexins*, which are formed *de novo* in response to pathogen attack.

One of the best-studied interactions where pre-formed plant toxins are known to determine host range is that between the root infecting pathogen *Gaeumannomyces graminis* and its cereal hosts (Osbourn, 1996). *G. graminis* is divided into oat- and wheat-infecting varieties with the oat variety being able to infect both wheat and oats. Oat roots contain a saponin toxin known as avenacin. The oat-infecting variety is resistant to avenacin as it possesses an enzyme avenacinase which can detoxify avenacin by removal of one or more sugar molecules (Figure 10.5). Mutants of the oat-attacking variety made by targeted gene disruption of the avenacinase gene can no longer infect oats but are still fully pathogenic to wheat showing that the host range of this pathogen is determined by the ability to detoxify avenacin.

The target of plant toxins must also be present in order for them to

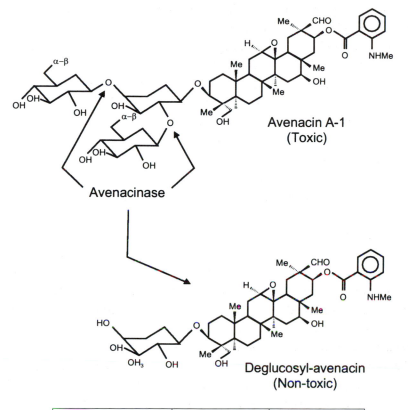

	Oat		Wheat	
Plant avenacin	+		−	
Fungal avenacinase	+	−	+	−
Pathogenicity	yes	no	yes	yes

Figure 10.5 Relationship between production of the saponin-detoxifying enzyme avenacinase and host range in the root-infecting pathogen *Gaeumannomyces graminis*. The fungal enzyme, avenacinase, can remove the two terminal glucose molecules from avenacin A1 resulting in the less toxic deglucosyl avenacin. The table shows pathogenicity of two different isolates of *G. graminis* to oats and wheat. Isolate 1 produces avenacinase and can infect oats which contain avenacin. Isolate 2 cannot produce avenacinase and is unable to infect oats which contain avenacin. This isolate is fully pathogenic to wheat and oat species that do not contain avenacin, showing that avenacinase is a host-range determinant rather than a pathogenicity determinant.

309

be effective. For example saponin toxins form complexes with fungal membrane sterols leading to ruptured membranes. In cases where the fungus lacks membrane sterols or has sufficiently different membrane sterols, there is no inhibition of fungal growth by saponins.

Recognition events

Pathogenic fungi do not constantly produce the enzymes and factors needed for pathogenicity. Rather they produce them in response to recognition of the fact that they are near or next to a plant. It is possible that the fungus can only recognize molecules from particular plants and will thus infect only those plants.

Plant responses to infection

Thus far we have considered plants to be merely the passive partner in the pathogenic process. In reality plants possess a range of responses to pathogen attack equal in range and complexity to the more familiar mammalian immune system. Plants are, however, generally more tolerant of losing parts of their anatomy in order to seal off an invading pathogen.

General responses

Plants express a range of responses to any type of wounding or damage including that caused by pathogens (Gebhardt, 1997). These include production of enzymes, and an 'oxidative burst' consisting of rapid evolution of superoxide radicals and production of salicylic acid. The enzymes are usually chitinases and glucanases; enzymes which might be involved in degrading the fungal cell wall. The oxidative burst produces H_2O_2, which is rapidly converted to the highly reactive and biologically damaging O_2^- radical. There is evidence that the oxidative burst is not just used to damage pathogens but may also have a role in triggering off later responses. Infection of one part of the plant causes reaction in the rest of the plant to the extent that later inoculation of the plant at different sites can be unsuccessful: the plant is 'ready for' the pathogen and can successfully ward off attack. This phenomenon is known as systemic acquired resistance (SAR) and implies the existence of a messenger molecule that can alert the plant to pathogen attack (Ryals *et al.*, 1996).

310

Salicylic acid (aspirin) is thought to be involved in this message though recent evidence suggests that it is not actually the messenger itself (Draper, 1997).

Plants often form dead layers of cells around the site of pathogen penetration. These cells are thought to block spread of the pathogen and to limit diffusion of any toxic compounds produced by the fungus. In certain cases the plant may kill a whole area of itself in order to limit pathogen spread. This *hypersensitive reaction* occurs by programmed cell death in tissues which receive a particular chemical signal from either plant or pathogen (Dangl *et al.*, 1996). Plants are capable of forming physical barriers to fungal growth inside individual cells, usually in the form of lignin deposition around invading hyphae. This is often a successful means of halting hyphal growth.

Fungal avirulence/plant resistance

The most striking example of a plant response to a perceived fungal signal is to be found in the *gene-for-gene* interaction first described by Flor in 1940. Flor noticed that certain isolates of flax rust caused the flax plant to produce areas of necrosis: the *hypersensitive response.* In a series of genetic experiments Flor showed that a single gene in the pathogen was responsible for eliciting this response and that a single gene in the plant was required for this response to occur. Absence of either the plant or the fungal gene meant that no *hypersensitive response* occurred and that the infection was successful (*compatible*). Flor proposed that the plant gene (the *resistance* gene) produced a product which recognized the fungal gene product (the *avirulence* gene). Many gene-for-gene reactions were discovered by genetic analysis in subsequent years until the application of molecular genetics began to unravel the nature of this recognition system. The plant–pathogen interaction in which much of the pioneering work was done is the *Cladosporium fulvum* – tomato system (De Wit, 1997; Gebhardt, 1997; Hammond-Kosack and Jones, 1996, Innes, 1995). Genetic analysis has shown that *C. fulvum* expresses a range of avirulence (*avr*) genes (*avr1–9*) recognized by corresponding tomato resistance genes (*Cf1–9*). Thus fungi carrying the *avr4* gene cause the hypersensitive reaction on plants which carry the *Cf4* gene but not the *Cf9* gene. Conversely fungi carrying the *avr9* gene elicit no response from plants

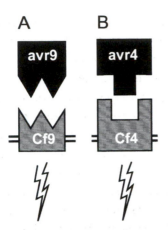
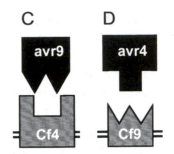

A   B   C   D

Figure 10.6 The gene-for-gene system: a model for specificity in plant–pathogen recognition. Fungi secrete a variety of factors during infection, some of which can be recognized by the plant. Such factors are termed avirulence proteins (avr). Plants possess many specific receptors which can recognise specific avr factors. These plant factors trigger a variety of responses in the plant and are termed resistance (R) factors. This figure shows examples from the *Cladosporium fulvum*–tomato interaction. Here avirulence gene products are shown interacting with plant R receptor factors (in this case termed Cf for resistance to *C. fulvum*). In A and B, avr and Cf proteins can interact and a defence response such as hypersensitivity is triggered. In C and D no interaction occurs although both avr and R factors are present and no plant defence is triggered.

carrying the *Cf4* gene (Figure 10.6). *C. fulvum* races which are pathogenic on *Cf9*-containing plants have been shown to have lost the entire *avr9* gene. The cost to the organism of such drastic evasive action is not clear but *C. fulvum* races lacking *avr9* are less virulent in the field.

Confusingly it has been proposed that *avr* genes may also be pathogenicity or virulence genes. To be recognized by a plant receptor gene, the fungal proteins must be extracellular and hence might be playing some role in feeding or breaking down plant defences; in other words such proteins might be pathogenicity factors. To date only the *NIP1* gene of *Rynchosporium secalis* has been shown to have the role of *virulence* factor (a toxin) and *avr* gene (Knogge, 1996). Interestingly the toxin activity and the avirulence activity reside on different parts of the NIP1 protein suggesting the presence in the plant of two separate NIP1 receptors.

Plant resistance genes have recently been cloned and shown to encode leucine-rich repeat type proteins, generally known to be involved in protein–protein interactions and receptor signal transduction. The obvious (but as yet unproven) hypothesis is that the avirulence protein binds to the resistance gene causing conformational change and triggering a signalling cascade resulting in the plant defence response. The functions of the other known avirulence genes have not been determined but there is clear pressure for the fungus to lose or mutate these genes to avoid recognition by the plant. This suggests that their continued maintenance may reflect an important role in pathogenesis. Targeted disruption of these genes causes no observable effects on *ex planta* growth or pathogenicity to non-resistant plants but these experiments have been carried out under laboratory conditions; the possibility remains that they may have important functions in the 'real' world.

Fungal toxins

Plant pathogenic fungi produce a range of toxins that can interfere with plant metabolism with results ranging from death of the whole plant to subtle effects on gene expression (Walton, 1996). Fungal toxins must interact with a plant enzyme or component and if the particular toxin target is missing or altered the toxin will have no effect, hence fungal toxins and their targets may be an important determinant of pathogen host range. Here we will consider three chemically diverse toxins produced by *Cochliobolus* species responsible for three new outbreaks of cereal disease caused by the ability of the particular *Cochliobolus* species to infect a new host.

T-toxin

Until 1969 only one pathogenic race (race O) of the maize pathogen *Cochliobolus heterostrophus* was known. Race O is of low virulence and does not pose a serious threat to maize farmers. In 1970 a devastating epidemic of *C. heterostrophus* swept across North America causing

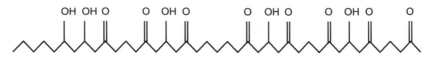

T-Toxin

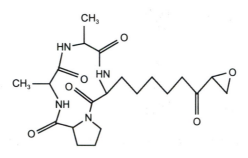

HC-Toxin

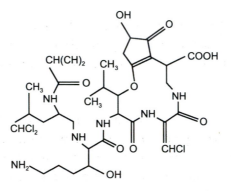

Victorin-C

Figure 10.7 Some host-specific toxins produced by fungi.

huge crop loss and bankrupting many farmers. This new form of the disease (known as race T) was distinguished by its extreme virulence to maize carrying Texas (T) male sterile cytoplasm, which happened to be the predominant maize variety at that time. Genetic analysis showed that race T pathogen had arisen from race O by mutation at a single genetic locus, Tox1. This locus is responsible for production of a polyketide toxin, T-toxin (Figure 10.7) which interacts with the T-cms protein in the maize mitochondrion causing cell death. Plants which have no T-cms

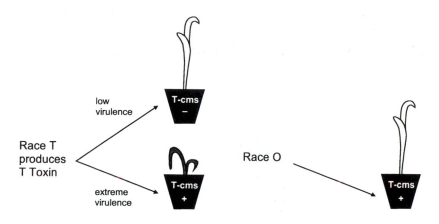

Figure 10.8 Specificity of the *Cochliobolus*–maize interaction produced by T-toxin.

gene are unaffected by T-toxin (Figure 10.8). Thus the production of T-toxin is a determinant of host range in this fungus in that it allows the fungus to successfully attack a particular variety of maize.

HC-toxin

Cochliobolus carbonum is responsible for the leaf spot disease of maize. This pathogen produces a cyclic peptide toxin, HC-toxin (Figure 10.7), which is able to suppress maize gene expression by inhibition of histone deacetylase, an enzyme involved in chromatin decondensation (Figure 10.9). This toxin does not directly cause cell death as does T-toxin but is rather thought to suppress maize defence gene activation to facilitate fungal spread through the plant. Maize plants that carry one or more copies of the Hm gene are resistant to *C. carbonum*. The Hm gene is now known to encode an enzyme, HC-toxin reductase, which can reduce a peptide carbonyl group of HC-toxin rendering it inactive. Plants which express this gene can therefore deactivate HC-toxin before it can stop defence gene activation and hence are resistant to this disease (Figure 10.9).

Victorin-C

In the 1940s Victoria blight devastated oat production in North America. This epidemic was caused by a new pathogen, *C. victoriae*,

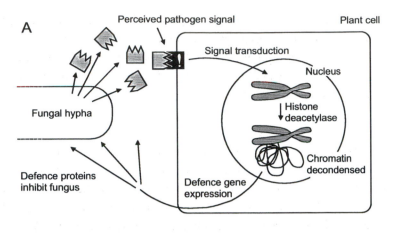

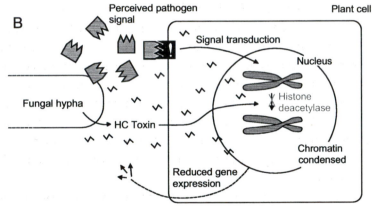

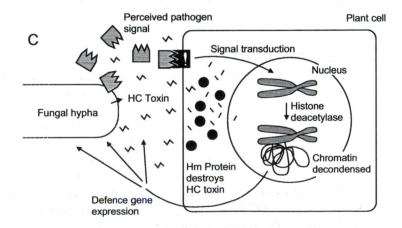

which produces victorin-C toxin (Figure 10.7), a compound specifically toxic to Victoria oats. Victorin-C is an unusual chlorinated peptide compound. It was subsequently found that oats carrying the Vb gene were susceptible to victorin-C. *C. victoriae* isolates which do not produce victorin-C have very low virulence on oats.

Toxin biosynthesis

Small and/or cyclic peptide toxins are often produced by enzymes known as peptide synthases. These enzymes are able to assemble stretches of amino acids without needing an RNA template to direct the synthesis. Peptide synthases are very large multi-domain enzymes of *c.* 500 kDa which are also involved in the production of penicillin by *Penicillium* spp. Gene cloning of the TOX2 locus from *C. carbonum* suggests that HC-toxin is synthesized by such an enzyme. The peptide nature of victorin-C also suggests the involvement of a peptide synthase although the genes and proteins responsible for victorin synthesis remain uncharacterized.

T-toxin is known to be synthesized by a polyketide synthase gene. Molecular and genetic evidence suggests that this gene may have arisen as the result of a chromosomal translocation which brought parts of two previously unconnected polyketide synthases together. The new hybrid gene produced T-toxin which serendipitously interacted with the T-cms protein in a lethal manner.

Non-host-specific toxins are produced by many other species. These toxins increase the extent of the disease but are not essential for it.

Control of plant diseases

Due to the impossibility of eradicating plant diseases from the environment, and the difficulty of curing a disease that has become

Figure 10.9 The mechanism whereby the Hm gene causes resistance to *Cochliobolus* in maize. **A** The presence of a pathogen is recognized by the plant resulting in defence gene expression and production of an inhibitory reaction. **B** When the fungus produces HC toxin histone deacetylase is inhibited resulting in decreased defence gene expression, suppressing the defence reaction and allowing fungal growth. **C** The Hm gene inactivates the HC toxin resulting in a normal resistance response.

established in a plant, most control measures involve attempts to prevent the pathogen reaching the plant. These measures include:

(a) Use of pathogen-free seed or propagating material. Such material may be obtained from areas where a pathogen is not present.
(b) Good farm hygiene, which may include separation of surplus unused crop from subsequent sowing or burning of infected stubble.
(c) Crop rotation. Soil-borne diseases are often unable to persist in the soil through more than one winter. Changing to a non-host crop for a season may effectively 'starve out' the soil inoculum.
(d) Legislation to prevent import or transport of diseased plant tissue.
(e) Biological control. The presence of antagonistic microorganisms in soil may destroy or suppress pathogen populations. Although this phenomenon holds great promise for control of plant disease little progress has so far been made in exploiting it for crop protection.
(f) Use of resistant plant varieties. As previously mentioned many plants carry resistance genes which are able specifically to recognize certain pathogens and trigger defence responses. These genes have been bred into commercial crop varieties to produce disease resist-ant crop plants. Such disease-resistance may be long lasting or may break down after a few years; the recent cloning of resistance genes from a variety of sources offers the possibility of manipulating these systems to provide stable or novel disease resistances.

Chemical control

One of the most common effective measures used to combat fungal pathogens is the spraying of fungicides onto crops. Most fungicides are used against foliar pathogens or for disinfestation of the stored harvest. Though the earliest fungicides, such as Bordeaux mixture, were first used in the nineteenth century (and are still used today!) the first group of specific chemicals, the organic sulphur compounds (dithiocarbamates), dates from the 1930s. New classes of fungicide were not discovered until the 1970s, but tough licensing regulations, high research and development costs and lack of understanding of pathogen biochemistry have limited the emergence of new compounds since then. Fungi are continually evolving resistance to the existing classes of fungicide and discovery of new classes of compound is becoming urgent. Such discovery may well depend on more detailed knowledge of the molecular genetics of fungal pathogenicity.

Recent research on systemic acquired resistance has yielded compounds which can trigger the plants defence response at very low concentration. Such compounds are currently marketed for use as disease prevention treatments and should provide a potent weapon in the fight against fungal disease.

The future

Although great advances have been made in the understanding of how a few model pathogens cause disease, the diversity of life styles among plant pathogens means that it is difficult to apply this knowledge to other fungi. Genes that are important in one organism may be present in other pathogens but be irrelevant as far as disease is concerned. Knowledge gained from pathogens that penetrate the leaf surface via appressoria may be meaningless when applied to pathogens that penetrate through the stomata and so forth. Additionally it is becoming clear that pathogenesis is not the result of a few key genes but of a large number of genes acting in a concerted manner. It is clear that more work needs to be done to further produce and characterize mutations which alter pathogenicity in a variety of organisms. A new prospect for gaining an overview of the genes involved in disease arises from the emerging discipline of genomics. This involves finding out a large amount of information about an organism's genes either by sequencing of the entire genome or by sequencing and automated analysis of expressed genes. Computer analysis is then used to find patterns in this information to determine what genes are present or absent in the organism or to show which genes are expressed at given stages in the life cycle. It should be possible to apply such techniques to finding out which genes are involved in pathogenicity and to find answers to the question 'what makes a pathogen different from a saprophyte?'

An improved understanding of the factors which make plants resistant to fungal attack will also allow manipulation of plant resistance and recognition systems to improve defence against disease. It is unlikely that crop modification programmes will always be able to counteract the appearance of new pathogenic races but it is likely that such techniques will significantly improve the efficiency of crop production in years to come.

319

Further reading
Recommended plant pathology textbooks

Agrios, G. N. (1997). *Plant Pathology*, fourth edition. Academic Press, London.

Lucas, J. A. (1998). *Plant Pathology and Plant Pathogens*, third edition. Blackwell Science, Oxford.

Recommended web sites and web starting points

British Society of Plant Pathology home page http://www.bspp.org.uk/

Plant pathology Internet guide http://194.247.68.33/ppigb/

References

Ballance, G. M., Lamari, L., Kowatsch, R. and Bernier, C. C. (1996). Cloning, expression and occurrence of the gene encoding the Ptr necrosis toxin from *Pyrenophora tritici-repentis*. *Molecular Plant Pathology On Line*: http://www.bspp.org.uk/mppol/1996/1209ballance.

Dangl, J. *et al.* (1996). Death don't have no mercy: cell death programs in plant–microbe interactions. *Plant Cell* 8:1793–1807.

de Jong, J. C., McCormack, B. J., Smirnoff, N. and Talbot, N. J. (1997). Glycerol generates turgor in rice blast. *Nature* 389:244–245.

De Wit, P. J. G. M. (1997). Pathogen avirulence and plant resistance: a key role for recognition. *Trends Plant Sci.* 2:452–458.

Draper, J. (1997). Salicylate, superoxide synthesis and cell suicide in plant defence. *Trends Plant Sci.* 2:162–165.

Ebbole, D. J. (1997). Hydrophobins and fungal infection of plants and animals. *Trends Microbiol.* 5:405–408.

Flor, H. H. (1955). Host parasite interactions in Flax rust: its genetics and other implications. *Phytopathology* 45:680–685.

Gebhardt, C. (1997). Plant genes for pathogen resistance: variations on a theme. *Trends Plant Sci.* 2:243–244.

Hammond-Kosack, K. E. and Jones, J. D. G. (1996). Resistance gene dependent plant defence responses. *Plant Cell* 8:1773–1791.

Innes, R. W. (1995). Plant–parasite interactions: has the gene-for-gene model become outdated. *Trends Microbiol.* 3:483–485.

Knogge, W. (1996). Fungal infections of plants. *Plant Cell* 8:1711–1722.

Kronstad, J. W. (1997). Virulence and cyclic AMP in smuts, blasts and blights. *Trends Plant Sci.* 2:193–199.

Lawrence, G. (1995). Do plant-pathogens produce inhibitors of the resistance reaction in plants. *Trends Microbiol.* 3:475–476

Mendgen, K. and Dessing, H. (1993). Infection structures in fungal plant patho-
    gens: a cytological and physiological evaluation. *New Phytol.* 124:192–213.
Newsham Fitter, A. and Watkinson, A. (1995). Multi-functionality and biodiver-
    sity in arbuscular mycorrhizas. *Trends Ecol. Evol.* 10:407–411.
Oliver, R. and Osbourn, A. (1995). Molecular dissection of fungal pathogenicity.
    Microbiology 141:1–9.
Osbourn, A. E. (1996). Saponins and plant defence: a soap story. *Trends Plant Sci.*
    1:4–9.
Ryals *et al.* (1996). Systemic acquired resistance. *Plant Cell* 8:1809–1819.
Smith, F. A. and Smith, S. E. (1997). Structural diversity in (vesicular)-arbuscu-
    lar mycorrhizal symbioses. Tansley review 96. *New Phytol.* 137:373–388.
Walton, J. D. (1996). Host selective toxins: agents of compatibility. *Plant Cell*
    8:1723–1733.

Recommended 'browsing' journals
Fungal Genetics and Biology
Molecular Plant Microbe Interactions
The Plant Cell
Plant Pathology
Trends in Microbiology
Trends in Plant Science

11 Fungi as animal pathogens

S. HOSKING

The fungi are admirably equipped with an array of degradative and lytic weaponry to colonize and enter the animal body. The host, of course, has an equally impressive armoury not only of defensive, but also aggressive machinery to prevent such an invasion. Very few fungi are obligate animal pathogens, but many are opportunists, able to take advantage of any chink in the host's defences.

Fungal infections of humans
The dermatophytes
The dermatophytes are a group of fungi which are particularly well adapted for the infection of the superficial keratinized structures of the body, including the nails, hair shafts and the stratum corneum (the cornified layer) of the skin. The true dermatophytes all belong to one of three genera of fungi, and are all able to use keratin as a source of nutrients. They are distinguished from a number of other fungi which may colonize the same superficial sites, including *Candida albicans*, *Malassezia furfur* and *Aspergillus* spp., by the fact that the dermatophytes are not opportunists. They are well adapted to their pathological niche and able to infect the immunocompetent individual.

The three genera of dermatophytes are *Trichophyton*, *Epidermophyton* and *Microsporum*, which together contain over 40 species. These organisms are relatively similar, but can be distinguished by colony morphology, macroscopic appearance and some biochemical tests. Recently, discovery of the perfect, or sexual, stage of several species of dermatophyte has led to their reclassification into two new genera, *Arthroderma*, containing the perfect *Trichophyton* species, and *Nannizzia*, containing the perfect *Microsporon* species. The two new genera are

322

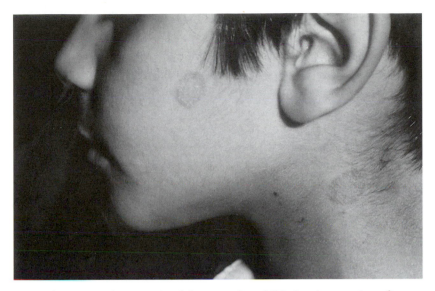

Figure 11.1 An example of ringworm in a child, showing two sites of disease, one on the cheek and one on the neck. There are many species of fungi that cause ringworm (tinea corporis) including *Trichophyton rubrum* and *Microsporum canis*.

included in the subdivision Ascomycotina. The original three genera are part of the subdivision Deuteromycotina.

Dermatophytosis, the disease caused by the dermatophytes, can be described by the site of infection, using the word 'tinea' followed by a term for the particular body site. The word 'tinea' comes from the Roman word for the clothes moth, and is used because of the similarity between the effect of the moth on wool clothing and the appearance of the fungal skin infection. For example, tinea pedis, or athlete's foot, is an infection of the foot, usually between the toes, and is probably the most common example of dermatophytosis. Tinea corporis refers to dermatophytosis occurring on one of the major surface areas of the body. Tinea corporis frequently appears as ring-shaped lesions, hence the misleading term 'ringworm' to describe this condition (Figure 11.1). Tinea unguium is a form of onychomycosis, or fungal infection of the nail, and is frequently caused by *Trichophyton rubrum*.

The skin represents a hostile environment to potential infectious organisms, because of the exposure to UV, drying and colonization by the normal bacterial flora. However, unlike most microorganisms, the dermatophytes produce keratinases which enable them to utilize keratin as a source of nutrients. Dead skin cells are continually shed from the stratum corneum, and therefore infection depends on the organism's ability to grow into the deeper layers of the skin. If it is unable to do so, it will be removed with the shed skin cells. The physical nature of the site of infection may protect the dermatophyte from contact with at least part of the host's immune system. This 'separation' apparently also works in the host's favour, for even in severely immunocompromised patients, dermatophytes rarely invade the deep tissues or cause systemic infections.

Superficial mycoses

The superficial mycoses are fungal infections of various superficial structures of the body, including the hair, nails, stratum corneum of the skin, the cornea, and the lining of the external ear canal. This list is also frequently extended to include infection of the mucosal surfaces of the body, including the buccal cavity (the mouth) and the vagina.

The causative agents of superficial mycoses can be split into two groups; those that are largely or exclusively associated only with such infections, and those that are opportunists and cause superficial infections as just part of their repertoire. The former group are largely restricted to infections of the hair, nails and skin, and are afforded the same protection from host defences by these environments as the dermatophytes. They are effectively partially isolated by the cornified layer from the effectors of the immune system. Such infections include tinea versicolor, which is an infection of the stratum corneum, and malassezia folliculitis, which is an infection of the hair follicles. The causative agent of both these diseases is *Malassezia furfur*, and in both cases the infection is largely confined to the trunk. Other superficial infections of this class occur mainly in warm climates and include tinea nigra, which is located most commonly on the palms of the hands and is caused by *Phaeoannellomyces werneckii*. In this case the lesions are dark or black in colour, hence the descriptor 'nigra'. White piedra is an asymptomatic infection of the hairshaft which results in light-coloured soft nodules that are attached to the hair shaft and may cause the hair shaft to break. The

324

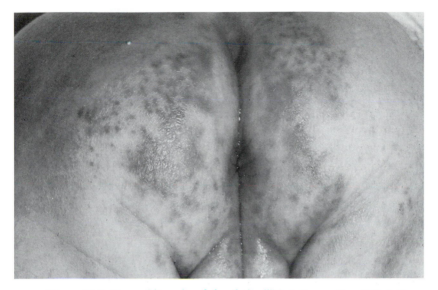

Figure 11.2 Nappy (diaper) rash in a baby. This is frequently coinfected with *Candida albicans* and improves with topical antifungal therapy and barrier creams.

causative agent is *Trichosporon beigelii*, and although the disease is found predominantly in warm climates, it has also been reported to occur with relatively common frequency in the United States.

Of the fungi which cause opportunistic superficial infections, the dimorphic fungus *Candida albicans* is by far the most prevalent (Figure 11.2). *C. albicans* is a normal commensal of the healthy individual, residing in the gastrointestinal tract and the vagina. Under normal circumstances, when it must compete with resident bacteria and when the mucosa is able to exhibit a normal non-specific barrier to infection, the fungus does not penetrate the mucosal surfaces. However under certain conditions, *C. albicans* can multiply on or invade through the mucosa to cause a localized or, in extreme cases, a systemic or widespread infection. Circumstances which promote this opportunistic invasion include the hormonal changes which result from pregnancy and the oral contraceptive pill, which typically results in vaginal candidiasis or thrush, and the use of broad spectrum antibiotics, which eliminate the protective capacity of the friendly flora of the alimentary tract. More severe forms of candidiasis are associated with immunocompromisation, including the aggressive chemotherapy associated with the treatment of certain cancers and HIV. Severe oral and/or oesophageal candidiasis are typical symptoms of the progression of a patient from HIV infection to full-blown AIDS. Although *C. albicans* is the major human pathogen of the genus *Candida*, several other species are also able to cause similar infections, the most common being *C. krusei*, *C. tropicalis*, *C. glabrata* and *C. parapsilosis*. The *Candida* species are true opportunists and are also able to invade the cornified surfaces of the body (skin, nails and hair) and also to cause blood stream infections and infect organs of the body cavity. These latter infections are known as deep-seated mycoses.

Deep-seated mycoses

The deep-seated mycoses are fungal infections which pervade into deeper regions of the body than those described above. They may be isolated to a single organ or site, or may occur in several disparate locations. The latter condition is known as a disseminated infection. Many fungi are able to cause deep-seated mycoses. The majority of such infections are opportunistic, but a few fungi are able to cause severe infection in the immunocompetent host.

Coccidioidomycosis is a systemic mycosis caused by the dimorphic fungus *Coccidioides immitis*. The fungus is a soil saprophyte which occurs naturally only in the Western hemisphere in a region extending from northern California to Argentina. Coccidioidomycoses can therefore be described as endemic to this region, and rarely occurs outside this area, with the exception of isolated cases brought about by travel of an infected individual from the endemic area. The usual route of infection is pulmonary. The vast majority of residents in endemic regions have experienced infection with *C. immitis*, but in most cases, the infection will be asymptomatic or manifested as a mild bronchopneumonia. Infection other than pulmonary is rare but serious and may occur in 1 per cent of cases and extends from the primary site of infection, the lungs. Disseminated coccidioidomycosis may occur in virtually any organ, but the skin, bones and joints, meninges and genitourinary system are common.

Paracoccidioides brasiliensis is another dimorphic pathogenic fungus which is found and causes disease only within a restricted, or

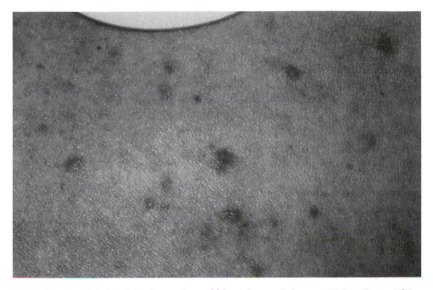

Figure 11.3 Rash in disseminated histoplasmosis in an AIDS patients. This is caused by *Histoplasma capsulatum.* Rash only occurs in 10 per cent of cases of histoplasmosis in AIDS. (Courtesy of Dr P. Moruzumi, Kaiser Permanente, Santa Cloras, California.)

endemic, area. It occurs only in Latin America from Mexico to Argentina. However, the disease apparently has a potentially extremely long period of latency, for paracoccidioidomycosis has been identified in individuals who left the endemic region 30 to 40 years prior to diagnosis. The primary infection site is the lungs, but dissemination to the mucous membranes, skin, lymph nodes, adrenals and other organs occurs frequently. Interestingly, the disease is most frequently seen in adult males. It has been shown that oestrogens inhibit the mycelial to yeast transition of *P. brasiliensis* which is necessary for the survival of the organism in the host tissues. Therefore, when the initial contact with the fungus takes place, the female individual may be protected by hormone levels which prevent the necessary dimorphic transition.

Histoplasma capsulatum is another fungal pathogen which may cause disease, histoplasmosis, in the otherwise healthy individual, although without doubt the immunocompromised individual is far more at risk of serious infection (Figure 11.3). The most common form of the disease is an acute pulmonary infection, which normally resolves itself spontaneously after 2 to 4 weeks. Primary pulmonary infections may be accompanied by spread of the infection site to other organs, the spleen, lymph nodes and liver being common sites. Fungaemia due to *Histoplasma* also occurs. Usually, in otherwise healthy individuals, immunity is acquired which results in combat of the disease, but in immunocompromised individuals acute disseminated histoplasmosis often results.

Opportunistic fungal infections

Few human fungal pathogens are dependent on the host for any phase of their life cycle or for the perpetuation of the species, with the exception of some dermatophytes. Several fungi are only able to cause severe infections in the immunocompromised or otherwise debilitated host. Immunocompromisation can arise through various means: through hereditary immune disorders, such as chronic granulomatous disease; through HIV infection and AIDS or through immunosuppressive chemotherapy used to treat transplant patients or cancer chemotherapy. Other predispositions to fungal infections include prolonged treatment with broad spectrum antibiotics which can deplete the body of its natural 'friendly' flora which would otherwise out-compete the fungal pathogen. Corticosteroids actively promote fungal growth and therefore patients receiving corticosteroid therapy can be susceptible to aggressive fungal infections. Surgical procedures and intravascular catheters can both provide a direct route for invasion of the body by the fungus by circumventing the body's primary barrier to infection, the skin.

As the use of modern clinical therapies and procedures has risen, and with the increase in incidence of HIV infection, the occurrence of opportunistic fungal infections has grown over the past few years. Immunocompromisation permits the commensal and environmentally prevalent fungus to become pathogenic. Not surprisingly, the nature of the immunocompromisation dictates what type of infection the patient will be susceptible to, to the extent that the infective organism and presentation of the infection can indicate the defect in the immune system. Many fungal species are removed from the circulation by phagocytosis by

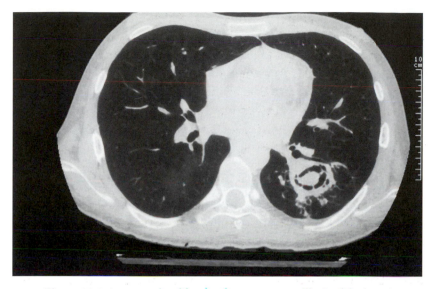

Figure 11.4 An example of focal pulmonary aspergillosis of the lung in a bone marrow transplant patient. This image is of a computer tomography scan of the lungs and heart showing a circular lesion in the superior part of the left lower lobe. The scan illustrates the phenomenon of cavitation which is likely to lead to major bleeding in the lung that is frequently fatal and so surgery is urgently indicated. Invasive pulmonary aspergillosis has a 60–90 per cent mortality, as current antifungals are relatively ineffective.

macrophages and/or neutrophils. Neutropenic (neutrophil-deficient) patients such as those undergoing chemotherapy for leukaemia and bone marrow transplants, are predisposed to disseminated fungal infections, especially candidiasis and invasive infection by an *Aspergillus* species (aspergillosis) (Figures 11.4, 11.5). T cell-deficient patients, such as AIDS patients, are predisposed to mucosal candidiasis, but disseminated candidiasis or aspergillosis is rare. Cryptococcosis (infection by *Cryptococcus neoformans*) is rare in patients other than those infected by HIV.

Candida species are responsible for the majority of fungal infections in the immunocompromised host. Oropharangeal candidiasis has been described above, and is extremely common in AIDS patients (Figure 11.6). In severely neutropenic patients, *Candida* fungaemia (blood stream infection) and/or systemic candidiasis can also occur. This can commonly manifest as hepatosplenic candidiasis (localised to the liver

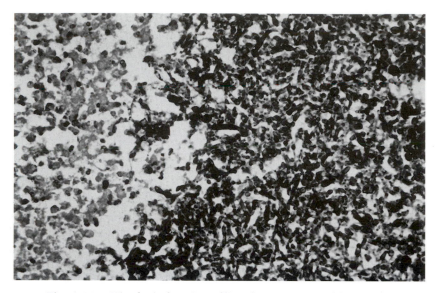

Figure 11.5 Histological section of liver from a 21-year-old woman with acute lymphoblastic leukaemia who developed chronic disseminated candidiasis, involving the liver and spleen. The section shows fungi as black and tissue as grey (Grocott staining). Both yeast and hyphal forms are visible, consistent with dimorphism. This disease is very difficult to treat, and despite amphotericin B therapy, it spread to the brain and her leukaemia relapsed. She died about 15 months after this biopsy was taken.

and spleen), or as disseminated candidiasis occurring at multiple sites throughout the body.

The *Aspergillus* species are widespread filamentous fungi found in soil, water, decaying vegetation and other waste matter. They have very light, easily dispersed conidia. The primary route of infection is via the respiratory tract, and the majority of infections are pulmonary (Figure 11.7). The infection can become disseminated, in which case virtually any organ may be involved. Outbreaks of *Aspergillus* infections have occurred in hospitals during construction work and when air conditioning systems have become heavily contaminated. *Aspergillus fumigatus* is responsible for most *Aspergillus* infections, although *A. flavus*, *A. niger* and *A. terreus* are also occasionally isolated.

Cryptococcosis is caused by *Cryptococcus neoformans*. The organism is distributed widely in nature, being common in pigeon droppings.

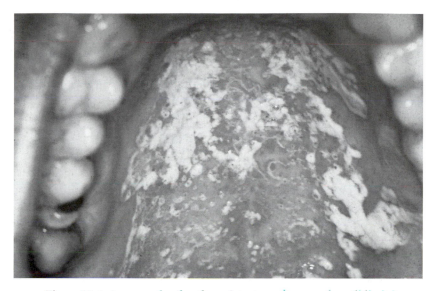

Figure 11.6 An example of azole-resistant oropharangeal candidiasis in AIDS. The causative agent was *Candida albicans*, and the fluconazole MIC (minimum inhibitory concentration) was 50 μg ml^{-1} (sensitive organisms have an MIC of less than 6.25 μg ml^{-1}). He responded to topical itraconazole. Resistance occurs late in AIDS in about 15–20 per cent of patients.

The principal route of infection is via inhalation of the basidiospores. However the most common clinical form of cryptococcosis is infection of the central nervous system, or meningitis. It has been estimated that up to 13 per cent of AIDS patients will develop cryptococcosis, and that up to 75 per cent of these will be cryptococcal meningitis. Disseminated cryptococcosis also occurs, and virtually any organ of the body may be involved, although the liver and spleen are most common.

Mechanisms of fungal pathogenesis

Most fungi are not able to survive in the environment provided by human tissue, and therefore cannot cause disease in humans. A few fungi are well adapted to be human pathogens and can cause disease in the otherwise healthy individual. The majority of fungal infections are caused by the opportunist pathogens, able to grow and invade human tissue, but only when the host's defences are compromised. Some factors must obviously

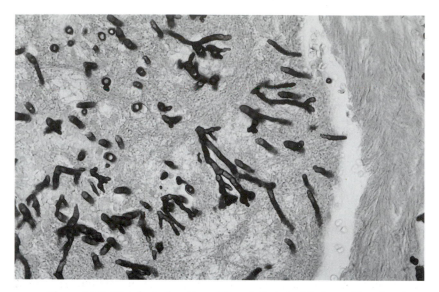

Figure 11.7 Histological section of lung infected with *Aspergillus*, showing typical branching pattern of *Aspergillus* hyphae (septate and branching at 45°).

differentiate the pathogen and opportunistic pathogen from the non-pathogen. Perhaps the most obvious is the ability to grow at body temperature. The majority of fungi have an optimum growth temperature of 25 to 30 °C, and may grow only weakly or not at all at 37 °C. Beyond this primary consideration, there is no single factor which confers pathogenesis for any single species. For most pathogens, the factors which contribute to their pathogenicity are unknown, and are at best suppositions derived from *in vitro* findings. However, the development of a fungal infection must satisfy several considerations, irrespective of the site of infection. The fungus must first be able to adhere to the host tissues. The fungus must colonize the host and invade the host tissues. Once the fungus has invaded the host tissue, it must be able to adapt to the tissue environment. Probably most importantly, the fungus must be able to avoid the host's cellular defences.

Adherence, colonization and invasion

Before colonisation and invasion can take place, the fungus must first adhere to the primary barrier to infection, generally the skin in the case of dermatophytes, or the mucosa in the case of most other pathogens. Occasionally an alternative portal of entry arises in the form of an intravenous line or deep-seated catheter. Adherence is dependent on a variety of factors, including surface glycoproteins, fungal cell surface hydrophobicity, pH, temperature, and of course, phenotype of the organism. It has been reported that *C. albicans* adheres better to epithelial cells under conditions which promote germ tube development. *C. albicans* is generally hydrophilic when grown at 37 °C, but can become hydrophobic under certain conditions. Hydrophobic *C. albicans* cells are better able to adhere to mucosal cells, and are less rapidly killed by neutrophils than hydrophilic cells. Attachment of *C. albicans* to host cells is minimal at acidic pH, but increases dramatically under neutral conditions. This is reflected in the worsening of the symptoms of thrush during menstruation, when the vaginal pH rises from acidic to neutral. *Candida* species express several cell surface proteins termed adhesins which actively promote binding to host cells. These include a lectin-like protein that recognizes sugar residues of epithelial cell surface glycoprotein, and a complement receptor-like protein, CR3, which may play in a role in adherence to endothelial cells.

Colonization can best be described as the presence of a pathogen about the host in significantly high numbers. Invasion occurs when the fungus physically enters the host tissue. Colonization is generally a prerequisite for invasion. *Candida albicans* is a normal commensal of the healthy individual, and therefore the majority of the population could be considered to be colonized by the organism, but infection only occurs under aberrant conditions when the fungus breaches the barriers to infection and invades the host tissue.

The development of deep mycoses demands penetration into the internal tissues of the body. Many infections arise through inhalation of spores or yeast cells into the lungs, where the moist environment promotes germination and multiplication and hence colonization. Deep-seated mycoses can also arise through direct invasion as a result of surgical procedures and indwelling catheters. In this case the catheter or wound act as the site of colonization. Direct penetration and passage through the epithelial mucosa by *C. albicans* germ tubes has been demonstrated. Invasion is probably aided by hydrolytic enzymes, such as proteinases and lipases, and in the case of dermatophytes, keratinases.

Adaption and avoidance of host defences

In general, the development of human mycoses is related primarily to the immunological status of the host and environmental exposure, rather than to the infecting organism. However, the invading organism must display a number of adaptive and avoidance mechanisms to avoid the host's defences.

Many fungal pathogens of man are dimorphic and can exist in either yeast or hyphal growth forms. Several fungi exist outside the body as filamentous fungi, but adopt a yeast (e.g. *Histoplasma capsulatum, Paracoccidioides brasiliensis*) or spherule-endospore growth form within the body (e.g. *Coccidioides immitis*). For these organisms, the shift from environmental to body temperature is the critical cue for transition to the yeast or spherule form. *Candida albicans* is able to adopt both a yeast and a hyphal growth form in the host tissue, but adopts predominantly a yeast form under normal laboratory growth conditions. Again, temperature is important for the dimorphic transition, as are pH and the presence of other factors including serum. The role of dimorphism in pathogenesis is unclear, but it does appear to be important. The yeast and hyphal forms of a fungus show dramatically different antigenic properties, having significant variation in cell wall composition and cell surface glycoprotein expression. The dimorphic transition stimulated by temperature shift is associated with the production of heat shock proteins, and it has been shown that these may be important in evoking an immunological response to both *Histoplasma capsulatum* and to *C. albicans*. The hyphal growth form of *C. albicans* has often been associated with pathogenesis, and it is easy to envisage the foraging hyphal form being more invasive than the budding yeast. However, this assumption has never been proved, and it has been demonstrated that a mutant strain incapable of producing hyphae was as capable of causing disease in an animal model as a wild-type dimorphic strain.

Perhaps of more importance to avoidance of the host immune system is the ability of some fungi, notably *C. albicans* and *H. capsulatum*, to undergo a process known as phenotypic switching. In both these organisms, switching can manifest *in vitro* as a spontaneous change in colony morphology from rough to smooth and *vice versa*. In *C. albicans*, the different phenotypes result in differences in hyphal formation and susceptibility to antifungal agents, and, although switching has not been demonstrated *in vivo*, it has been proposed that it may promote invasion and proliferation in entirely different tissue types and evasion of host defences by alterations in surface antigenicity. Furthermore, switching may enhance adhesion to mucosal surfaces, tissue penetration and secretion of lytic enzymes.

The secretion of lytic and degradative enzymes is of obvious importance to the invasion of host tissues. One family of such enzymes, the secreted aspartyl proteinases (Sap), has been extensively studied in *C. albicans*. Sap enzyme activity has been correlated with virulence. A family of at least eight SAP genes can be used to produce Sap enzyme activity. The SAP genes have been shown to be differentially expressed, according to the growth phase and phenotype of the organism. *SAP2* mRNA was the dominant transcript in the yeast phase organism; *SAP4*, *SAP5* and *SAP6* transcripts were observed only at neutral pH during serum-induced yeast to hyphal transition. In the *C. albicans* strain WO-1 which undergoes overt phenotypic switching between white and opaque colony phenotypes, *SAP1* and *SAP3* are differentially expressed in the white and opaque switch types. This data indicates that members of the SAP gene family may have distinct roles in the colonization and invasion of the host.

Cryptococcus neoformans is unique among the fungal pathogens of man in being an encapsulated yeast. Phagocytosis by macrophages is an important factor in the resistance to *C. neoformans* infection. The polysaccharide cryptococcal capsule is a potent inhibitor of phagocytosis. Non-encapsulated strains are quite sensitive to phagocytosis, whilst encapsulated isolates are engulfed poorly. In variants displaying different sizes of capsule, a direct relationship has been observed between capsule size and resistance to phagocytosis. The mechanism for inhibition of phagocytosis by the capsule is not known, although studies have shown that it is unlikely to be due to a direct effect of the capsular polysaccharide on macrophages. It is more probable that the capsule prevents efficient opsonization of the organism, and hence prevents recognition of the yeast by the host macrophages.

Treatment of fungal infections

Like humans, fungi are eukaryotic, and not prokaryotic like bacteria. Thus it is not surprising that antifungal agents are associated with a

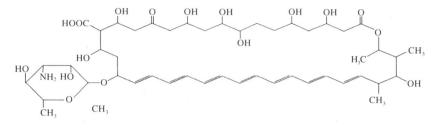

Figure 11.8 Structure of amphotericin B. Amphotericin B is considered to be the 'gold standard' antifungal drug currently available, despite significant associated toxicity.

higher toxicity than antibacterial agents. It might be envisaged that anti-fungal agents could be developed which target fungus-specific enzymes such as those involved in chitin, mannan and glucan biosynthesis. However, despite extensive effort by the pharmaceutical industry, such approaches have not yet proved fruitful, and the few systemic antifungal agents that are available have mechanisms of action that are directed against sterols in the cell membrane or against enzymes involved in nucleic acid biosynthesis.

There are two major classes of antifungal agents used to treat the variety of fungal infections found in humans. These are the polyenes, which include amphotericin B and nystatin, and the azoles. The azole antifungals can be further subdivided into the imidazoles, such as micon-azole and ketoconazole, and the newer triazole compounds itraconazole and fluconazole. Other compounds, such as griseofulvin which selec-tively inhibits fungal mitosis, are occasionally used to treat superficial and dermatophyte infections.

Amphotericin B is often considered to be the 'gold standard' anti-fungal (Figure 11.8). It is extremely antifungal and has a broad spectrum of action including most *Candida* species and some *Aspergillus* species. Amphotericin B is a naturally occurring antibiotic, produced by the acti-nomycete *Streptomyces nodosus*. Amphotericin B acts by binding to sterol components of the cell membrane and inducing permeability changes which results in leakage of the cell contents through the membrane. It binds preferentially to ergosterol, the major sterol component of the fungal cell membrane, and less so to cholesterol, the principal sterol of the

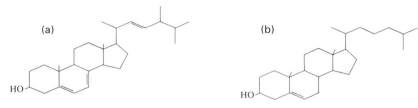

Figure 11.9 Comparison of the structures of (a) ergosterol and (b) cholesterol. Ergosterol is the major sterol component of fungal cell membranes,
whilst cholesterol is the principal sterol component of mammalian cells.

human cell membrane (Figure 11.9). Fungal resistance to amphotericin B
is largely due to the substitution of ergosterol in the cell membrane with
other components which do not bind to amphotericin B.

Despite the selectivity of amphotericin B for ergosterol over cholesterol, the drug is associated with severe and wide-ranging toxic side
effects, most serious of which is nephrotoxicity (toxicity to the kidneys).
Virtually every patient treated with amphotericin B develops some degree
of renal damage, either temporary or permanent. The toxicity is so great
that it is often the toxicity rather than the response to therapy which dictates the dose and schedule of therapy. In addition, amphotericin B is
only suitable for intravenous administration, and therefore treatment
generally requires hospitalization. Not surprisingly, despite its excep-
tional antifungal properties, amphotericin B is reserved for treatment of
life-threatening fungal infections which cannot be treated by any other
means. In an attempt to reduce the toxic effects of amphotericin B, some
alternative formulations of the drug have recently been developed which
alter the pharmacological properties and therefore toxicity of the drug.
These include administration in the form of lipid complexes or lipo-
somes. A significant reduction in toxicity has been achieved, although
this is counter-balanced by a significant (some hundred-fold) increase in
the price of the drug.

Flucytosine (5-fluorocytosine) is a fluorinated pyrimidine which
interferes with pyrimidine metabolism in fungi, and therefore RNA and
protein synthesis. In the fungal cell, flucytosine is deaminated to 5-
fluorouracil which is incorporated into RNA. Mammalian cells do not

convert large amounts of flucytosine to 5-fluorouracil, so toxicity is selective against fungi. Flucytosine is active against most yeast pathogens, but resistance is frequently seen, particularly by *C. albicans*. Because of the potential for resistance, flucytosine is rarely used alone, but in combination with amphotericin B. Such combination therapy permits effective therapy with lower doses of amphotericin B, and hence less risk of toxicity.

The principal mechanism of action of the azole antifungals is to inhibit cytochrome P-450 enzymes. The main enzyme which is inhibited in fungi is the 14 α-demethylase which is responsible for the conversion of lanosterol to ergosterol. The azoles also interfere to some extent with the mammalian enzyme which converts lanosterol to cholesterol, therefore toxicity can be a problem. The azole antifungals are generally considered to be fungistatic rather than fungicidal. That is, they suppress growth rather than kill the organism. Clearance of the organism from the body is dependent on host defences. The spectrum of activity of the azoles varies slightly amongst the group. Most are active against most species of *Candida*, but poorly active against filamentous fungi. All are active against *C. albicans*. Fluconazole is the newest and best tolerated of the azoles, and for this reason has frequently been used in a prophylactic (preventative) capacity within patient groups at risk of fungal infection. Concomitantly, an increase in the incidence of *Candida* infections resistant to fluconazole has been observed. Despite this potential for resistance, fluconazole is currently the drug of choice for treatment of mucosal *Candida* infections. A particularly advantageous property of fluconazole is that it is administered orally, and can therefore be used to treat patients on an out-patient basis; an important consideration when considering the quality of life of terminally ill patients.

The need for new classes of novel antifungal drugs

It is clear that the incidence of serious fungal infections is on the increase, and that this increase is due largely to the increase in AIDS patients and the advent of modern medicine. Current antifungal therapies are subject to toxicity and resistance problems. The incidence of acquired resistance, particularly of fluconazole-resistant *C. albicans*, is increasing, as is the incidence of inherently fluconazole-resistant *Candida* species such as *C. glabrata*. Current leading therapies, the polyenes and azoles, target sterols within the cell membrane and sterol biosynthesis, and cross-react to some extent with the mammalian homologues. It is unquestionable that a need exists for new antifungal drugs, and that these should have novel mechanisms of action which are specific for fungal cellular components to reduce mammalian toxicity.

Fungal infections of fish

In recent years, the commercial farming of several species of fish in crowded environments has highlighted the occurrence of fungal infections in fish. The superficial infections seen in farmed fish are frequently caused by oomycetes, and in particular, *Saprolegnia diclina*. External infections caused by oomycetes are therefore termed saprolegniasis, although saprolegniasis may be caused by species of oomycete other than the Saprolegniales. Infection of the fish is believed to be aided by the motile zoospores. The zoospore cysts have hairs and hooks which aid attachment to the fish surfaces.

Systemic infections of fish have been noted both in farmed fish and among free-living fish populations. The principal causative agent of internal mycosis of fish is *Ichthyosporidium hoferi*. *I. hoferi* is an obligate pathogen which was first described in 1893 as a disease of trout. The organism is found in both fresh and salt waters, and has an extensive host range. Catastrophic mortalities of more than 80 species of fish around the world have been reported, including mackerel in England, trout in Germany, and the herring and mackerel populations of Canada. Mass mortalities occur in the herring population of the North Atlantic with periodic frequency. The route of infection is generally accepted to be ingestion of infected materials.

The growing need to cultivate fish, both marine and fresh water, for the economic benefit of society means that much research effort will be directed towards fungal infections of fish in the future.

Acknowledgement

With great thanks to David Denning, Department of Infectious Disease and Tropical Medicine (Monsall Unit), North Manchester General Hospital, for provision of clinical photographs.

Recommended reading

Calderone, R. A. (1993). Recognition between *Candida albicans* and host cells. *Trends Microbiol.* 1:55–58.

Cox, G. M. and Perfect, J. R. (1993). Fungal infections. *Curr. Opin. Infect. Dis.* 6:422–426.

Cutler, J. E. (1991). Putative virulence factors at *Candida albicans*. *Annu. Rev. Microbiol.* 45:187–218.

Kwon-Chung, K. J. and Bennett, J. E. (1992). *Medical Mycology*. Lea & Febinger, Philadelphia.

Vanden Bossche, H. V., Marichal, P. and Odds, F. C. (1994). Molecular mechanisms of drug resistance in fungi. *Trends Microbiol.* 2:393–400.

Further reading

Arora, D. K., Ajello, L. and Mukerji, K. G. (1991). *Humans, Animals and Insects*. Handbook of applied mycology, vol. 2. Marcel Dekker, New York.

Calderone, R. A. and Braun, P. C. (1991). Adherence and receptor relationships of *Candida albicans*. *Microbiol. Rev.* 55:1–20.

Kozel, T. R. (1996). Activation of the complement system by pathogenic fungi. *Clin. Microbiol. Rev.* 9:34–46.

Murphy, J. W., Friedman, H. and Bendinelli, M. (eds.) (1993). *Fungal Infections and Immune Responses*. Plenum Press, New York.

Prassad, R. (ed.) (1991). Candida albicans: *Cellular and Molecular Biology*. Springer-Verlag, Heidelberg.

Rinaldi, M. G. and Dixon, D. M. (1994). The evolving etiologies of invasive mycoses. *Infect. Dis. Clin. Prac.* 3(suppl. 2).

Young, D. W. (1994). Superficial fungal infections of the skin. *Curr. Ther.* 35:63–73.

12 Biotechnology of filamentous fungi: applications of molecular biology

D. B. ARCHER

Introduction

Scope

Biotechnology is defined here as the exploitation by man of biological systems for manufacture of biomass or derived products. Filamentous fungi have been used by man for centuries but the technology is now advancing at an unprecedented rate raising the potential for fungal applications to new heights. A successful product-based biotechnology relies upon the combination of several factors but, principally, being able to supply a desirable product at the right price. Although such economic realities pervade all biotechnology, they will concern us little in this chapter. Rather, we will concentrate on fungal products already available commercially and some which could become so. The aim of the chapter is to discuss the biological aspects of the formation of such products by fungi and to discuss the impact that modern molecular biology plays in improving their yields and creating novel products.

History

The long history of using filamentous fungi in the food industry indicates that many species are safe either for consumption or for the production of food components. Several species of edible mushrooms have long been cultivated but, in addition, others that do not form large fruiting bodies are also grown for human consumption. The other established technology using filamentous fungi is the production of fermented foods, e.g. soy sauce, where fungi are used as sources of enzymes that degrade complex substrates not otherwise available for use by bacteria and yeasts. The ability of many filamentous fungi to secrete degradative enzymes is exploited in the production of enzymes for a large range of commercial applications and is a reflection of the natural habitats of such species where survival depends on their protein secretory capacity. As will be seen later in this chapter, a wide range of metabolites are produced by fungi; some of these are exploited by man (e.g. antibiotics, food components) whilst others present a problem (e.g. toxins). In all cases there is an opportunity to explore and manipulate the mechanisms by which a target metabolite is made and the contribution to this exploration by the application of molecular biology techniques is profound.

Technology for the future

The development of transformation in the yeast *Saccharomyces cerevisiae* rapidly led to similar procedures being applied to the filamentous fungi so that by the early 1980s a number of species could be transformed, selection of transformants made possible by using a small range of different markers. The range of fungal species that can be transformed has increased steadily since, as has the variety of options available in terms of selection procedures and transformation protocols. Transforming DNA is ordinarily integrated into the fungal genome although a vector that supports autonomous replication, albeit unstable, can give very high transformation frequencies and has even been used to create instant gene banks. Transformation allows the introduction of trans genes for heterologous expression purposes, is an essential part of functional analysis of isolated genes and provides the means for altered expression of cloned genes. Thus, only since the advent of transformation procedures with filamentous fungi has the ability to exploit molecular biology in technological applications been possible.

This chapter concentrates on fungal products (small metabolites and enzymes) and the impact that molecular biology has on their variety and yield. Fungi are put to a far wider range of uses than can be discussed here and suitable reference is made for further reading at the end of the chapter.

Fungal metabolites

Filamentous fungi are used in commercial production of a variety of organic acids, edible oils, β-lactam antibiotics and polyketides. In

342

Table 12.1. *Fatty acids and their nomenclature*

ω 3 2 1	
$H_3C\text{-}(CH_2)_n\text{-}CH_2\text{-}CH_2\text{-}COOH$	
Palmitic acid	16:0
Stearic acid	18:0
Oleic acid	18:1 ($\omega-9$)
Linoleic acid	18:2 ($\omega-6$)
α-linolenic acid	18:3 ($\omega-3$)
γ-linolenic acid	18:3 ($\omega-6$)
Arachidonic acid (ARA)	20:4 ($\omega-6$)
Eicosapentaenoic acid (EPA)	20:5 ($\omega-3$)
Docosahexaenoic acid (DHA)	22:6 ($\omega-3$)

addition, a host of other products have the potential for commercial applications and are being seriously examined at some scale. Three case studies are presented and have been chosen because there is current research activity into their production that exploits, or is aiming to exploit, molecular biology techniques in order to enhance their yields or diversity.

Fungal oils

A wide range of oils can be extracted from filamentous fungi and, in some cases, to extremely high yields because oils accumulate in some species (termed 'oleaginous'). Such systems have been exploited commercially, e.g. in the production of γ-linolenic acid (GLA 18:3 ω-6). That particular process lost out commercially in the battle with competing plant-derived sources of GLA (e.g. borage and evening primrose) and serves as a reminder that although cost competitiveness is vitally important so is public perception, i.e. marketing a plant-derived product can be easier than a fungal product. The future of fungal oils most probably lies with those oils that plants do not produce, e.g. glycerol esters containing the very long chain polyunsaturated fatty acids (PUFAs). Some of the terminology for describing fatty acids is given in Table 12.1.

Fatty acid synthesis is catalysed by fatty acid synthase (FAS) in the

Table 12.2. *Domain order and size of fungal fatty acid synthase (FAS) and polyketide synthase (PKS) proteins*

	Common domain order					Size (amino acids)[b]
FAS 1 (β-subunit)	AT	ER	DH	MT/PT		2037–2076
FAS 2 (α-subunit)	ACP	KR	KS			1857–1894
PKS	KS	AT/MT	KR	ACP[a]	TE	1774–2187

Notes:
[a] Or two core ACP domains
[b] Of those fungal proteins reported
Domain abbreviations are as in the legend to Fig. 12.1.
Dr D. A. MacKenzie is thanked for compiling the table from the published sources.

cytoplasm. In yeasts and filamentous fungi there are two multifunctional FAS subunits encoded by unlinked genes. In the yeast *S. cerevisiae*, the *FAS1* gene encodes the pentafunctional β subunit and the *FAS2* gene encodes the trifunctional α subunit. The active FAS enzyme is $\alpha_6\beta_6$ in all fungi examined to date and the domain structures of the subunits from *S. cerevisiae*, *Y. lipolytica* and *Penicillium patulum* are summarized in Table 12.2. The synthesis of fatty acids by FAS and polyketides by polyketide synthase (PKS) has similarities and is summarized in Figure 12.1. Synthesis of even chain length fatty acids involves the addition of C2 units from malonyl-CoA, derived from acetyl-CoA by carboxylation. Synthesis of odd chain length or branched chain fatty acids is achieved by using a different starter unit, e.g. methylmalonate as extender, rather than malonate. Thus, supply of acetyl-CoA from the mitochondria and the activity of acetyl-CoA carboxylase are important factors and targets for genetic manipulation of fungi when overproduction of fatty acids is sought. It is important to acknowledge, however, that while listing targets for manipulation is relatively simple once the metabolic steps leading to a product are known, it is not trivial to predict the outcome of manipulating one step in a complicated metabolic pathway.

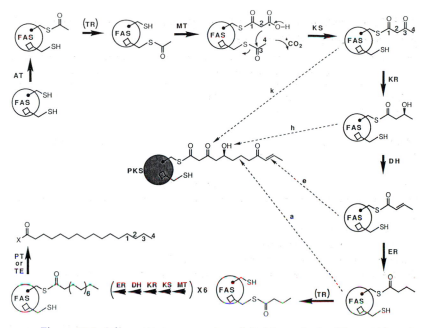

Figure 12.1 Schematic representation of the biosynthesis of fatty acids and polyketides. The circle labelled FAS or PKS represents the fatty acid or polyketide synthetase, carrying two thiol groups, one on the β-ketoacyl synthetase (condensing enzyme) (*open rectangle*) and the other on the acyl carrier protein (*filled circle*). The reaction steps are labelled: AT, acetyl transferase; TR, acyl transfer reaction; MT, malonyl transferase; KS, β-ketoacyl synthetase; KR, ketoreductase; DH, dehydrase; ER, enoyl reductase; PT or TE, palmityl transferase or thioesterase, involved in chain termination to produce palmityl CoA (X = CoA) or free palmitic acid (X = OH) respectively. 1, 2, 3, 4 designate carbon atoms of malonate and acetate that contribute to chain building, while the asterisk labels the carbon of malonate that is eliminated as CO_2. k, h, e, a represent the various possibilities that follow each condensation step to give keto, hydroxyl, enoyl or alkyl functionality at specific points in the product of a PKS. (From Hopwood and Sherman, 1990, with permission.)

345

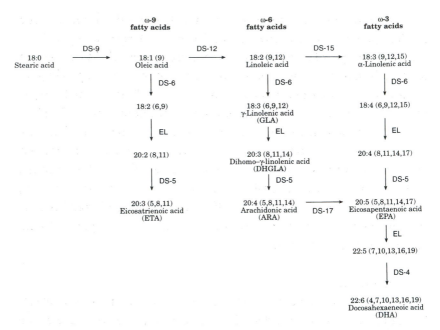

Figure 12.2 Biosynthetic pathway of polyunsaturated fatty acids in fungi. DS-n, desaturase functioning at the designated carbon; EL, elongase. (From Ratledge, 1994.)

The processes of fatty acid elongation and desaturation that lead to the synthesis of PUFAs are illustrated in Figure 12.2. By analogy with better studies of mammalian systems and because some enzyme activities have been found in fungi to be associated with microsomes, these reactions probably occur on the endoplasmic reticulum. The fatty acids are found in fungi either as components of the membrane, mainly phospholipids, or as accumulated oils existing as droplets of triacylglycerols. The esterification of glycerol 3 phosphate to form the triacylglycerols is also associated with the endoplasmic reticulum. The details of the biochemistry of these reactions are beyond the scope of this chapter and can be found in biochemistry texts. The interesting feature for fungal biotechnologists is that some fungi do accumulate large amounts of PUFA-containing oils which offer a potential alternative source of these nutritionally and developmentally important fatty acids. The zygomycetous genera *Mucor* and *Mortierella* contain species which look most

346

promising for the production of PUFAs at high yields although some marine species may also prove useful. Of these, *Mortierella alpina* is an excellent example of an oleaginous fungus producing PUFAs which has a record of safety in use of its oil as a human food additive. Depending on the strain, up to 44 per cent of its dry weight can be lipid and up to 70 per cent of the fatty acids in extracted oil can be arachidonic acid (C20:4 ω-6). Several studies have shown that the oil composition can be altered by changes in culture conditions (e.g. growth temperature, growth on different media) and that supplied oils can be modified. Although initial molecular biological studies with *M. alpina* are only now being reported, mutagenesis has led to strains with different fatty acid profiles and genetic manipulation of fungi to produce strains modified in the type and yield of fatty acids will soon follow.

Polyketides

Polyketides are mainly formed by some species of bacteria, plants and fungi and have a wide diversity of structures, a consequence of the detailed 'programming' of their biosynthesis. Thus, many of the keto groups, which give rise to the term polyketide, might be reduced to hydroxyls or removed. The fungal polyketides are significant in biotechnology either because they are exploited as antibiotics (e.g. griseofulvin from *Penicillium griseofulvum*), as drugs (e.g. lovastatin from *Aspergillus terreus*) or because they are toxins which can contaminate crops and derived foods (e.g. aflatoxins from *Aspergillus flavus*). In all cases, application of molecular biology approaches has begun to explore the biosynthesis with a view to producing new antibiotics and drugs with enhanced yields and, in the case of toxins, either to investigate the presence and activities of toxin 'pseudogenes' in food-grade fungi or to engineer non-toxin-producing strains for use in bio-control in agriculture. Most published information on fungal polyketides pertains to the genetics of aflatoxin biosynthesis and this will therefore be taken as the example of polyketide production by fungi.

Aflatoxins are produced by some strains of *Aspergillus flavus*, *A. parasiticus* and *A. nomius* although a much wider group of fungi are capable of producing sterigmatocystin, a toxic intermediate in the aflatoxin biosynthetic pathway. One such species is *Aspergillus nidulans*,

347

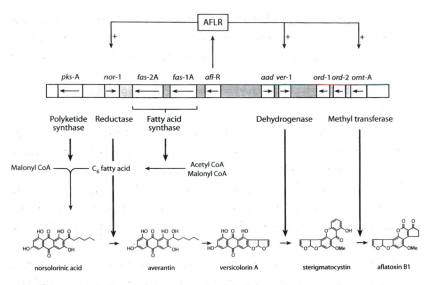

Figure 12.3 Genes and enzymes in the aflatoxin B1 biosynthetic pathway.

which is widely used as a model fungus for genetic studies. The biosynthetic pathway of aflatoxins, based on characterization of intermediates and purification of some of the enzymes involved is shown in Figure 12.3 together with some details of the gene cluster in *A. parasiticus*. The genes encoding enzymes of the biosynthetic pathways leading to sterigmatocystin in *A. nidulans* and aflatoxins in *A. flavus* and *A. parasiticus* have been studied in recent years. Although not all the genes have yet been fully characterized, it appears that most of the genes are clustered within the genome. The clusters have been sequenced and studies of gene regulation have begun with particular attention being paid to the regulatory genes that are known. Physiological conditions which favour aflatoxin production (e.g. high C/N ratio, low oxygen levels) have been known for some time and the gene studies will explain those effects in terms of the regulation of gene expression. A regulatory gene *afl*R located in the clusters encodes a DNA-binding protein with a zinc binuclear cluster motif that relieves the nitrate inhibition of aflatoxin biosynthesis. Molecular analysis of mutant strains deficient in aflatoxin synthesis is revealing other regulatory elements. The synthesis of polyketides requires a large number of genes and a level of regulation beyond our present understanding. The

348

effort devoted to unravelling the mechanisms of aflatoxin synthesis is being rewarded and similar approaches can be expected to be successfully applied to the production of novel polyketides intended for therapeutic use.

β-lactams

The 'β-lactam' antibiotics, penicillins and cephalosporins, are produced following condensation of three amino acids: L-cysteine, D-valine (from epimerization of L-valine by ACV synthetase; see below) and α-aminoadipic acid. The β-lactam thialozidine (penam) ring of the penicillins and the β-lactam dihydrothiazine (cepham) ring of the cephalosporins are shown in Figure 12.4. Commercially, penicillins are produced primarily from *Penicillium chrysogenum* and the cephalosporins by *Cephalosporium acremonium* (syn. *Acremonium chrysogenum*). *Aspergillus nidulans* also produces penicillins, albeit in low quantities, and is used as a model organism for many studies. Commercial production of these antibiotics is almost exclusively achieved with strains developed through strain improvement programmes involving mutagenesis. The economics of commercial operations dictate that robust, high-yielding fungal strains are used that achieve a 'high in-tank productivity' and antibiotic yields have increased with time as a result of strain and technology improvements without recourse to using strains derived from recombinant DNA technology. It is, however, in the last decade that it has been possible to introduce DNA into the main antibiotic-producing organisms and that the genes encoding the enzymes involved in the metabolic pathways depicted in Figure 12.4 have been cloned. Studies of their regulation of expression are currently in progress. So, the ability to produce recombinant strains that either overproduce the antibiotics or produce variant structures is recent; even then, the suitability of the strains for commercial scale production plants has to be established. So, while it is not surprising that recombinant strains are not yet prominent in commercial antibiotic production, their use is likely to expand.

The penicillin biosynthetic genes in both *P. chrysogenum, P. notatum* and *A. nidulans* are clustered (Figure 12.5). The *acv*A (*pcb*AB) and *ipn*A (*pcb*C) genes of *C. acremonium* are also clustered and the *cef* EF and *cef*G genes form a separate cluster. The *cef* D gene has not been

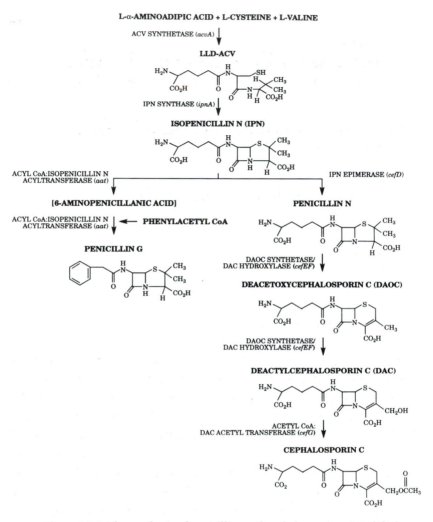

Figure 12.4 The synthesis of penicillins and cephalosporins. (Modified from Brakhage and Turner, 1995.)

cloned from a fungus despite its cloning from cephamycin-producing bacteria, e.g. *Streptomyces clavuligerus*. Note that the *acv*A and *ipn*A genes are contiguous but divergently transcribed. Thus, the intergenic region (872 bp in *A. nidulans* and 1.16 kb in *P. chrysogenum*) houses the transcription regulatory regions for both genes. The *acv*A gene encodes the ACV (α-aminoadipyl-cysteinyl-valine) synthetase enzyme. Both the

350

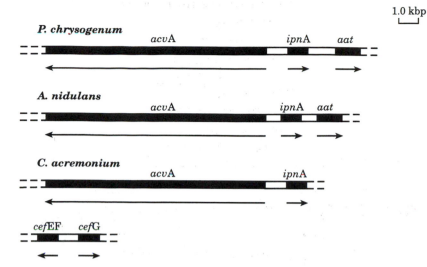

Figure 12.5 Clustered genes encoding enzymes for biosynthesis of penicillins and cephalosporins. Gene designations are explained in Fig. 12.4. (Modified from Brakhage and Turner, 1995.)

gene (coding region of over 11 kbp with no introns found) and the protein (about 426 kDa in *P. chrysogenum* and 415 kDa in *C. acremonium*) are very large. Transcription of the genes encoding the penicillin and cephalosporin synthesis enzymes is regulated by nutritional and developmental factors. The biosynthesis of the β-lactam antibiotics can be, depending on the fungal species, negatively regulated by carbon source, nitrogen and, to some extent, phosphorus sources. In addition, gene expression is subject to regulation by pH, oxygen levels and some amino acids. The use of reporter constructs and promoter analysis is now yielding detailed information on the regulation of gene expression in particular species. Analysis of proteins that bind to the promoters is also helping to provide a complete picture of the regulatory factors involved in the biosynthesis of β-lactam antibiotics.

Alongside detailed studies of the regulation of gene expression are studies designed to alter the biosynthesis of β-lactam antibiotics in order to achieve higher yields or modified structures. Driving expression of the *ipn*A and *aat* (*pen*DE) genes with the strong *alc*A promoter in *A. nidulans* resulted in enhanced enzyme levels without making much difference to the yields of penicillin although it must be remembered that altering metabolic flux through a pathway is not necessarily achieved by altering a single step (depending on the contribution of that step to flux control). The *acv*A gene and pathways that supply the necessary amino acids for ACV synthetase are now the focus of attention. Overexpression of the *acv*A gene in *A. nidulans* leads to enhanced penicillin production indicating that the step catalysed by ACV synthetase is a major factor in flux limitation to penicillin biosynthesis in that species. Isopenicillin N can be secreted by *C. acremonium* during production of cephalosporins. Introduction of extra copies of the linked *cef*G and *cef*EF reduced the levels of isopenicillin N and significantly increased the levels of cephalosporin C. Introduction of the *cef*EF gene with and without *cef*G into *P. chrysogenum* leads to production of cephalosporin intermediates which can be varied by feeding adipic acid as a side chain precursor. Such studies show that the high production capacity of *P. chrysogenum*, relative to *C. acremonium*, might be harnessed to produce novel cephalosporins. Aside from manipulating the pathway structural genes and their promoters, synthesis of the amino acid precursors and the transport of pathway intermediates between subcellular compartments are also aspects of the biosynthesis of β-lactam antibiotics that are being studied and the results are expected to find application in the achievement of enhanced antibiotic yields.

Enzymes

The saprophytic life style of many fungi is made possible by their ability to secrete enzymes that degrade polymeric material to be used subsequently as nutrients (see Chapter 9). This capacity for enzyme secretion enables fungi to be the primary colonizers of many habitats and to be the principal group of microorganisms responsible for nutrient recycling. The main polymeric components of organic matter found in nature are derived from plant cell walls and include cellulose, hemicellulose, lignin and pectin. Other polymers available are starch, protein and lipid. Fungi secrete a wide range of enzymes capable of degrading each of these polymers even when the polymers are naturally found in association with each other. Fungal genomes can encode a very large range of different degradative enzymes and many of the enzymes are harnessed in biotechnology. The enzymes used are either cell associated, i.e. the fungus is grown on a material which is degraded during growth, or cell free, i.e. the enzymes are separated from the fungal biomass and then used in an application. This exploitation of fungal enzymes is increasingly dependent on the application of molecular techniques.

Cell-associated enzymes are exploited in the commercial production of edible mushrooms (where the enzymes facilitate growth of the fungus on a particular substrate) or in the production of various fermented foods and beverages (where the enzymes degrade an otherwise recalcitrant material such as rice or soy bean and permit growth of other microorganisms such as yeasts and lactic acid bacteria). A diverse array of foods and beverages are produced, mainly in China, Japan and other far-eastern countries, using fungi. Rice (and other cereals), soybeans, cassava, meat, fish all serve as starting substrates and the fungi most commonly used depend on local practice. In Japan, *Aspergillus oryzae* and *A. sojae* are mainly used for the manufacture of sake and soy sauce whereas *A. niger, A. awamori* and other black-spored fungi are used in southern Japan for saccharification of steamed rice and sweet potato. *Rhizopus* spp. are predominant in China although a number of other fungi are also used. The key to choice of organism is its process suitability based on experience but, in relation to enzyme secretion by the fungus, the main enzymic activities sought are amylases and glucoamylases. The same cell-free activities are used commercially in starch processing for production of syrups rich in maltose or glucose. The commercial value and broad division of uses of industrial enzymes are shown in Figure 12.6. In this chapter attention is focused on commercially important enzymes for which there is detailed information on regulation of transcription and where genetic manipulation has been used to produce fungal strains with different enzyme profiles. Regulation of gene expression is a critical aspect to be considered but the secretion process itself can impose a bottleneck on secreted yields of target enzymes, so the molecular basis of protein secretion is also considered.

Amylases and glucoamylases

The best studied starch-degrading enzymes from fungi are α-amylase and glucoamylase. α-amylase cleaves internal α-1,4 glucoside bonds to

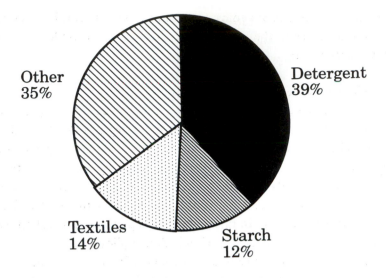

Total Market £0.9 bn

Figure 12.6 Global market for industrial enzymes in 1994. (Adapted from the 1994 Annual Report of Novo Nordisk, Denmark.)

release maltodextrins whereas glucoamylase releases β-D-glucose from the non-reducing ends of starch and maltodextrin. In addition, α-glucosidase genes have been cloned from *Aspergillus niger* and *A. oryzae*. Several genes encoding amylases and glucoamylases have been cloned from fungi and current studies are aimed at dissecting the elements that regulate expression of these genes. Production of the amylase family genes is subject to control by culture conditions. For example, their production is generally favoured by the presence of starch or maltose but not by glucose. The molecular basis of this regulation is slowly being revealed and is summarized below in the section on regulation of gene expression.

One of the strategies for maximized production of an enzyme, such as glucoamylase, is the introduction of multiple gene copies. Although this approach does work, the complexity of DNA–protein interactions in expression of *gla*A encoding glucoamylase dictates that the availability of transcription factors can become limiting at very high ($> c.$ 20) *gla*A copies. Commercial strains for glucoamylase production are generally the result of mutation and screening programmes. Such strains have several genetic differences to the wild-type strains that may include enhancement of gene copies.

Plant cell wall-degrading enzymes

The complexity of plant cell wall structures and compositions requires a diverse array of enzymic activities for degradation. In particular, the cellulases, hemicellulases (especially the xylanases), pectinases, ferulic acid esterases and ligninases are necessary. Several of these enzymic activities and encoding genes have been described from different fungal species and the degradative mechanisms for lignocellulose have been described elsewhere in this volume (Chapter 9). This section will therefore be restricted to the biotechnological applications.

Trichoderma reesei is exploited commercially for the production of cellulases and hemicellulases used, for example, in paper manufacture, vegetable processing and for cotton polishing in detergents. The enzymes can be secreted to very high yields at industrial scale, and strains have been developed using mutagenesis and screening programmes. Different applications require enzyme mixtures tailored to the purpose and using fungal strains that have been engineered to produce either a single cellulase, for example, or a defined mixture is one approach that can be undertaken now that several of the relevant genes are cloned. Deletion of genes that encode unwanted enzymes or upregulation of expression of desired genes are typical approaches adopted for altering the cellulase profiles in *T. reesei*. Cellulase-encoding genes from other cellulolytic fungi continue to be cloned and some of these might find biotechnological applications, e.g. from *Phanerochaete chrysosporium*. As described for the amylase family genes, another route for biotechnology with the cellulases is to employ engineered proteins based on known structures.

Other plant cell wall-degrading enzymes are secreted from fungi and many are used in commercial quantities, e.g. in fruit and vegetable processing. Many components of the plant cell walls, e.g. 'pectin' and 'lignin', are varied and complex structures which need several different enzyme activities for their degradation. Several genes encoding enzymes that degrade pectins, lignins and hemicellulose have been cloned from a variety of different fungal species. Studies in these systems are leading to the point where fungi such as *A. niger*, *T. reesei*, *P. chrysosporium* and others will be engineered to produce tailored mixtures of enzyme activities for commercial applications.

Other enzymes

A wide variety of fungal enzymes find commercial use and several of the genes involved have been cloned, including lipases, nucleases and proteases. The proteases are a large group of exploited enzymes and a number of protease-encoding genes have been isolated from several different fungal species. Aspartic proteases from *Rhizomucor* spp. are used as substitutes for rennin in cheese manufacture. The gene encoding an aspartyl protease from *R. miehei* has been cloned and the cDNA has also been expressed in *A. oryzae* to produce the recombinant product at high yield ($c.$ 3 gl^{-1}). The capacity of industrial fungi to secrete proteases can be a problem if the desired enzyme preparation is degraded by co-secreted proteases. Therefore, protease-deficient strains obtained by mutagenesis have been used as part of strain development programmes. More recently, protease-deficient mutants have been analysed further and the mutations mapped. Although some of the mutations cause a reduction in a single protease being produced, others affect more than one protease suggesting that the mutations affect the regulation of protease gene expression. Some of the fungal proteases have been characterized structurally raising the potential for producing engineered proteins. This type of approach, together with the increasing understanding of gene regulation will provide versatility in using filamentous fungi for protease production.

Production of heterologous proteins from fungi

The natural capacity of fungi to secrete a variety of proteins, together with the ability to manipulate fungi genetically makes them obvious targets as hosts for heterologous protein production. This is particularly so for those fungi with a record of safety and commercial cultivation, e.g. *A. niger*, *A. oryzae*, *Fusarium graminearum* (*renenatum*) and *T. reesei*. The reason for wanting heterologous production of certain proteins is because some proteins derived from their natural sources are very expensive, either because the source is difficult to obtain, the yield is low

356

or the extraction procedure is difficult (and, consequently, expensive). Alternatively, there may be ethical objections to using some sources of useful enzymes: it is not only the use of recombinant products which raises ethical questions! Considerable success has been obtained in the production of some heterologous proteins from fungi, i.e. the yields have been high and the proteins have authentic structures and functions. Similarly, there have been many disappointments which have led to the realization that more fundamental knowledge on the issues relating to protein production by fungi is required. In order to help in prioritizing the areas for research, identification of bottlenecks in protein production has led to a focus of studies in gene expression and protein secretion. However, it has become clear that more knowledge of fungal biology in general is required and that improvements in our knowledge will find application in all aspects of fungal biotechnology.

In the main, the yields of heterologous proteins from fungi have been much lower than the yields of secreted homologous proteins, even when expression is driven by a strong homologous fungal promoter. Exceptions are proteins encoded by genes from other fungi, e.g. *R. miehei* protease from *A. oryzae*. A few non-fungal proteins have been secreted from fungi at yields in shake flasks exceeding 1 gl^{-1}. The key strategy employed to achieve such high yields has been to use translational fusions with a homologous fungal protein and, in some cases, using protease-deficient host strains. For example, chymosin, lactoferrin, antibodies and some other proteins have been secreted using the fusion approach. The rationale that underlies the use of a fusion carrier protein to enhance secreted yields is not fully explained although current data favour an improvement to mRNA stability and the passage of protein during secretion being brought about by the homologous fusion carrier protein. Separation of the target protein from the carrier protein can be achieved in more than one way. A glucoamylase-prochymosin fusion protein was cleaved at the pro-sequence boundary to yield mature chymosin, probably by chymosin itself. Alternatively, an endoproteolytic cleavage site can be included when making the fusion construct at the DNA level. Many eukaryotic cells have an enzyme which cleaves preferentially at exposed LysArg, and some other dibasic residues, and this has been used in production of heterologous proteins from fungi. The endoprotease is physically located in the late secretory pathway so that the mature target protein is secreted following cleavage from the fusion protein. The processing of a fusion protein during secretion is illustrated in Figure 12.7. Even when using strong homologous promoters to drive expression, the fusion strategy and protease-deficient host strains, there are still some target proteins which have proved difficult to produce from fungi. Bottlenecks to the complete process need to be identified before remedies can be applied; one known bottleneck is the secretory pathway and this is discussed below.

Protein secretion from fungi

Considerably less is known of the protein secretory pathway in filamentous fungi than in the yeast *Saccharomyces cerevisiae* or in some mammalian cells. While the knowledge available with these systems affords an advantage to investigations with the filamentous fungi, differences between various systems must not be overlooked because they will explain why one cell type is a better secretor of proteins than another. The key features of protein secretion are common between eukaryotic cells: entry of proteins into the endoplasmic reticulum (ER), folding of the protein into its 3D structure, glycosylation in the ER and post-ER compartments, targeting to the cell exterior and fusion of secretory vesicles that enclose the secretory proteins with the plasma membrane releasing the protein to the exterior of the cell membrane, albeit with the cell wall still to negotiate. Protein secretion is an active process (requiring energy) and catalysed by a variety of other proteins. The secretory proteins themselves contain signals that determine their destination; for example entry to the ER is mediated by the secretion signal sequence. Sequences for targeting to particular cell compartments are contained within the proteins, as are sequences for glycosylation.

A large number of genes have been associated with protein secretion in *S. cerevisiae* and some of the cloned yeast genes have been used as heterologous probes for equivalent filamentous fungal sequences. Some of the genes that encode ER-resident proteins involved in protein folding (the so-called foldases and folding chaperone proteins) have now been cloned from *A. niger*. In the yeasts *S. cerevisiae* and *Schizosaccharomyces pombe*, the secreted yields of two heterologous proteins were increased by

1. Translocation of nascent protein to the ER lumen

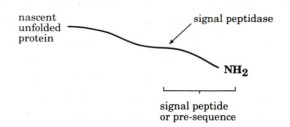

nascent unfolded protein

signal peptidase

NH$_2$

signal peptide or pre-sequence

2. Late secretory pathway/Golgi

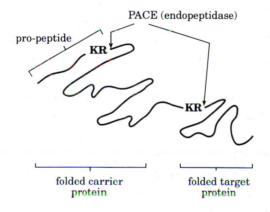

PACE (endopeptidase)

pro-peptide

KR

KR

folded carrier protein

folded target protein

3. Secretion of carrier and target proteins

Figure 12.7 Proteolytic processing of a fusion protein during secretion. PACE, paired basic amino acid cleaving enzyme (e.g. kexin of *S. cerevisiae*). The signal sequence directs the protein during translation across the membrane of the endoplasmic reticulum (ER). The signal sequence is removed by signal peptidase. Protein folding occurs within the lumen of the ER. Endoproteolytic cleavage at the dibasic amino acid pair (KR) occurs later in the secretory pathway. K, lysine; R, arginine.

overexpressing a foldase although it cannot be concluded either that this approach will benefit the secretion of all heterologous proteins from yeast or that the same approach will be successful in filamentous fungi. Equivalent studies are now being conducted in filamentous fungi.

Regulation of gene expression in fungi

In applying molecular biology to filamentous fungi for biotechnological purposes the capacity to manipulate gene expresion is an important tool. For this reason, much of what has been written in this chapter is devoted to the study of relevant genes. This work leads to investigations of gene expression. In the context of biotechnology, process optimization during fungal culture is a key feature, i.e. the yield:cost ratio should be maximized. A number of process variables can be adjusted, e.g. oxygen levels, shear in the reactor, pH, salt levels, medium composition and temperature. Other factors such as bioreactor characteristics of the organism and strain stability are also important. Many of these factors are outside the scope of this chapter which will focus briefly on the impact of nitrogen and carbon source regulation of gene expression, pH and chromatin structure.

Gene expression in fungi is subject to specific regulatory effects and wide-domain controls (e.g. by nitrogen and carbon sources; see Chapter 4). While most of the detailed studies have been with 'model' fungi such as *A. nidulans* and *N. crassa*, parallels can be drawn with the industrial fungi. Nitrogen metabolite repression leads to repression of the expression of genes necessary for the utilization of less-favoured nitrogen sources by the preferred sources ammonia or L-glutamine. The *trans*-acting regulatory protein AREA (encoded by *areA*) in *A. nidulans* is a DNA-binding protein necessary for alleviation of nitrogen metabolite repression. The equivalent protein, NRE, has been identified in *P. chrysogenum* to bind to the intergenic regions of *niiA–niaD* and *acvA–ipnA* in the penicillin gene cluster. A large number of genes in *A. nidulans* are known to be affected by nitrogen metabolite repression but to different extents. This indicates that the response to AREA depends upon the DNA recognition sequence and interaction of AREA with other regulatory proteins. Although other wide-domain controls act, in the main,

360

independently of nitrogen metabolite repression, promoter-binding sites for regulatory proteins can be physically close, suggesting some interaction. This must also be a factor when designing hybrid promoters for use in biotechnology that have altered regulatory characteristics.

In the presence of glucose, a readily metabolized carbon and energy source, the expression of genes required for utilization of alternative carbon sources is repressed. This is termed carbon catabolite repression and is mediated by DNA-binding transcription factors. In *A. nidulans* and *A. niger* the mediating protein is termed CREA and in *T. reesei* CRE1. The consensus recognition sequence for CREA binding has been found in the promoter regions of several genes and functional analysis of genes known to be regulated by carbon catabolite repression is an active area of study. For example, the promoter region of *gla*A (encoding glucoamylase) has been subject to deletion analysis in order to identify the important CREA-binding sites as well as sites responsible for upregulation of expression. A hypercellulolytic strain of *T. reesei* (Rut-C30) was obtained by screening for carbon catabolite derepression and it has since been shown that the *cre*1 gene is truncated, lacking the coding sequences for 80 per cent of the CRE1 protein, including one of the zinc finger regions. Although mutant strains might carry other mutations which contribute to the phenotype observed, this work demonstrates that alterations to regulatory proteins can affect gene expression and produce strains advantageous to biotechnology.

Extracellular pH is sensed by fungi and alters the expression of some genes. Some fungi secrete acids during growth making their environment particularly acidic. In *A. nidulans pac*C encodes a regulatory protein with three putative zinc finger domains that activates transcription of genes expressed during growth at alkaline conditions and prevents transcription of genes expressed at acid pH. The *pac*C gene from *A. niger* has also been described. A number of other genes originally found by analysis of mutants have now been identified as mediating pH regulation and some of these genes have been cloned. In the design of novel promoters it seems likely that pH regulation will be a significant factor, and improved understanding of pH sensing and regulation of gene expression will play an important part in gene manipulation for biotechnology.

361

The eukaryotic nuclear membrane envelops DNA, RNA and protein which is both densely packed and ordered. In fungi, as with other eukaryotes, we know that the genomic site of integration of introduced genes can affect expression, i.e. some genomic loci are better suited to expression than others. In a practical sense, methods have been developed for targeting introduced genes to loci of known transcriptional activity. However, the impact of chromatin structure on gene expression will need to be considered before a complete picture of gene expression in fungi is gained. Studies with other eukaryotes have shown that regions of DNA that bind to the proteinaceous nuclear matrix (also called the nuclear scaffold) can be used to insulate against the effects of chromatin structure on gene expression. Such matrix attachment regions (MARs) have been isolated from *T. reesei* but it is too early to say if they can be used in expression constructs. However, gene manipulation to improve the biotechnological applications of filamentous fungi will ultimately take into account genome structure as well as the direct interactions of DNA with *trans*-acting regulatory proteins.

Concluding remarks

Filamentous fungi are already widely used in biotechnology. A far wider range of potentially useful metabolites are produced from fungi than are currently used commercially. Whether such fungal metabolites are developed commercially will depend on the acceptability of the product, the market size and the profit margin in its manufacture and sale. For some products fungi will have an edge over competing sources whilst, with others, fungi will not be competitive. However, these positions are not static, being subject to consumer preferences and regulatory concerns as well as developments in the process itself. Applications of molecular biology in fungal biotechnology are very recent and while this technology offers unprecedented potential in developing the biotechnology of fungi its impact is only just being felt. This is due in part to ethical and regulatory issues but also because we need to know more about the biology of fungi at the molecular level. Genome sequencing projects for filamentous fungi are under way and will contribute invaluable information to the development of fungi as biotechnological organisms. The *S. cerevisiae* genome sequencing project is completed, and functional analysis of open reading frames will bring an ever more complete understanding of its biology. The molecular biology of filamentous fungi will benefit from continuation of present studies with individual fungi and from the information gained through the various genome sequencing projects. It is likely that most advances in our understanding of fungal molecular biology will find applications in fungal biotechnology.

Further reading

Archer, D. B. and Peberdy, J. F. (1997). The molecular biology of secreted enzyme production by fungi. *CRC Crit. Rev. Biotechnol.* 17:273–306.

Archer, D. B., Jeenes, D. J. and MacKenzie, D. A. (1994). Strategies for improving heterologous protein production from filamentous fungi. *Antonie van Leeuwenhoek* 65:245–250.

Brakhage, A. A. (1998). Molecular regulation of β-lactam biosynthesis in filamentous fungi. *Microbiol. Mol. Biol. Rev.* 62:547–585.

Brakhage, A. A. and Turner, G. (1995). Biotechnical genetics of antibiotic biosynthesis. Pp. 263–285 in *The Mycota.* II. *Genetics and Biotechnology* (U. Kück, ed.). Springer-Verlag, Berlin.

Brown, D. W., Yu, J.-H., Kelkar, H. S. *et al.* (1996). Twenty-five coregulated transcripts define a sterigmatocystin gene cluster in *Aspergillus nidulans. Proc. Natl. Acad. Sci. USA* 93:1418–1422.

Brown, D. W., Adams, T. H. and Keller, N. P. (1996). *Aspergillus* has distinct fatty acid synthases for primary and secondary metabolism. *Pro. Natl. Acad. Sci. Sci. USA* 93:14873–14877.

Crawford, L., Stepan, A. M., McAda, P. C. *et al.* (1995). Production of cephalosporin intermediates by feeding adipic acid to recombinant *Penicillium chrysogenum* strains expressing ring expansion activity. *BioTechnology* 13:58–62.

Cullen, D. and Kersten, P. (1992) Fungal enzymes for lignocellulose degradation. Pp. 100–131 in *Applied Molecular Genetics of Filamentous Fungi* (J. R. Kinghorn and G. Turner, eds.) Blackie Academic and Professional, London.

Dunn-Coleman, N. S., Bloebaum, P., Berka, R. M. *et al.* (1991). Commercial levels of chymosin production by *Aspergillus. BioTechnology* 9:976–982.

Gouka, R. J., Punt, P. J. and van den Houdel, C. A. M. J. J. (1997). Efficient production of secreted proteins by *Aspergillus*: progress, limitations and
prospects. *Appl. Microbiol. Biotechnol.* 47:1–11.

Hopwood, D. A. and Sherman, D. H. (1990). Molecular genetics of polyketides
and its comparison to fatty acid biosynthesis. *Annu. Rev. Genet.* 24:37–66.

Lowe, D. A. (1992). Fungal enzymes. Pp. 681–706 in *Handbook of Applied
Mycology, Fungal Biotechnology*, vol. 4 (D. K. Arora, R. P. Elander and K. G.
Mukerji, eds.). Marcel Dekker, New York.

MacKenzie, D. A., Jeenes, D. J., Belshaw, N. J. and Archer, D. B. (1993). Regulation
of secreted protein production by filamentous fungi: recent developments
and perspectives. *J. Gen. Microbiol.* 139:2295–2307.

Ratledge, C. (1994). Yeasts, moulds, algae and bacteria as sources of lipids. Pp.
235–291 in *Technological Advances in Improved and Alternative Sources of
Lipids* (B. S. Kamel and Y. Kakuda, eds.). Blackie Academic and
Professional, London.

Sakaguchi, K., Takagi, M., Horiuchi, H. and Gomi, K. (1992). Fungal enzymes
used in oriental food and beverage industries. Pp. 54–99 in *Applied
Molecular Genetics of Filamentous Fungi* (J. R. Kinghorn and G. Turner,
eds.). Blackie Academic and Professional, London.

Skatrud, P. L. (1992). Genetic engineering of β-lactam antibiotic biosynthetic
pathways in filamentous fungi. *Trends Biotechnol.* 10:324–329.

Trail, F., Mahanti, N. and Linz, J. (1995). Molecular biology of aflatoxin biosynthesis. *Microbiology* 141:755–765.

Unkles, S. E. (1992). Gene organization in industrial filamentous fungi. Pp.
28–53 in *Applied Molecular Genetics of Filamentous Fungi* (J. R. Kinghorn
and G. Turner, eds.). Blackie Academic and Professional, London.

van den Hombergh, J. P. T. W., van de Vondervoort, P. J. I., Fraissinet-Tachet, L.
and Visser, J. (1997). *Aspergillus* as a host for heterologous protein production: the problem of proteases. *Trends Biotechnol.* 15:256–263.

Index

acetyl-CoA carboxylase 344
Acrasiomycota 22–3, 35
actin 173
activator protein
    QA-1F 142, 155, 157
    QUTA 140–4, 146, 155, 157–9
adenylate cyclase 183
adherence 332–3
aflatoxins 347–8
Agaricales 31
Agaricus bisporus 12, 235, 279–80
algae
    brown 24, 37
    golden brown 24, 37
    green 24
    red 24, 51
Allomyces 27
allosteric changes 155, 157
Alternaria 8
alveolates 24
American Type Culture Collection (ATCC)
    81–2
α-aminoadipyl-cysteinyl-valine (ACV)
    synthetase 349–52
amoeboflagellates 23, 24
amphotericin *see* antifungal drugs
amylases 12, 290, 353–5
anaerobes
    obligate 27
anaphase 222, 226
    inhibitors 222
    promoting complex (APC) 224–5
anastomosis 232–3
anthrax 3
antibiotics 3, 5, 347
    cephalosporin C 13, 350–1
    β-lactam 13, 342, 349–52

penicillin 3, 13
    regulation of biosynthesis of 351
    streptomycin 5
anti-fungal drugs
    amphotericin B (amphotericin) 180,
        336–7
    aureobasidin A 39
    azoles 180, 336–8
    echinocandins 178–9
    fluconazole 336, 338
    griseofulvin 336, 347
    imidazoles 336
    itraconazole 336
    ketoconazole 336
    miconazole 336
    nystatin 6, 180, 336
    pentamidine 39,
    polyenes 180, 336
Aphyllophorales 31
apothecium *see* fruiting body
appressorium 301–4
Arabidopsis thaliana 119
arabinose 274, 290
Archea 22–3, 35, 47
AROM protein 137, 139, 151, 153, 154,
    155, 156–9
Arthroderma 322
ascogonium 253, 262
ascomata 260, 262–3
ascomycetes 6, 207, 260
Ascomycota (Ascomycotina) 28–35, 37,
    176, 323
ascorbate oxidase 287–8
ascospores *see* spores
ascus 6, 50
Ashbya gossipii 120
aspergillosis 329

Index

Index

bleaching 292
Botrytis cinerea 295
Buchner 3
budding yeast *see Saccharomyces cerevisiae*
Buller's drop 268

Ca^{2+} ion channels 173, 182
Caenorhabditis elegans 88, 90, 119
cAMP 183, 291
   dependent protein kinase (PKA) 183,
     303
Candida 2, 12, 329
Candida albicans 10, 32, 35, 39, 175–6,
     322, 325–6, 333, 335, 338
   secreted aspartyl proteinases in 335
Candida glabrata 326, 338
Candida krusei 326
Candida parapsilosis 326
Candida tropicalis 326
Candida utilis 5
candidiasis 325–6, 329
carbon catabolite repression 136
carbon sources
   fermentative 3
   glucose 360
   non-fermentable 8
   quinate 137, 140
cell cycle
   meiotic 210
   mitotic 210– 28
cell cycle control 8, 11, 14, 15
   progression 216–7
cell division cycle 8, 11, 213–13, 215
   mutants (cdc) 212
   phases
     G0 217
     G1 216,225, 236,240
     G2 216, 221,225
     M 216
     S 216
     START 217–9, 223
   proteins (*A. nidulans*)
     BIMA 225
     BIMC 222
     BIME 225
     BIMG 225

proteins (*S. cerevisiae*)
   Cdc16p 225
   Cdc23p 225
   Cdc25p 226
   Cdc27p 225
   Cdc28 217
proteins (*S.pombe*)
   cdc2$^+$-encoded 217
   Dis2 225
cellulases 12
cellulose 273–5
   binding domain (CBD) 279–81
cell wall 176–9
   chitin 25, 176–8
   inner (primary) 176
   outer (secondary) 176
   polysaccharides 22
Cephalosporium spp. 13
Cephalosporium acremonium 349
chaperones 358
checkpoint controls 11, 225–228
chiasmata 258
chitin 25, 176
chitin synthase 171, 178
chitinases 310
chitosomes 171
chlorophylls 24
choanoflagellates 25
cholesterol 180, 336–7
Chromocrea spinulosa 240
chromosome disjunction 258
chromosomes 80
chymosin 12, 357
Chytridiomycota 24–6, 37, 180
chytrids 43
cis-acting elements
   ARE198
   BRE 195
   MCB 219
   SCB element 218
   UAS 99
   URS 99
citric acid 13
Cladosporium fulvum 8, 304, 311–12
clamp connections 246, 248, 255
classification of fungi 49–50

Index

Index

Index

Index

Index

Index

Index

Index

Index